Withdrawn!

PRAISE FOR *FOOD FORENSICS*

"*Food Forensics* definitely is a must-have book for serious-minded researchers, healthcare practitioners, and consumers, plus everyone who eats! I highly recommend it! It's a virtual 'encyclopedia' about toxins we don't know about in our foods."

—Activist Post

"I emphatically agree that this book is an important public record of scientific truth. An amazing job on so many levels and an absolute must-read."

—Michael T. Murray, ND, coauthor of
The Encyclopedia of Natural Medicine

"*Food Forensics* is an incredible, groundbreaking book. Just buy it. And read it. You won't be sorry. It might be the best decision of your life."

—Ty Bollinger, author and documentary
film producer, TheTruthAboutCancer.com

"Mike Adams has done a great service for all those who are concerned about the quality of what goes into their body. *Food Forensics* is an essential reality guide to food and water in the 21st century. Mike reveals why relying on federal bureaucrats to protect us from toxic additives and contaminants is a dangerous proposition. While the FDA and EPA are derelict in their duty even with multibillion-dollar budgets, the Health Ranger references and democratizes food safety science for the people so that practical strategies for clean living and safe detoxification are available to us all, right now!"

—Robert Scott Bell, D.A. Hom.

"Radical, irreverent, always provocative. Mike Adams's new book, *Food Forensics*, should be required reading for all the Big Food, Big Biotech apologists who continue to poison and mislead us, as well as for activists and concerned consumers."

—Ronnie Cummins, cofounder and international
director of the Organic Consumers Association

"It's unfortunate, that just as Americans are beginning to wake up to the importance of natural health approaches and a clean diet, Big Food and others are trying to capitalize on this new market by deceiving consumers into buying expensive, heavily contaminated 'health' foods. Mike Adams has done a real public service by exposing the deceit and arming consumers with the knowledge to make informed decisions about the food they eat. *Food Forensics* is a must read for anyone wanting not only to avoid disease, but also to achieve optimal health and wellbeing."

—Gretchen DuBeau, executive and legal director of the Alliance for Natural Health

"It's not enough to just read the labels these days. In this book, Mike exposes some of the most prevalent, hidden dangers in your food that he's identified through years of research and scientific testing—you won't find any of these listed in the ingredients. But he doesn't just expose these toxic chemicals, he gives you a roadmap to help you and your family avoid them to confidently live a healthy and happy life."

—Kevin Gianni, author of *Kale and Coffee: A Renegade's Approach to Health, Happiness, and Longevity*

Withdrawn/ABCL

FOOD FORENSICS

FOOD FORENSICS

The Hidden Toxins Lurking in Your Food and How You Can Avoid Them for Lifelong Health

MIKE ADAMS

BenBella Books, Inc.
Dallas, Texas

3907505172353 1

This book is for informational purposes only. It is not intended to serve as a substitute for professional medical advice. The author and publisher specifically disclaim any and all liability arising directly or indirectly from the use of any information contained in this book. A health care professional should be consulted regarding your specific medical situation. Any product mentioned in this book does not imply endorsement of that product by the author or publisher.

Copyright © 2016 by Mike Adams

All rights reserved. No part of this book may be used or reproduced in any manner whatsoever without written permission except in the case of brief quotations embodied in critical articles or reviews.

BENBELLA

BenBella Books, Inc.
PO Box 572028
Dallas, TX 75357-2028
www.benbellabooks.com
Send feedback to feedback@benbellabooks.com

Printed in the United States of America
10 9 8 7 6 5 4 3 2 1

Library of Congress Cataloging-in-Publication Data
Names: Adams, Mike.
Title: Food forensics : the hidden toxins lurking in your food and how you
   can avoid them for lifelong health / Mike Adams.
Description: Dallas, TX : BenBella Books, Inc., [2016]
Identifiers: LCCN 2016004389| ISBN 9781940363288 (trade paper) | ISBN
   9781940363462 (electronic)
Subjects: LCSH: Nutrition. | Diet. | BISAC: HEALTH & FITNESS / Food Content
   Guides. | HEALTH & FITNESS / Nutrition. | SCIENCE / Research & Methodology.
Classification: LCC RA784 .A33 2016 | DDC 613.2—dc23 LC record available at https://lccn.loc.
gov/2016004389

Editing by Leah Wilson   Text design and composition by Aaron Edmiston
Copyediting by Elizabeth Degenhard   Cover design by Jason Gabbert
Proofreading by Amy Zarkos and Kim Broderick   Printed by Lake Book Manufacturing
Indexing by WordCo Indexing Services, Inc.

Distributed by Perseus Distribution
www.perseusdistribution.com

To place orders through Perseus Distribution:
Tel: (800) 343-4499
Fax: (800) 351-5073
E-mail: orderentry@perseusbooks.com

Special discounts for bulk sales (minimum of 25 copies) are available.
Please contact Aida Herrera at aida@benbellabooks.com.

CONTENTS

INTRODUCTION

To pursue scientific research into food forensics, I oversaw the construction of a food forensics laboratory in central Texas. The lab's central feature was an inductively coupled plasma mass spectrometry instrument, called ICP-MS for short. It has the unique ability to detect metals and elements including nickel, lead, mercury, or magnesium at very low concentrations—in almost any sample you might want tested. I call it "Star Trek technology" because it seems to function almost as if by magic. But it isn't magic. It's just "sufficiently advanced technology," as Arthur C. Clarke once explained.

In the months after its installation and calibration by expert chemists and instrumentation engineers, the ICP-MS instrument began to lift the veil on what was really present in all sorts of foods: junk foods, fast foods, superfoods, herbal supplements, vitamins, and more.

That's when things began to get weird.

When the instrument identified very high levels of lead and cadmium in popular vegan protein products, I contacted the manufacturers of these products to suggest they pursue a voluntary recall of their products. A recall wasn't an option, I was informed, and I was urged to be careful about releasing anything publicly that would "impact sales revenues" of these companies.

When I discovered that popular ginkgo herbs grown in China contained a whopping 5 parts per million (ppm) of toxic lead—an element proven to cause cancer and brain damage—I was told that the lead contamination was "naturally occurring" and therefore didn't matter. Yet when I tested ginkgo herbs grown on U.S. soil, they tested remarkably clean, showing near-zero levels of heavy metals. It turns out that when ginkgo is grown in contaminated soils, it accumulates heavy metals in the herb. (This should not be surprising to anyone.)

When I found very high levels of tungsten (greater than 10,000 parts per billion, or ppb) in superfoods imported from China and Southeast Asia, I was told that tungsten was of no concern because "the U.S. Food and Drug Administration (FDA) has no limits on tungsten," and that therefore everyone should ignore the presence of this heavy metal in popular superfood products.

When I discovered an astonishing 11 ppm of lead in mangosteen superfood powder imported from Thailand, I went public with the finding and warned people not to eat mangosteen powder unless it had been tested. In response, I was blacklisted from several importers and not allowed to purchase their raw materials anymore. (My company purchases raw materials to manufacture certified organic foods and superfoods in Texas, and we meticulously test each material before purchasing it in volume for manufacturing.)

Over and over again, as I began to find alarming levels of lead, aluminum, tungsten, mercury, arsenic, and other toxic elements in everyday foods, superfoods, pet treats, and even certified organic foods, the response I got from manufacturers of these products was, *"Don't tell anyone!"*

Before disclosing some of my results, it's important to understand the thresholds at which heavy metals begin to affect human health.

Mercury: According to the World Health Organization (WHO), mercury, even in small amounts, may cause serious health problems, earning it a spot on the top ten list of the most dangerous chemicals to humans. The Environmental Protection Agency's (EPA) maximum containment level goals for mercury in drinking water is 2 ppb.[1]

Tungsten: Cases of acute poisoning by this heavy metal can be caused by just 5 mg/L, or approximately 5 ppm. Exposure to high levels of tungsten has been linked to an increase in strokes.[2]

Lead: While there is no safe blood lead level in children, the U.S. Centers for Disease Control and Prevention (CDC) recommends the threshold at which a child is deemed to have lead poisoning is 5 micrograms per deciliter of blood, or 50 ppb.[3]

Arsenic: Long-term exposure to this heavy metal through drinking water and food may cause neurotoxicity, cancer, developmental issues, cardiovascular disease, and diabetes, according to the WHO. The EPA has set the arsenic standard for drinking water at .010 ppb.[4]

Cadmium: When ingested in high doses, this heavy metal can cause nausea, vomiting, diarrhea, abdominal cramping, and severe gastroenteritis, according to the Agency for Toxic Substances and Disease Registry (ATSDR). The reference dose for dietary exposure to cadmium is 0.001 mg/kg/d.[5]

In just the first few months of ICP-MS research on samples of foods, vitamins, and consumer products, I discovered

- Over 500 ppb mercury in cat treats and fish-based dog treats
- Over 10 ppm tungsten in rice protein products
- Over 5 ppm lead in ginkgo herb products
- Over 11 ppm lead in mangosteen powder
- Over 400 ppb lead in cacao powders
- Over 500 ppb lead and more than 2,000 ppb cadmium in rice proteins
- Over 6 ppm arsenic and more than 1 ppm lead in some spirulina products
- Over 500 ppb mercury in dog treats
- Over 200 ppb lead in brand-name mascara products

(Note: 1,000 ppb = 1 ppm)

In nearly every case, when I contacted the manufacturer of the product to warn them about the high levels of heavy metals found in their products, they insisted their products were perfectly safe while urging me to remain silent and keep their secret from the public.

A real-life conspiracy of silence

Conspiracies really do exist, of course. New York Attorney General Eric Schneiderman said that pharmaceutical companies conspired to set artificially high drug prices in that state. U.S. federal trade authorities say the Chinese government conspires to dump cheap solar panels on the U.S. market to drive U.S. solar manufacturers out of business. And many food companies, I've discovered, actively conspire to keep their own customers ignorant of the toxic substances routinely found in their products.

The point of this book is to break that conspiracy of silence and reveal what's lurking in your favorite foods, superfoods, organic foods, dietary supplements, vitamins, and even pet foods. The information in this book is precisely the information these companies desperately hope you never see.

Recent experience has taught me some valuable lessons in how these companies operate:

Step 1: **Deny** the existence of heavy metals or other harmful substances in their products.

Step 2: **Attack** the source of the information. Try to create doubt about the motives of the researcher (me) or the accuracy of the findings.

Step 3: Should denials and attacks fail, **twist** scientific facts to claim that all heavy metals are "naturally occurring" and therefore don't count, even if they are found in high levels due to heavy industrial contamination of the farms where the food is grown.

Step 4: If steps 1 through 3 are unsuccessful, **lie** to customers by telling them that heavy metals are *good* for them! This strategy has already been invoked by one company whose products tested at high levels of lead and cadmium. Instead of announcing they would reduce the level of these metals in their products, they posted an article that claimed heavy metals were good for you and people shouldn't be concerned about eating them.

Sheer deception and consumer fraud

The process of denial and obfuscation I'm describing here is routinely pursued by companies of all sizes, including some companies catering to organic consumers, raw foodies, vegans, vegetarians, detox patients, and health-conscious buyers.

The deceptions are quite incredible. One company that imports nearly 100 percent of the rice protein used by all the vegan protein manufacturers in the United States is fully aware that their product contains high concentrations of toxic lead, cadmium, tungsten, and mercury. On their website, however, they claim their material is "Prop 65 compliant," referring to Proposition 65 in California.

Prop 65 says that if your product exceeds 0.5 micrograms of lead per serving, then you must put a cancer warning on your product label. The rice protein material being imported by this company delivers over 16 micrograms of lead per serving, which is 34 times higher than the Prop 65 lead limit. So how is that "compliant"? Because companies using the material place a small cancer warning on their product labels to "comply" with Prop 65. So even though this material contains 34 times more lead than is allowed under Prop 65, the importer claims the material is "compliant" with Prop 65, thereby grossly misleading buyers into thinking the material has low lead composition.

This sort of deception and consumer fraud, I've found, is routinely carried out across organic foods, natural products, superfoods, and dietary supplements companies. Many companies that sell products emblazoned with phrases like "better than organic" or "high raw" are actually poisoning their own customers with toxic heavy metals. And they almost never test their own products for heavy metals, which is why they are so surprised when I confront them with the truth about what's found in their products. Even then, when made aware of the heavy metals concentrations found in their products, they invoke denial and obfuscation rather than transparency. Just like drug companies, weapons manufacturers, or Wall Street investment houses, many natural products companies seem to be run by people who place profits over consumer safety . . . almost by default.

That's why this book is such an important public record of scientific truth. This book documents the heavy metals that are really found in these products, mapping out the actual metals composition of products that were acquired in 2013 through 2016, then analyzed via atomic spectroscopy for their elemental composition. In early 2016, we expanded our laboratory to include liquid chromatography–mass spectrometry (LC-MS) instrumentation for the detection of pesticides, herbicides, and other organic molecules.

We hope to report on those findings in subsequent books and website reports. (See labs.naturalnews.com for the latest analysis reports.)

This book will spur widespread denials and possibly even a few lawsuit threats. It will enrage unethical product manufacturers but empower consumers with a new source of information that should appear on Nutrition Facts labels but doesn't. This book will not only indict dishonest companies selling contaminated products, but it will also celebrate those many companies whose products are remarkably clean of toxic heavy metals (and yes, they do exist).

Substantial efforts to silence this work

I have been offered money not to publish this book. I've been offered large advertising contracts to leave certain products out of this book. I've been threatened with lawsuits for publishing laboratory results on the Internet. One of the largest natural products retailers in the United States, a $12 billion company, deliberately trained its employees to lie about me in very specific terms by telling customers that "Mike Adams doesn't have a lab" and that all the laboratory results I've been publishing are fictional.

Substantial efforts have been made to discredit me and silence this work, and yet the fact that you hold this book in your hands is proof that all of those efforts failed. No matter how much I am threatened, I refuse to remain silent on this crucial issue for public health and food transparency.

We live in a world that's heavily contaminated with industrial waste. Much of our organic food now comes from China, where the term "organic" is a cruel joke. Air quality in Beijing was recently recorded as being 1,100 percent higher than the maximum air pollution limits set by the WHO, reaching the astonishing pollution concentration of 268 micrograms per cubic meter.[6]

Much of our food is now grown on lands where industrial waste is intentionally dumped and used as "fertilizer." As a result, many foods are heavily contaminated with toxic substances. The environmental science cannot be denied, and the scientific findings of this book can be replicated by any competent laboratory running ICP-MS instrumentation.

Please value what you now hold in your hands and understand how incredibly rare it is for this information to have finally been made public, despite all the threats and intimidation attempts that were unleashed in a desperate effort to keep this information hidden. Ask yourself this question, too: "Why isn't the FDA conducting this research and publishing the results for the public to see?" I ask myself that same question every time I step into my lab. If the FDA really cared about food safety and public health, it would never have left this task to private citizen scientists like myself. The only reason I've taken up this task is because everybody else refuses to do it. The FDA, food manufacturers, and mainstream media outlets funded by food advertising are all colluding to ignore this science and prevent the public from learning the truths you'll read here.

To stay up to date on the latest findings in this realm, visit the website of which I am the editor, www.naturalnews.com.

Laboratory methodologies and accuracy

Can you trust the data presented in this book? My laboratory is accredited by the International Organization for Standardization (ISO) under its global analytical accuracy standards program known as ISO 17025. This is the gold standard for internationally recognized analytical laboratory accuracy, and it means we operate under a strict set of rules, guidelines, and procedures that are enforced by a third-party audit.

The scientific methodologies we use for testing food and water are universally recognized by the scientific community and are sourced from organizations such as the Association of Analytical Communities (AOAC), the EPA, and the FDA. For example, we use a minor variation of AOAC 2013.06 for testing heavy metals in foods.[7]

For testing water samples, we use methodology EPA 200.8.

My lab was accredited in 2016 after two years of preparation involving analytical repeatability determinations, validation of analytical methods, and exhaustive documentation of our laboratory quality control procedures and error-correction processes. Because of this extensive experience in ICP-MS analysis and laboratory protocols, I even plan to announce my availability as a science consultant to food manufacturers or retailers who wish to set up similar testing for their own operations.

But what is ICP-MS? How are heavy metals really tested in foods and beverages?

To help understand analytical accuracy a bit further, it's important to understand the nature of ICP-MS testing.

ICP-MS results across competent laboratories can and do vary by as much as 20 percent due to differences in methodologies and instrument sensitivities. Within the same lab, variation in results from different samples of the same product may vary as much as 10 percent due to several reasons, but competent laboratories demonstrate strong repeatability within a range of plus or minus 10 percent.

From lab to lab, analytical results of the same substance may vary slightly. So if two different labs test the exact same protein powder, for example, it is perfectly reasonable that one lab might report lead at 450 ppb while a second lab reports lead at 500 ppb.

However, you won't find orders of magnitude differences. No competent lab would report lead at just 45 ppb or at 4500 ppb for the same sample, in other words.

In summary, it's important to understand that ICP-MS laboratory results do have some natural variability within a reasonable range. Metals composition will also vary from gram to gram and lot to lot. Every production lot of a commercial product has a different metals composition from previous lots. Because of these simple truths, **all the numbers in this book should only be used as a general guide** to help you decide what to eat and what to avoid. They do not describe absolute concentrations that are consistent across all products of the same name.

It's also true that because of the efforts already made by myself and the launch of the Natural News Forensic Food Lab, some companies are making tremendous efforts to clean up their raw materials and produce cleaner products. That's why products sold on the market at the time you read this may be substantially cleaner than the products tested in this book. A book takes at least a year to go from manuscript to store shelves, so what you are seeing in this book is actually a snapshot of products that were available in the three years prior to publication. If you'd like to see more up-to-date results, you'll find them at labs.naturalnews.com.

Many commercial labs deliberately produce artificially low results

Another important thing to keep in mind here is that many commercial labs that cater to food companies are in the business of producing artificially low metals test results because that's precisely what their customers want to see. Producing artificially low results is very easy to accomplish by various means that are readily accessible to anyone who wants to commit such violations of ethics.

At the Natural News Forensic Food Lab, we use a slow digestion method that prevents the nitric acid from boiling. This retains nearly all heavy metals found in the original food sample. As a result, our metals tests are typically slightly higher than what most commercial labs produce, but they are also more accurate. Our open-block digestion cycle typically takes two hours, not the forty-five minutes often used by other labs. We also use closed cell (microwave) digestion for difficult samples to ensure complete digestion.

Any competent university lab can easily reproduce our results within plus or minus 10 percent by using appropriate digestion equipment and procedures.

How we assure scientific accuracy at our lab

In the interests of full disclosure, here are some of the methods and safeguards we've used in the Natural News Forensic Food Lab to ensure the best possible accuracy:

- As noted above, our lab is ISO 17025 accredited, having achieved the global standard for trusted analytical accuracy in laboratories.
- All instrumentation is calibrated and certified accurate by its original manufacturer.
- All analytical methodologies we use are derived from globally accepted methodologies published by the AOAC International or other similar scientific organizations.
- All external standard solutions are traceable to National Institute of Standards and Technology (NIST) or other standards bodies. Custom standards are formulated and validated by highly competent, experienced custom formulations companies.
- We do not re-use sample digestion vessels or autosampler vessels. Our laboratory process relies on disposable vessels that eliminate vessel contamination concerns.
- After every tenth sample is run via ICP-MS, a blank vial and a calibration vial are run to ensure the ICP-MS instrumentation

remains well calibrated. If significant analytical drift is detected (i.e., results of the midrange calibration checks begin shifting), the run is halted, the instrument is cleaned (or consumable parts are changed out), and the run is repeated from the start. Analytical drift during our ICP-MS testing has been nearly eliminated through the use of the Niagara Plus sample injection system manufactured by Glass Expansion.

- ICP-MS instrumentation is routinely maintained in accordance with manufacturer recommendations. For example, sample cones and skimmer cones are routinely cleaned. Sample uptake tubing in the peri-pump is routinely changed. Argon air is in-line filtered, as is our helium source.

- For each food sample tested, three separate samples from the same product lot are run. Results are then averaged across the three to help eliminate variability and improve reliability.

- All sample test vials are archived for a period of one year so that any challenged result can be re-validated if needed.

- The validity of digestion methods and ICP-MS analysis methods are further validated through the frequent use of Certified Reference Materials (CRMs) with known concentrations of elements verified by more than a dozen other laboratories.

- Outside labs are used to further validate and spot-check in-house laboratory results. We have at times used a third party commercial laboratory as well as a university laboratory, both of which have confirmed our findings on multiple occasions.

- The dilution water used in sample preparation is laboratory-grade deionized water produced by a high-end Thermo Scientific water filtration system specifically designed for laboratories.

- Oxidation acids used for sample digestion are trace-grade acids and are routinely tested for their purity. The very small concentrations of elements (parts per trillion) found in these acids are measured at the beginning of each sample run, then subtracted from the results of all subsequent samples.

- Samples that show curiously high results are re-analyzed a fourth or fifth time to make sure the results are accurate.

- All raw sample data for each run is archived on multiple backup servers residing at two different physical locations.

U.S. Department of Agriculture (USDA) and Food and Drug Administration (FDA) have no heavy metals limits

Neither the FDA nor USDA has any official, universal limit on heavy metals in foods, beverages, and dietary supplements sold to U.S. consumers.

This fact is, of course, astonishing. Most consumers of USDA-certified organic foods automatically assume those foods are substantially free of heavy metals because they are labeled organic. But in our lab, we've found USDA-certified organic foods to consistently contain far higher levels of heavy metals than many conventional foods (which tend to be aggressively processed, removing minerals and heavy metals alike).

So why don't the USDA or FDA set heavy metals limits for the U.S. food supply? Surely they have their own explanations, but my view as a food researcher and investigative journalist is that both the USDA and FDA are far too intertwined with the interests of the industries they claim to regulate. Most of the top people at the USDA, for example, have a revolving-door history with the cattle industry or herbicide companies such as Monsanto and DuPont. Top FDA people, similarly, are far too cozy with drug companies and processed food manufacturers to make reliable decisions in the public interest.

Rather than regulating these industries for the benefit of the public, both the FDA and USDA seem far more interested in protecting these industries from public scrutiny. As a result, there is no real incentive to disclose the heavy metals contamination of agricultural products, or canned soup or beef jerky, for that matter. Because the truth of all this might "cause alarm" among consumers, government regulators essentially play along with the conspiracy of silence preferred by food manufacturers.

This is why I strongly support the establishing of heavy metals limits in foods, beverages, and dietary supplements. Without such limits, food manufacturers can get away with essentially any amount of toxic elements in their products.

It is noteworthy that, in February 2016, the nation was outraged over the discovery of 1–2 ppm of lead in the water supply of Flint, Michigan. Yet I have personally found food products with far higher levels of lead that are consumed by a consumer cross section of the entire nation. Strangely, there is so far no outcry over high lead levels in food products, even though lead in water is widely recognized as so dangerous to children that many citizens of Flint, Michigan, called for the criminal prosecution of those responsible.

Moving toward a low heavy metals industry standard

Until the USDA and FDA come around to establishing heavy metals limits for foods, superfoods, and dietary supplements, we've created our own limits, which have been published online and embraced by several companies.

The website lowheavymetalsverified.org provides a voluntary heavy metals guide for manufacturers of foods, superfoods, and dietary supplements. The site describes a letter-grade self-certification system ranging from A+++ on the super clean side down to F for foods that are more heavily contaminated with heavy metals. (This grading system is printed in full on page 212 of Part 3: The Data near the end of this book.)

Because these standards may be revised from time to time as more information is learned about the impact of heavy metals on human health, please refer to lowheavymetalsverified.org to view the latest numbers. In particular, we hope to begin the speciation of arsenic so that we can distinguish organic arsenic from inorganic. Once that is accomplished, we plan to alter this standard to consider solely *inorganic* arsenic (the dangerous variety).

Most food products available in the marketplace today fall between A and D on the grading scale. This scale sets a voluntary standard by which food products can be easily compared on their heavy metals composition. It also allows consumers to more easily shop for products that are cleaner than others. For example, almost every health-conscious consumer would prefer to eat grade-A chocolate rather than grade-B chocolate, assuming all other properties of the chocolate are equal.

The downside of this system is that it is purely voluntary and, as you might have already guessed, many companies will flat-out lie to their customers and claim lower heavy metals concentrations than really exist in their products.

For this reason, Natural News will be policing the industry by randomly purchasing products from companies who claim these heavy metals limits and testing those products for compliance. Products that do not comply with the claims levels will be published on naturalnews.com.

Our hope is that both the USDA and FDA will eventually take over this function and establish their own procedures for heavy metals limits and industry spot-checking. Until that day comes, Natural News is the only organization on the planet that will be fulfilling this important role in the interests of public safety.

Some observers find it quite curious—perhaps even bizarre—that a private sector company is doing a better job of policing the U.S. food supply for heavy metals than the entire federal government, with a seemingly infinite budget.

I find it bizarre, too.

PART 1

EVERYTHING YOU NEED TO KNOW

ABOUT TOXIC ELEMENTS

The next section of this book is scientific in structure as it discusses the origins of toxic heavy metals and other chemical contaminants such as pesticides, fertilizers, and preservatives. It also explains the way in which humans absorb these contaminants and the resulting health effects.

If you're really only interested in the heavy metals test results for your favorite foods and superfoods, you can skip ahead to the charts section of this book, beginning on page 214.

But for those who want more in-depth research and explanations about how heavy metals and other contaminants harm biology and why they are so difficult to get rid of, this section documents the harm of heavy metals with a considerable amount of scientific explanation and research citations.

Just a warning, though: This section can get a bit technical. (Doctors, scientists, and biologists, however, will find it familiar reading.)

HEAVY METALS

Where do heavy metals come from?

Life on Earth in its rawest natural form is fraught with countless dangers and immediate threats to your existence. Numerous toxic metals and compounds are found almost everywhere on this planet in some concentration. However, potentially poisonous forms of mercury, lead, cadmium, arsenic, aluminum, copper, tin, tungsten, chromium, beryllium, and other elements are increasingly found in our post-industrial environment.

As elements, they can be transmuted through nuclear fusion in exploding stars, but they are not destroyed in mundane Earthly environments. Until the industrial revolution accelerated mining and pollution operations across the planet, most toxic heavy metals were buried deep underground, far from the concerns of simple human civilizations. As human industry expanded in the nineteenth and twentieth centuries, toxic heavy metals were mined, smelted, and added to any number of products that released those metals directly into the environment. Leaded gasoline, for example, released lead directly into the air with every stroke of the combustion engine. Mercury fillings resulted in thousands of tons of mercury being expelled into the atmosphere as the bodies of those who passed away were cremated. Lead arsenate was also widely used as a pesticide on orchards and food crops across North America for much of the nineteenth century.

Once expelled into the open environment, heavy metals may be inhaled, ingested, or absorbed into humans, animals, plants, and fungi, or they may

2

be transformed and combined with other substances to create new compounds . . . but they cannot simply vanish. They persist.

It is the industrial exploitation and expelling of these elements—which were originally sparse and spread out at relatively low levels—that has turned vague primordial threats into everyday dangers. As by-products of smelting, ore extraction, energy production, and commercial goods, heavy metals and refined chemical compounds have poured into our air, water, soils, foods, ecosystems, and bodies.

In September 2013, the CDC issued its updated fourth National Report on Human Exposure to Environmental Chemicals, detailing more than 201 chemical substances that have been identified in blood serum and urine levels throughout the U.S. population.[8] These can be ingested, absorbed, stored, excreted, metabolized, or bound to other compounds—potentially interacting with, blocking, or amplifying reactions within the body.

While many elements, including trace levels of certain minerals, are essential nutrients for catalytic conversions and biological functions, alarming concentrations of toxic forms of these elements have found their way into our lives at a pace that's wildly out of balance with nature and hazardous to our health and longevity.

A few dozen key contaminants may be taking a crucial but yet uncalculated toll on the well-being of everyone around the world—with increased levels of toxins in everyday foods contributing to a general rise in inflammation, immunological and digestive disorders, neurological damage, organ failure, heart and lung ailments, cancer, and other serious diseases and conditions.

When most people think of being poisoned, they typically imagine ingesting a large, concentrated dose that quickly induces acute toxicity, often followed by a swift and horrible death. In reality, the real danger to health comes from long-term exposure to low-level doses of toxins *over time*, including heavy metals.

Science now recognizes that these detrimental health effects are triggered by gradually accumulating, minuscule concentrations of toxins through repeated dietary or environmental exposure.

The tidal wash of pesticides, herbicides, insecticides, fungicides, rodenticides, fertilizers, preservatives, emulsifiers, and additives across the agricultural practices of the entire Western world—and increasingly the developing world—has contributed to the introduction of known toxins into the environment at apocalyptic levels. They interact with and are absorbed by soils, bodies of water, vegetation, fish, and wildlife. They are absorbed and integrated into plant and animal tissues. As humans, we breathe in these compounds, eat them, drink them, and accumulate them in our bodies. We also excrete them, or their metabolized by-products, back into the environment, furthering the cycle of death and destruction brought on by these toxins. While further research is needed to expand our understanding of exactly how these toxins interact to produce disease and death, there is little debate about the importance of limiting environmental and dietary exposure to these toxins in the first place.

Dietary exposure to toxic heavy metals through foods is a far greater problem than most people suppose. Even USDA-certified organic foods are not tested for heavy metals like cadmium, lead, arsenic, or mercury. Thus, there are no limits on heavy metal levels in these foods, including those sold in upscale healthy food retailers such as Whole Foods. The organic label simply describes the process through which the food was grown and that a farmer hasn't used additional pesticides, herbicides, or other petrochemicals during that process. "Certified organic" in no way requires any heavy metals testing of soils, irrigation water, or even the final food product.

The reality is that one farmer's "organic" food can differ widely from another farmer's food simply because the air, water, and soil in which the food is grown is overwhelmingly contaminated with heavy metals.

Toxic heavy metals and other elemental poisons—whether they circulate around us or are absorbed into our bodies—definitively remain in the biosphere in one form or another *in perpetuity*. They are part of a vicious and deadly cycle that modern life has exponentially accelerated through the industrial mining, concentration, and dispersing of toxic elements that would have been far better left alone, buried in the Earth's crust.

Some of the worst offenders, including metals like lead, mercury, cadmium, and arsenic, have long since thoroughly infiltrated our lifestyles, and each poses its own significant hazards. Because the functions of the body are complex, many of the harmful effects are still being discovered and documented to this day. The scientific work on understanding the effects of toxic elements on biological systems, in fact, has only just begun.

Already, there is ample evidence of heavy metals disrupting chemical reactions throughout the body and blocking important nutrient absorptions. Toxic metals often compete with nutritional elements in metabolic processes; poisonous metals can imitate essential, or "good," trace metals, rendering elements the body needs unavailable as chemical catalysts. Even when heavy metals don't interfere with key metabolic functions, they still cling to cell walls, interfering with other cellular functions such as waste excretion, immune defense, healing, and adaptation.

Scientists have spent a considerable amount of time and effort researching the processes by which heavy metals undermine and destroy the body over time. Oxidation is one such process, whereby cells are disrupted and damaged, often leading to disease or weakened organ vitality. This is one reason why antioxidants are essential for good health: They protect cells from dangerous and deadly exposure to free radicals.

Emerging science reveals that toxic elements, including heavy metals, have a greater propensity than previously thought for damaging DNA and disrupting cellular processes. Not only are these metals shown to cause cancer, but there is increasing evidence now confirming their potential roles as co-carcinogens that increase mutations and disruptions when combined in the body with other types of toxins.[9]

Heavy metals poisoning is trans-generational

An even more important—and destructive—role may be played by toxic heavy metals in interfering with the process of DNA methylation, which transforms cytosine and adenosine nucleotide bases in the DNA sequence. This interference can cause inheritable changes in what is known as the epigenome, a genetic roadmap parallel to DNA that records changes to gene expression that are passed on to the next generation.

The process of DNA methylation plays a role in gene regulation and is a vital process during early fetal development when methylation during cell division directs specific tissue formation and other processes. Approximately 70 percent of human DNA is naturally methylated when the attachment methyl groups switch a gene on or off, but when toxic metals attach to these methyl bonds, they can interfere with vital cellular functions or even block them altogether.[10]

Through the still-emerging understanding of epigenetics, science has uncovered the specific process by which environmental factors, diet, stress, and exposure to toxins rewrite the intended gene expression and alter DNA. This, in turn, influences an individual's chances of contracting disease—and of passing along those risks to their children. Despite the fact that epigenetic influences are not hard-coded into DNA, epigenetic effects influence traits that appear to be inherited by offspring.

A full understanding of this phenomenon should cause immediate alarm in the mind of anyone reading this. Epigenetic inheritance of toxic side effects from dietary exposure to heavy metals means that *toxicity is trans-generational.* This means that the toxic environment in which we live today will negatively impact future generations for an unknown number of generations *even if we eliminate all exposure starting tomorrow.*

For example, studies have shown an inverse relationship between a mother's cumulative cord blood lead levels and the epigenome of her developing fetus, strongly suggesting that toxins interfere with "long-term epigenetic programming and disease susceptibility."[11,12] Arsenic exposure was likewise found to affect DNA methylation in fetal development, damaging DNA and disrupting gene regulation.[13]

In many ways, we are already too late to save future generations from the harmful effects of exposure to toxic elements. And because exposure is only getting worse, not better, trans-generational negative effects are likely to significantly worsen with each subsequent generation. This cycle may place the very sustainability of the human race in a precarious situation, with its effects only becoming more widely apparent in the coming years. Broadly speaking, we may already have doomed ourselves to global increases in infertility, devastating cancer rates, and a planet-wide decline in cognitive function due to heavy metals exposure in modern-day foods.

In other words, we may have already set out on a path by which the great-grandchildren of today's young adults will be increasingly mentally challenged, infertile, and possibly incapable of surviving without significant medical assistance. The destruction of sustainable human life on our planet, in other words, may have already been set into motion, only to play out through several generations of suffering and bewilderment as government regulators and food companies continue to push their conspiracy of silence about the actual underlying causes.

Heavy metals interfere with your biology

There are many ways in which heavy metals interfere with and distort healthy biological functions. As just one example, heavy metals may interfere with normal cellular methylation cycles. When lead builds up in bones, it can negatively distort DNA methylation processes in white blood cells, which of course originate in bone marrow.[14] White blood cells are essential to a healthy immune system as they help the body fight infection by attacking foreign invaders such as viruses, bacteria, and germs.

Many metal toxins are classified as electrophiles, meaning these molecules are driven to steal electrons and bind to chemical compounds in the body in processes similar to methylation. Lead, mercury, arsenic, and cadmium are biochemical vampires, latching onto and interfering with vital molecular groups, disrupting their immunological and metabolic contributions to healthy biology. Even after they are expelled from the body, these heavy metals can go on to cause damage in downstream biological systems such as fish, amphibians, and ocean ecosystems.

Natural chelation and the removal of heavy metals from the body

Health-conscious consumers naturally want to find ways to remove heavy metals from their bodies. The most important method for accomplishing that is to eliminate dietary exposure to toxic heavy metals. Once sources of exposure are eliminated, the body's natural elimination processes will automatically and over time remove toxic heavy metal buildup in organs and tissues.

But even the process of removing heavy metals from your body can be toxic. One of the most common methods for this is called chelation, or the binding of metal ions. Chelation therapy involves the administration of chelating agents to bind to metals so they can be more easily excreted and removed from the body through detoxification.

When heavy metals are chelated out of the body's organs and tissues, they are dumped into the blood supply, which can have toxic effects on the body.

If this process is too rapid, the levels of heavy metals in the blood supply can increase so rapidly that they become acute and toxic on their own. This danger is why any heavy metals detoxification program must be pursued under the guidance of a clinically qualified chelation expert, naturopathic physician, or other holistic practitioner with years of experience in removing heavy metals from the body.

If you are looking for a chelation expert, my recommendation is to contact the American College for Advancement in Medicine (ACAM), a nonprofit organization dedicated to educating physicians and other health professionals about the efficacy of using integrative medicine, or medicine that treats the whole body, including the mind and spirit. ACAM's healthcare model focuses primarily on preventing illness, rather than masking symptoms with pharmaceutical drugs. At Acam.org, you can locate a qualified heavy metals removal clinician in your area.

The topic of heavy metals removal from your body is covered in more detail in the sections on specific heavy metals. As you read through these sections, however, keep in mind that removing the sources of exposure is the single most important principle of detoxification. Failure to remove the sources of exposure—even while undergoing aggressive detoxification therapies—will net you very few overall gains.

Chelation strategies are based on a metal element's natural affinity for molecules with a certain chemical charge. Chemical binding properties provide a pathway for removing damaging heavy metals from the body. Even the best chelators, however, are limited in their abilities. No chelation strategy offers 100 percent removal of any heavy metal from the body. Specifically, beware of dietary supplements that claim to rapidly pull heavy metals out of your brain or body tissues. Although certain supplements (such as oral ethylenediaminetetraacetic acid [EDTA]) may offer benefits if used over a very long period of time, many dietary supplements are presently sold with dubious detox claims backed by nothing other than wishful thinking. Some of them may even pose real dangers to your health.

In my view, no detox regimen should be pursued without first consulting with a naturopathic physician.

ZEOLITES AND HEAVY METALS

Beware of powdered zeolites sold alongside claims that they remove heavy metals from your body. All powdered zeolites contain very high concentrations of lead—typically 50,000 ppb and sometimes more—and their aluminum levels are many times higher. I recently discovered that powdered zeolites were being dishonestly marketed as a daily dietary supplement, pushed by unscrupulous companies that claimed you should "detox daily" by consuming these finely ground rocks containing very high levels of lead and aluminum. Remarkably, one of the primary claims of zeolite marketers was that it removed lead and aluminum from your body. To "prove" this, one of the companies commissioned a small-scale clinical trial in which the presence of toxic metals was measured in the urine of people consuming zeolites daily. Sure enough, people who consumed zeolites were found to urinate out higher levels of lead and aluminum (two elements found in powdered zeolites). From this, the study author "concluded" that zeolites remove heavy metals from the body.[1]

People who eat lead and aluminum, in other words, were found to urinate out lead and aluminum. Should we be surprised?

Talk about junk science!

In a clinical setting, common chelating agents for lead, arsenic, and other metals include meso-dimercaptosuccinic acid (DMSA), dimercaptopropane-sulfonic acid (DMPS), and 2,3-dimercaprol (BAL).[15] These chelates—termed after *chela*, or "claw," a Greek-derived Latin word—are often used in combination with vitamins and other antioxidants structured to bind more effectively with the metal while enhancing metabolic pathways for the metals' removal. While DMSA and DMPS are the most widely used chelates for lead and arsenic, studies have found them incompatible with mercury removal, where more custom chelates are typically used.[16]

There are many foods that naturally have some limited chelation properties. Cilantro,[17] chlorella,[18] and lemons[19] have all been identified as agents with some effectiveness for reducing heavy metal toxicity, while foods like garlic[20] can reduce levels of oxidative stress. It has also been found that citrate, cysteine, glutamate, ethylenediaminetetraacetic acid (EDTA), and yeast extract (particularly effective against copper toxicity) bind and remove certain metals.[21] (Although it must be noted that yeast extract is a common form of MSG, an excitotoxin with its own health concerns.)

In research conducted in 2010, Taiwan researchers found that lemon and orange peel could aid in the removal of heavy metal ions, particularly copper and nickel, which highlights the importance of consuming fruits and vegetables daily, as their benefits are extraordinary. Activated carbon (charcoal) is also very effective at neutralizing and removing metal toxins.[22]

Lack of exercise and sweating causes heavy metals to accumulate over time

The body's mechanisms for excretion also play an important role in detoxification; in studies, sweating in particular has been shown to remove heavy metals in vastly higher quantities than are expelled through urination. Endurance exercises and use of infrared saunas have been successfully used to sweat out toxins, in many cases surpassing the level of toxins removed through urination.[23,24]

The fact that more and more Americans pursue sedentary lifestyles lacking almost all vigorous exercise—and therefore lacking sweating—helps explain why metals so rapidly accumulate in the bodies of the obese. A 2014 study published in the *Mayo Clinic Proceedings* and conducted at the University of South Carolina's Arnold School of Public Health found that obese Americans spend less than one minute per day engaged in vigorous exercise.

Yes, that's one minute per day. The study found that obese women were far worse off than men, engaging in less than one hour of vigorous exercise *per year*.[25] With that near-zero level of exercise and sweat excretion of heavy metals, it's only a matter of time before the accumulation of heavy metals reaches a crisis point in the body, contributing to dysfunction and symptoms that are often diagnosed as diabetes, cancer, heart disease, or Alzheimer's and dementia.

Modern humans are, in a very real sense, walking time bombs of toxic metals and chemicals, all accumulated through the routine consumption of contaminated foods, personal care products, and environmental exposure. It is irrational to expect that a nation can protect the health of its people—or even control its health care costs—unless this trend is sharply reversed by cleaning up the food supply and reducing heavy metals in personal care products and dietary supplements.

The FDA, in other words, should be doing exactly what I'm doing here. A nationwide effort needs to be undertaken to test all the popular foods and other items that might contain toxic metals such as lead, cadmium, and mercury. Reasonable concentration limits need to be standardized at the national level, and the diligent efforts of people like myself and others who are attempting to lift the veil on food industry contamination should be celebrated, not vilified.

There aren't many of us who genuinely care about the quality of food our fellow Americans are routinely swallowing. We, the few who dare to spend our time, money, and effort examining the food contamination that's contributing to the disease epidemics now devastating our world, are the pioneers of the clean food movement. Through tools of modern science, we effectively give consumers a kind of X-ray vision into what they're eating, drinking, and putting on their skin. It is precisely this clarity that the food industry fears, because the more closely people are allowed to look at what they're really eating, the more persistently they may begin to ask the really important questions like, "Hey, why *isn't* anybody testing these protein powders for lead?"

HEAVY METALS: INTERNATIONAL LIMITS CHART

	EPA[a]	FDA[b]	USP[c]	WHO[d]/ FAO[e]
	Drinking Water		Oral Limit	Food
ELEMENTS	mg/L (unless specified)		ppm	
Aluminum	50–200 µg/L	—	5,000	—
Arsenic (inorganic)	0.01	apple juice: 10 ppb	1.5	provisional tolerable weekly intake (PTWI) 15 µg/kg body weight
Cadmium	0.005	food color additives: 15 ppm	2.5	provisional tolerable monthly intake (PTMI) 25 µg/kg body weight
Copper	1.3	—	50	—
Lead	0.015	total daily intake (TDI) 75 mcg/day	1	Previous limit withdrawn in 2011.
	0.015	bottled water: 5 µg/L	1	
	0.015	candy: 0.1 ppm	1	
	0.015	fruit juices: 50 ppb	1	
Mercury	0.002	elemental: 1 ppm	1.5	PTWI 1.6 µg/kg per bw
	0.002		1.5	
Tin	—	—	3,000	PTWI 14 mg/kg per bw

	WHO[d]/ FAO[e]	EU[f]	EU[f]	CA Prop 65[g]
	Water	EU Directive 1881/2006	EFSA[h]/ CONTAM[i]	
ELEMENTS		mg/kg wet weight	total weekly intake (TWI)	
Aluminum	100–200 µg/L	—	—	—
Arsenic (inorganic)	10 µg/L	—	No limit; panel says it needs more data.	10 mcg daily intake
Cadmium	3 µg/L	.05–3.0	2.5 µg/kg per bw	4.1 mcg daily intake
Copper	2,000 µg/L	—	—	—
Lead	10 µg/L	.02–3.0	Previous limit withdrawn in 2013.	0.5 mcg daily intake
Mercury	inorganic: 6 µg/L	0.–1.0	inorganic mercury: 4 µg/kg bw	0.3 mcg daily intake
			methylmercury: 1.3 µg/kg bw	
Tin	—	50–200	—	—

[a] Environmental Protection Agency
[b] Food and Drug Administration
[c] U.S. Pharmacopeial Convention
[d] World Health Organization
[e] Food and Agriculture Organization
[f] European Union
[g] California Proposition 65
[h] European Food Safety Authority
[i] The Panel on Contaminants in the Food Chain

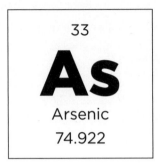

33

As

Arsenic

74.922

ARSENIC (As)

ATOMIC NUMBER: 33

GROUP 15: NITROGEN AND PHOSPHOROUS

The very mention of the element arsenic evokes thoughts of its notorious role as a poison in the commission of murder, often incited by passion, jealousy, or the quest for power. This use, long captured in literature and the infamous crimes of centuries past, continues today.

Yet in modern times, the broader impact of arsenic as a chronic, cumulative contaminate in water, food, and the air eclipses the significance of acute, deliberate poisoning. Arsenic does not always kill so quickly. It is a known carcinogen that has been linked to tumors formed in the skin, lungs, bladder, kidneys, and digestive tract[26] as well as the lymphatic and hematopoietic systems[27] in both humans and animals. Arsenic's numerous detrimental health effects have been well documented to include diabetes, heart disease, cardiovascular issues, respiratory distress, impaired neurological development, and even depression. Arsenic toxicity has also been linked to increased infant mortality and early developmental issues.

Notably, arsenic comes in two forms: organic and inorganic. Defined by their bonds with carbon and hydrogen, the organic forms of arsenic are largely considered harmless. The inorganic forms of arsenic widely used in industrial applications, which are typically bound to elements such as oxygen, sulfur, or chloride, are the varieties associated with arsenic's poisonous and carcinogenic effects. Common inorganic forms of arsenic include arsenic trioxide (a common industrial by-product also used in some medical treatments), chromate copper arsenate (widely used as a wood preservative that also acts as an insecticide), and pesticides. Lead arsenate, calcium arsenate, "Paris Green" (copper acetoarsenite), and sodium arsenate are all pesticides derived from inorganic arsenic.

Arsenic in drinking water

The tainting of well-water supplies across the globe with arsenic trioxide is a mounting catastrophic problem affecting more than 137 million people who have been exposed to levels exceeding 10 ppb in drinking water, the standard set by both the United Nations WHO and the EPA. A geological study conducted by Peter Ravenscroft at the University of Cambridge further discovered that some 57 million people are drinking water at peak contamination rates of more than 50 ppb—putting them at a serious risk for cancer and other health effects.[28]

This problem with arsenic contamination in water is most concentrated in Bangladesh and the neighboring Indian state of West Bengal, where nearly half the population drinks from contaminated sources after decades of Western aid directed the construction of tube wells that tapped directly into arsenic-tainted water reservoirs.[29] Because of this, Bangladesh has 27 million people drinking from sources that contain greater than 50 ppb of arsenic, while West Bengal and a few other areas of India have a combined 11 million people exposed to carcinogenic levels of arsenic-tainted drinking water.

An astounding 80 million people in this region drink water containing more than 10 ppb of arsenic. Bangladesh is considered the "biggest arsenic catastrophe in the world,"[30] where fifty-nine out of sixty-four districts are affected, and more than half the total population is at risk of arsenic contamination. This repeated exposure to arsenic is known as *arsenicosis*, which is typically diagnosed via visible skin lesions, although symptoms can also include dehydration, abdominal pain, vomiting, diarrhea, dark urine, delirium, vertigo, shock, and eventually death.

A study carried out in Bangladesh also confirmed a link between high arsenic exposure and anemia, a condition in which a person lacks healthy red blood cells and suffers from inadequate oxygen delivery to the body's cells and tissues.[31]

Other parts of the world face significant arsenic levels in drinking water as well. Another 5.6 million people in China and an astonishing 3 million in the United States also drink water that's heavily contaminated with arsenic. Several millions more across Southeast Asia and the Pacific Region, Russia, the Middle East, South America, and other pockets of the world are exposed to arsenic in their drinking water.[32]

While lakes, streams, and groundwater remain unregulated for arsenic, the EPA has limited public drinking water sources to 10 ppb. Despite this, several thousand water districts across the United States continue to contain dangerously high levels of arsenic.

Arsenic in the food chain and biosphere

Arsenic has thoroughly contaminated our food chain and the environment. Chronic exposure to arsenic compounds in food—even in low doses over time—has been definitively linked with the development of cancers, especially in the skin, liver, bladder, and lungs.[33]

The ability of inorganic arsenic to destroy and kill has also made it an important and widespread element in a cocktail of pesticides as well as an important wood preservative that doubles as an insecticide. As a result of the widespread use of agricultural and industrial arsenic compounds, arsenic has entered the soil and our surrounding environment at nearly every conceivable point—ultimately tainting the world's food supply.

In addition to organic arsenic compounds that are frequently found in small amounts in many foods, a number of inorganic arsenic varieties have contaminated production crops that feed America and the world. The real sources of concern are those accumulated from widespread pesticide and fertilizer use, runoff from industrial production, and—a factor of greater importance than most people realize —from pressure-treated wood.

Arsenic as a pesticide

Before the development of dichloro-diphenyl-trichloroethane (DDT), lead arsenate—a deadly cocktail of the heavy metals lead and arsenic—was one of the most widely used pesticides, dominating agriculture in the first half of the twentieth century. Along with other arsenic-based pesticides like calcium arsenate and "Paris Green," arsenic was used to control moths and other pests, especially in apple orchards and other fruit trees as well as cotton crops—despite the fact health concerns over arsenic residues had been officially acknowledged as far back as 1919.[34] Other inorganic varieties and a few organic varieties of arsenic were used for mosquito control and as insecticides, rodenticides, and herbicides sprayed on everything from curbs to sidewalks to road perimeters.

In addition to pesticide applications, a number of phosphate and micronutrient fertilizers—even those meant for organic food production—have been found to contain elevated arsenic and heavy metal levels, further contaminating many soils.[35]

The EPA's first comprehensive report on arsenic pesticides in 1972 listed numerous compounds and their known uses and hazards.[36] They include lead arsenate, "Paris Green,"calcium arsenate, basic copper arsenate, ammonium arsenate, arsenic acid, arsenic pentoxide, arsenic trioxide, sodium pyroarsenate, sodium arsenate, and potassium arsenate, as well as several harmful "arsenic-containing organic compounds used in formulating pesticides," including cacodylic acid—just to name a few.

According to the EPA, although DDT replaced much of the use of lead arsenate in the post-war period, that later reversed after federal regulations severely limited the use of DDT and other organochlorine insecticides. Subsequently, the use of some arsenicals as pesticide resumed by the late 1960s. By 1969, annual production of arsenic trioxide had increased to 66,000 tons. Meanwhile, more than 4 million pounds of lead arsenate and some 2 million pounds of calcium arsenate were also produced for industrial purposes.

These varieties of pesticides were useful in controlling moths, beetles, and other pests, particularly in orchards during the period spanning from 1890 to 1940, where lead arsenate was sprayed directly onto fruits, including apples, apricots, cherries, peaches, pears, plums, prunes, nectarines, quinces, and grapes.

Calcium arsenate was also frequently used as a pesticide on a wide range of agriculture crops, including asparagus, beans, blackberries, blueberries, boysenberries, broccoli, Brussels sprouts, cabbage, carrots, cauliflower, celery, collards, corn, cucumbers, dewberries, eggplant, kale, kohlrabi, loganberries, melons, peppers, pumpkins, raspberries, rutabagas, spinach, and squash—until the EPA canceled registration for its use in 1988. The registration was canceled after it was found that these pesticides posed "cancer risks to workers and acute toxicity to the general public."[37]

Not only were edible crops treated with calcium arsenate, but cotton crops spanning millions of acres in states including Texas and Oklahoma were annually sprayed with arsenic acid, leaving soils contaminated at levels that measured as high as 830 ppm.[38]

According to the EPA, many farmers who had been interviewed claimed their orchard trees lived shorter lives and that their fields were unsuitable for various forage crops typically grown during alternating years, giving support to the case for the negative effects presented by widespread arsenic soil contamination. The heaviest scheduled uses were in repelling Syneta beetles in apricots, peaches, and quince. Five to six pounds of arsenic-laced pesticide were used per 100 gallons of water, a mixture used on these crops for decades. Grapes were also subjected to some of the heaviest doses of arsenic, with sodium arsenate fungicide registered for use at an average rate of 3 to 9 pounds per acre in an effort to stop black measles and crown gall.

While arsenic pesticides have been found to metabolize into secondary forms with the aid of microorganisms, researchers have discovered that about 20 percent of the toxins remained in the soil *in their original form* decades later, even on fields that received only a single topical soil application. Researchers also found that 55 percent of croplands sprayed with pesticides containing arsenic trioxide back in the 1950s were irreversibly leaching into both groundwater and soils over time.[39]

Thus, repeated and widespread applications of lead arsenate and other pesticides have contributed to significant accumulations of lead and arsenic in soils, and these toxins can still be found even decades after their use declined or was banned, with horrible health implications that continue to this day.[40]

Ken Rudo, who has worked as the state toxicologist for North Carolina's Division of Public Health for more than twenty-four years, confirmed that arsenic compounds bind tightly to the soil, presenting a multitude of potential issues. "These chemicals have just tremendously long half-lives in the ground,"

Rudo stated in an EPA report.[41] The extensive spread of lead arsenate has made remediation of soils difficult, particularly as arsenic moves to the subsoil layers much more quickly and pervasively than other metals such as lead.

Soil analysis studies in the arsenic- and lead-tainted orchards of Massachusetts have revealed that the two metals "Pb and As bind 'tightly' to soil HA [humic acids] molar mass fractions."[42]

A study in Taiwan found an important relationship between the geographical concentrations of leading heavy metals, including arsenic and nickel, and the prevalence of oral cancer in patients who smoked or chewed betel quid (a combination of betel leaf, areca nut, and slaked lime). That is, cancer and other malignancies predominated in areas where the soil was contaminated with those elements.[43]

TOXIC ELEMENTS IN FERTILIZERS

The prevalence of heavy metal compounds in most fertilizers used in agriculture today poses ongoing problems for the bioaccumulation of toxins in crops, animals, humans, and the rest of the food chain.[1]

Naturally occurring elements and heavy metals (including mercury, lead, cadmium, and arsenic) are frequently found in combination with some of the world's leading industrial ores. This means that mining and processing those ores brings to the surface of the planet toxic elements that would have otherwise stayed buried.

Such is the case with phosphorous, which, alongside nitrogen and potassium, is one of the most important macronutrient constituents used in the creation of fertilizers. Phosphate ore typically contains cadmium in concentrations as high as 300 mg/kg, with sedimentary rock containing the highest concentrations. Other hazardous metals such as lead, nickel, and copper are also abundant in phosphate ores.[2,3]

As the primary application of phosphate ore is in the creation of fertilizers, its contamination by cadmium means a significant amount is added to the soil, creating abundant opportunities for human exposure to the known carcinogen and environmental toxin, especially through dietary uptake of foods and the inhalation of tobacco smoke.[4]

However, while phosphate fertilizers contribute a significant quantity of metals—particularly cadmium—to the soil, it is not the number one contributor. It may surprise many to know that industrial waste and sewage sludge is also exploited as a source of fertilizer, and contributes vastly higher quantities of heavy metals and other toxins to soils, and ultimately human intake, than nonwaste fertilizers ever could.[5]

EPA Okays Selling of Sewage Sludge

The wet, solid cake that remains after wastewater treatment plants process industrial and residential waste has long been referred to as sewage sludge. Decades ago, it was common practice for many municipalities—particularly very large urban areas—to haul the sludge and dump it into oceans and waterways, until the practice was banned by the EPA in 1992.[6]

In the mid-1990s, two lobbying groups—the U.S. Composting Council (USCC) and the Water Environment Federation (WEF)—joined forces with the EPA to promote the use of sewage sludge as a safe, effective, and cheap fertilizer under the rebranded name "biosolids." It was actively promoted by many agencies as an effective way to dispose of human waste, while creating a viable by-product market.[7]

In 1997, the EPA said their "longstanding policy encourages the beneficial reuse and recycling of industrial wastes, including hazardous wastes, when such wastes can be used as safe and effective substitutes for virgin raw materials."[8]

A study on the bioavailability of cadmium and its accumulation in soils found that while continued phosphate fertilization raised cadmium levels, the increase was much lower than those observed from the application of sewage sludge as fertilizer, both in overall accumulation as well as in bioavailability to Swiss chard and other plants.[9]

Heavy metals in biosolids can be a particularly worrisome issue, as the toxic elements frequently found in drinking water, food, and medicine tend to concentrate in the biosolids that are routinely applied to soils as fertilizer. There, they accumulate in the soil, leading to a persistent rise in toxic elements taken up by food crops.

Biosolids from sewage waste can contain especially high levels of accumulated metals—from lead, to cadmium, to mercury, to arsenic, or others such as nickel, copper, aluminum, or tin.[10]

In February of 2016, I acquired a bag of "Dillo Dirt" from the city of Austin, Texas, and I tested it for heavy metals via ICP-MS instrumentation. Dillo Dirt is composted human sewage that's purchased by landscapers and home gardeners for use on lawns and gardens. Even though the bag says, in small print, that it's not sold for use on edible vegetable gardens, it is positioned on retail shelves as a garden compost product (and no one reads the small print on a bag of compost anyway).

As you might expect, my ICP-MS analysis showed that Dillo Dirt was heavily contaminated with every toxic element tested, including lead, mercury, cadmium, arsenic, and copper. An organic chemistry analysis conducted by my colleague via LC/MS also revealed shockingly high levels of a chemical fungicide in the compost product.

Mercury used in dental amalgams poses a particularly significant source of concentrated metal exposure and environmental pollution through biosolids, as most dental practices have, for decades, used municipal water for waste disposal, and have been recognized as a significant contamination source.[11,12]

Estimates by the World Health Organization found that about one-third of mercury waste collected in sewage sludge substrate is derived from dumping amalgam fillings and related occupational implements. Moreover, many of the methods that have been implemented to separate dental mercury from wastewater were found to be inadequate.[13]

Once elemental mercury, used in dentistry, reaches waterways from direct dumping into groundwater, lakes, and streams, or indirectly from runoff on land tilled with biosolid fertilizer inputs, microbes readily convert it to methylmercury, which infamously bioaccumulates up the food chain in many fish and seafood, eventually reaching humans and others near the top of the food chain (see section on "Methylmercury in fish" on page 49 for more information).[14]

Biosolids from sewage sludge are now being increasingly produced and sold by most larger cities in the United States, and are increasingly used as a cheap and readily available source of fertilizer for crops intended for human and animal consumption. This poses numerous problems, including introducing a source of concentrated heavy metals as well as pharmaceutical, antibiotic, industrial, and medical waste, plus a multitude of pathogens, bacteria, viruses, and superbugs into the food chain.[15]

Cornell University conducted a 1981 report titled "Organic toxicants and pathogens in sewage sludge and their environmental effects," which found more than 60,000 toxic substances and chemical compounds of concern in sewage remains. In 1988, the U.S. Environmental Protection Agency conducted a National Sewage Sludge Survey, identifying 400 pollutants commonly concentrated in sludge that posed the greatest hazards for large cities; later, in 2001, the EPA followed up with monitoring the levels of carcinogenic dioxins and dioxin-like compounds commonly found in sludge. The possibilities of interaction and further amplification by any or all of these toxic elements and compounds is understudied and unknown, but they present a clear and present risk to public health and safety.[16]

Industrial waste from animal feeding operations, and livestock manure in general, is also a source of metals contamination.[17]

Arsenic has for many decades been added to the diets of broiler chickens, as well as pigs, turkeys, and other animals, to promote growth. The resulting litter of chickens and other livestock, rich in arsenic compounds, is frequently used as a cheap and readily available fertilizer that the industry would otherwise have to dispose of at great cost.[18]

Cow and pig manure from factory farms used as biofertilizers contains concentrated metals and toxic elements. In China, this situation has become especially severe, with copper, arsenic, and zinc bioaccumulating through animals, manure, and soils. Chicken waste is the most significant source of metal pollution from manure in China, as in the United States, due to the deliberate addition of arsenic.[19,20,21]

Reusing excrement from both livestock and human populations is an age-old practice, but never before in history have these by-products included so many hazards in one application.

Cattle sludge from Concentrated Animal Feeding Operations (CAFO) add to the soil other pollutants such as antibiotics, pharmaceutical compounds, hormone mimickers, and hundreds of types of bacteria, which carry their own potential risks (see the "Animal Feed Contaminants" section on page 185 for more information). Many critics of CAFO practices believe this sludge by-product to be a potential culprit in recent E. coli outbreaks in the nation's produce.[22]

Arsenic-treated wood

About 90 percent of the arsenic produced for industrial purposes is ultimately used in wood preservation in the form of chromated copper arsenic (CCA). While CCA has now been phased out, it still permeates much of the existing infrastructure. This arsenic compound has been used in lumber treatment to both prevent rotting and to act as an insecticide that kills termites, ants, and other unwanted pests.

This arsenic-treated wood has been almost universally used in utility poles and for fencing and wooden decks around businesses and residences.[44] The Federal Insecticide, Fungicide, and Rodenticide Act now prohibits the use of CCA-treated wood in residential areas, but decades of nearly ubiquitous use has left an enormous exposure footprint on the environment.

The EPA has warned parents not to allow their children to play anywhere on, under, or even near patios and decks that were built with arsenic-treated wood, as the highly toxic arsenic compound is known to leach into the surrounding dirt or soil, as well as the surrounding landscape and any water sources.

Even worse, CCA-treated wood also contains chromium VI, better known as hexavalent chromium, the element that caused so many people in Hinkley, California, to get sick after industrial contamination (as portrayed in the based-on-a-true-story film *Erin Brockovich* starring Julia Roberts). Hexavalent chromium leaches into the environment at greater levels than arsenic and is considered a genotoxic carcinogen, meaning that it is linked with both cancer and damage to the DNA structure itself.

In addition to these concerns are neighborhood fences, electric poles, picnic tables, and playgrounds. In conjunction with its facilitation of the lumber industry's voluntary "phasing out" of what was once widespread CCA treatment, the EPA has provided oversight for "focusing on children" by assessing "the potential exposure of children to playground equipment built with CCA-treated wood" since 2001, while considering ways to deal with the countless structures in society that were built with components saturated in this harmful compound.[45]

Testing performed in areas around utility poles that had been heavily coated with a CCA treatment has confirmed that significant levels of both arsenite and arsenate had leached into soils and groundwater in the area.[46]

Some mitigation treatments have successfully converted the toxic inorganic arsenic trioxide to a less harmful pentavalent arsenate form; however, this form readily competes with phosphorous inside the body and thus has been known to impair essential bodily functions.

As far back as 1972, the EPA knew of the toxicity issues with arsenic-based pressure treatments and injection treatments including arsenic acid, arsenic pentoxide dehydrate, sodium arsenate, sodium hydroarsenate, and disodium arsenate, but the agency considered the implications of the loss of use to be a "national disaster" and thus downplayed the real environmental implications.

Arsenic in food

More than a century ago, it was arsenic that helped pave the way for modern reforms to clean up the food supply. In a case in Bradford, England, in 1858, which later spurred the Pharmacy Act of 1868, a sweetshop worker misidentified and then accidentally mixed some 12 pounds of arsenic trioxide into delicacies. Even though several of the experienced workers thought the sweets looked odd, they were still sold, prompting one vendor to demand a discount. Subsequently, twenty people were ultimately killed and at least two hundred others were sickened.[47] This haphazard poisoning opened the door to regulations that took on food adulteration as a major issue.

Though subsequent regulation has banned the use of many arsenic-based pesticides and curbed some of the chemical's industrial use, arsenic accumulation in the soil has thoroughly contaminated many areas throughout the world, thus severely affecting the food supply. Even low levels have shown carcinogenic effects through chronic exposure, raising serious concerns about staple food crops.

This problem is compounded by the volume of food exports coming from China and other countries where environmental standards are often lax.

By far the biggest source of total arsenic in foods comes from seafood, including fish, crustaceans, and seaweed. The CDC reports that the "biological half-life of [organic] ingested fish arsenic in humans is estimated to be less than 20 hours, with total urinary clearance in approximately 48 hours."[48] Most researchers have dismissed the role of organic sources of arsenic in causing any harm, but inorganic forms are widely recognized as being harmful to human biology. This difference is why a key question we're examining in our forensic laboratory concerns the ratio of organic versus inorganic arsenic in ocean-derived products. Many seaweeds sold for human consumption, for example, contain very high levels of arsenic. If most of that arsenic is organic arsenic, however, it likely poses no real long-term health risk to those who consume it.

CHINA'S TOXIC POLLUTION CATASTROPHE: IT'S "IMPOSSIBLE TO GROW TRULY ORGANIC FOOD" IN CHINA

China, the world's largest exporter, is also officially the world's largest carbon pollution emitter. While pollution is discussed by government organizations and on the news as an abstract but important environmental issue in America, the poor condition of the environment in China is so severe that toxic smog has from time to time closed down everything from roads and bridges to public schools.

In December 2013, emergency health warnings were prompted when record levels of severe air pollution descended over Shanghai, reducing visibility within the city to a mere 60 feet. Hazardous particulate matter in the air reached levels so high, it was well above even the highest warning level of the United States, prompting officials to cancel public school classes for seven consecutive days and ground hundreds of flights.[1] That same month, a deputy minister of China's Ministry of Land and Resources declared that 3.3 million hectares of Chinese farmland was too polluted to grow crops.[2]

Sadly, daily life-altering air-pollution levels are a common occurrence in China. The media has actually dubbed these events "Airpocalypse."[3] Pollution has even caused the blooding of rivers in China. Residents in northern China's Henan province panicked in December 2011 when they awoke to find the Jian River running blood red. The horrifying sight was later attributed to an illegal workshop that had been dumping red dye into the city's storm water drains. When China's Yangtze River, the world's third longest river, dubbed the "golden waterway," turned a murky red in 2012, the dumping of artificial coloring was thought to be the cause.[4]

In early 2013, Beijing's Environmental Protection Chief Bao Zhenming was offered more than £20,000 to take a 20-minute swim in a local river completely polluted with all manner of toxic industrial waste; he refused.[5] A recent Chinese government study admitted that a whopping 90 percent of the groundwater in China's cities is polluted.[6] Furthermore, after decades of persistent pollution, China has also admitted the existence of "cancer villages," where every other household contains someone dying of cancer, dotting the countryside.[7] In May 2013, government tests confirmed that almost half of all rice for sale in the southern China city of Guangzhou was tainted with toxic heavy metal cadmium, thought to be due to pollution.[8]

China's pollution problem has actually become so dire that the country's government has attempted to order all foreign embassies to stop releasing data regarding pollution and air quality in the nation's large cities in an attempt to censor the severity of this situation from the rest of the world.[9]

The fact that Chinese people have to suffer this environment is horrible, but with the globalization of the world's food supply, China's pollution issue and the resultant detriment to human health that comes with it is steadily spreading across the globe. Most people do not realize that a large portion of the world's food is grown in China's poisonous environment. China is the third largest source of U.S. food imports according to the USDA.[10] For example, according to the consumer watchdog organization Food & Water Watch, an astounding 78 percent of the tilapia and 70 percent of the apple juice Americans ate and drank in 2009 was imported from China.[11]

The USDA released a report that same year regarding safety issues with Chinese food imports. The agency noted that the FDA has repeatedly refused these food imports not just on the consideration of environmental pollution, but also due to lax safety standards, unsafe food additives and labeling, drug residue contamination, and "recurring problems with 'filth.'"[12] However, as Food & Water Watch observed, the FDA inspects less than 2 percent of the food imported to America from China for safety. Of the imports that actually do get inspected, many fail to meet quality standards and are rejected. In 2012 alone, the FDA stopped 260 shipments of imported Chinese food coming into the United States because of heavy contamination with pesticides, bacteria, and/or filth.[13]

This perpetual lack of oversight, safety inspection, and regulation enforcement in China, America, and countries around the world has resulted in notable outbreaks of foodborne illness and death in both humans and animals. Perhaps most well-known in recent history, China's 2008 melamine milk contamination scandal resulted in 300,000 Chinese children suffering urinary problems—54,000 were hospitalized and six infants eventually died.[14] Melamine is an industrial chemical material used to make shatter-proof plates and other durable items. It is extremely toxic to the kidneys. But because its powder resembles powdered milk in both color and texture, powdered milk producers in China decided to simply substitute melamine for powdered milk and sell it to everyone.

Before long, melamine-tainted dairy began turning up around the world, and the European Union extended its Chinese dairy ban to include a total ban on all products for children containing any percentage of milk whatsoever, including chocolate and biscuits. Melamine was also found in other Chinese foods, including eggs from Chinese chickens who had ingested it in their feed. The year before, melamine-tainted vegetable protein in pet and farm animal food from China resulted in thousands of sick animals and dead pets in the United States, and a hog farm in North Carolina had to be quarantined when the chemical was found present in all of its hogs. Even though China banned melamine in 2007, it should not have been in milk or pet food to begin with.

Melamine is just one instance of the chemical tainting of foods coming out of China—a microcosm of a larger, systematic problem with China's agricultural and food industry standards. Other Chinese food scandals run the gamut from utterly disgusting to nightmare inducing: pork laden with a phosphorescent bacteria that caused it to actually glow iridescent blue in the dark, garnering it the nickname "Avatar meat"; large portions of rice crops contaminated with aluminum and cadmium; tons of beans thoroughly drenched in poisonous pesticide; milk produced with leather-hydrolyzed protein; counterfeit jellyfish slices made out of sodium benzoate and calcium chloride; recycled cooking oil made from a medley of discarded animal parts or "edible" oil concocted out of chicken and duck feathers and even fox hair.[15,16]

The list goes on and on. A Chinese professor's undercover investigation in 2010 found that an estimated 10 percent of all meals in China were being cooked with "recycled" cooking oil, the majority of which was being scavenged from drains underneath restaurants. His findings prompted the Chinese Food and Drug Administration to respond to the aptly named "sewer oil" scandal. Despite all of this, the Chinese food imports to the United States only continue to grow. The USDA even ever-so-quietly lifted an import ban on Chinese poultry in August 2013.

The issue in China isn't just about a worsening breakdown of confidence in the global food supply, but also a pervasive problem with far- and wide-reaching consequences on the health of billions of people. China's regulations and safety oversight are lax. Further, more and more foods labeled "organic" are being exported from China these days, even though there are absolutely no real guarantees that the Chinese organic guidelines are as stringent as they are in other countries; if China's abysmally lax agricultural regulations are any indication, there is little reason to put any faith into anything coming from China with "organic" printed on it. The USDA's own reports have admitted that food oversight in China is nothing like that of the United States.[17] A comparative assessment of organic foods produced in both the United States and China published in the summer 2011 issue of the *Stanford Journal of International Law* concluded the "USDA Organic" label is ultimately misleading because, "the current regulatory framework is not only inadequate to the task of regulating domestic organics, but also incapable of ensuring the integrity of imported organics."[18] While China traditionally did use organic farming techniques, decades of heavy pesticide use followed the country's socialization in the 1960s, prompting Senior USDA Economist Fred Gale to declare it is now "almost impossible to grow truly organic food in China."[19]

There's a reason the phrase "Product of China" is printed in such a tiny font on the food products that are labeled with it.

Arsenic in apple juice

Controversies surrounding the arsenic content in juices and rice have made their way into the mainstream media over the last few years. The prominent TV show host Dr. Mehmet Oz created a significant stir after releasing test results that showed what his team considered dangerous levels of arsenic in apple juices[49]—many were top brand name products typically found in grocery stores across the United States. Many established voices tried to discredit the claims made by Dr. Oz by preying on the public-at-large's ignorance, focusing on the lack of differentiation between arsenic's organic and inorganic speciation.

However, watchdog *Consumer Reports* followed up with confirmation that many juices—including those of the ever-popular apple and grape varieties—were indeed found to contain arsenic levels higher than the federal standard for drinking water, and the majority of this arsenic was *inorganic* and linked to potentially deadly health effects, including cancer.[50] Approximately 10 percent of the eighty-eight samples, which included a variety of name brands, showed arsenic levels above the 10 ppb threshold.

Consumer Reports identified Denise Wilson, PhD, a professor at the University of Washington, as having conducted her own testing of apple juices in which she discovered high levels of arsenic, even in brands labeled as organic. Wilson stated, "We are finding problems with some Washington state apples, not because of irresponsible farming practices now, but because lead arsenate pesticides that were used here decades ago *are still in the soil.* Heavy metals like lead and arsenic just don't go away."

Concern was further elevated by the fact that more than 60 percent of juice imports come from China, where the use of arsenic-based pesticides may still be ongoing and regulations for foods are even shadier than those in the United States.

After significant public pressure, the FDA was forced to consider new rules and finally conducted its own tests. After the results were released in July 2013, essentially confirming the arsenic tainting that it had previously attempted to sweep under the rug, the agency established a new proposed limit of 10 ppb for inorganic arsenic levels in apple juice, the same as EPA standards for drinking water. While maintaining that no specific danger was posed by the arsenic levels it found in juice, the FDA did acknowledge that "the arsenic in these samples was predominantly the inorganic form"—a form that is a Class A known human carcinogen.[51]

The agency claims there is no "short-term risk" from arsenic levels in food. However, the data backing this up primarily consist of measurements of total arsenic (as opposed to inorganic arsenic) and set aside altogether any consideration of risk potential from long-term, chronic, bioaccumulated exposure. Prior to this, the FDA had few limits on how much arsenic was tolerated in specific foods and no general limit, even though it set up a Total Diet Study program back in 1991, supposedly to monitor food safety.

The European Food Safety Authority (EFSA) also has no hard limits on arsenic in food, but concluded that the "possibility of a risk to some consumers cannot be excluded," revising and lowering its provisional tolerable weekly intake (PTWI) levels in 2009 after acknowledging that previous data had not properly considered the levels of inorganic arsenic or its propensity to cause cancer in the lungs, bladder, and skin.[52]

The Joint FAO/WHO Expert Committee on Food Additives (JECFA), which set the Codex Alimentarius International Food Standards, has since laid down limits on inorganic arsenic, setting the provisional tolerable daily intake (PTDI) at 0.002 mg/kg bodyweight, which is approximated for the average-sized person as 0.12 mg/day (for a 60kg adult). There is no U.S. federal limit for inorganic arsenic levels in food.[53]

Arsenic in rice and vegetables

Rice is known for its higher arsenic absorption levels. The food staple found itself surrounded by controversy when laboratory tests in 2012 revealed high levels of arsenic in numerous commercial rice products across nearly every variety.

After playing a significant role in exposing arsenic levels in popular juice brands, Consumer Reports turned its spotlight on rice in November that same year.[54] Testing more than 200 samples, the organization determined that the daily limit of 5 ppb arsenic (the original limit proposed by the EPA for drinking water that was not adopted) was frequently exceeded by double and triple those amounts—including in brands specifically marketed toward gluten-free and health-conscious niches. Brown rice was also found to have more arsenic overall than white rice in every sample Consumer Reports tested.

Some attribute the elevated arsenic levels in rice to paddies like those in the southern United States, which are generally found near areas where arsenic pesticides for cotton or other crops were traditionally used on a wide scale and subsequently absorbed by rice plants through tainted soil and water.

A bigger offender than even rice and apple juice, which received significant negative press, is the consumption of arsenic in vegetables, which also absorb trace amounts of arsenic from contaminated soils and water. Studies estimate that about a quarter, or 24 percent, of the average arsenic-laced foods ingested are vegetables; this is more than the approximate 18 percent of dietary arsenic derived from fruits and their juices, and the 17 percent of dietary arsenic contributed by rice, according to Consumer Reports' findings.

The big secret: arsenic in chicken

While the alarm has been sounded on foods like fruit juices, rice, and even vegetables grown in soils contaminated by pesticides tainted with dangerous arsenic compounds, little has been said about the effects of arsenic in poultry and swine.[55,56]

Drugs used in animal feed for chickens to control internal parasites and promote growth during factory farm confinement have long contained high levels of inorganic arsenic, and humans have been ingesting significant quantities of these compounds for decades. Alarming concentrations of these arsenic compounds in the livers and muscles of young chickens have been discovered at levels far exceeding anything found in rice, grains, fruits, or vegetables.

A 2004 study conducted by the USDA used monitoring data for the Food Safety and Inspection Service National Residue Program to determine average consumption levels for people who ate significant quantities of poultry between 1989 and 2000.[57] Researchers discovered mean concentration levels of .39 ppm, or 390 ppb arsenic, levels three to four times higher than in other meats. The report concluded, "At mean levels of chicken consumption (60 g/person/day), people may ingest 1.38–5.24 µg[micrograms]/day of inorganic arsenic *from chicken alone*" (emphasis added). When vegetables, fruits, and rice consumption are factored into the mix, people are likely eating much more arsenic in a day than previously thought possible.

Revelations about these high levels of toxic, inorganic arsenic led to pressure on the poultry industry and resulted in the voluntary withdrawal of Pfizer's arsenic-based animal drug roxarsone[58] from the market in 2013.[59] The FDA states that roxarsone is used for "growth promotion, feed efficiency, and improved pigmentation."

Unfortunately, other agricultural arsenic drugs are still being used every day all over the world. One example, nitarsone, a chemically similar arsenical drug to roxarsone, is still being used in mass quantities today on turkeys destined for human consumption throughout the United States, where turkey consumption is only going up.[60]

A study published in May 2013 and conducted by the Johns Hopkins Center for a Livable Future examined samples of conventional, antibiotic-free and organic chickens purchased when roxarsone was still widely available on the market. These researchers discovered that levels of inorganic arsenic—again, a known carcinogen—in conventional chicken were *four times higher* than what they found in organic chicken.[61] The authors of the study found the industry boasting about the use of roxarsone in 88 percent of some 9 billion birds raised in the United States, and recommended the FDA ban the use of all arsenicals based on these results.

Further, fertilizers created with poultry waste tainted by inorganic arsenic could be leaching even more toxins back into the soil, which in turn accumulate in crops and humans.

Burning coal and airborne arsenate trioxide

Another source of widespread environmental arsenic contamination comes from burning coal. Scientists estimate that 80,000 tons of arsenic are released into the air each year through the burning of fossil fuels. In the southwest Guizhou region of China, for example, at least 3,000 arsenic-contaminated patients have been diagnosed with skin lesions and elevated urinary levels due to exposure to inorganic arsenic emitted from coal-burning power plants. Among this group, the Center for Disease Control and Prevention in China has noted that high cancer and mortality rates in the area are far more prevalent than even those found in areas with heavily contaminated drinking water.[62]

In the United States, even though it was known that coal plants were spewing more toxic pollutants into the air than any other industrial source— some 386,000 tons of 84 unique hazardous air pollutants including arsenic, lead, and mercury are released from over 400 U.S. plants each year alone— the EPA did not even formally introduce standards to limit this type of toxic pollution from power plants until December 2011.[63]

Arsenic interference in the body

Central to the issue of heavy metals in the body is their propensity to compete with essential nutrients. Phosphate, for example, is required by the body to build healthy bones and teeth; phosphate also makes muscles contract and helps nerves function properly. Both arsenic and phosphorus are in the same group on the periodic table, and both have five electrons on their outer shells; thus, they biochemically compete inside the body for binding and absorption.[64] Because of this, arsenic can block the production of necessary enzymes and proteins by binding in places where phosphorus would normally go.

As with other toxic heavy metals such as mercury, arsenic has also been shown to inhibit thiol compounds including glutathione, which is one of the body's key detoxification agents and mandatory for a properly functioning immune system and warding off disease. Arsenic compounds also alter the body's ability to use pyruvate properly.[65] This deficiency allows lactic acid to build up to toxic levels, leading to neurological problems including seizures, intellectual deficits, and problems with even basic motor skills like walking. Most children suffering from pyruvate dehydrogenase deficiency don't live very long past childhood, and those who do suffer developmental disabilities.[66]

Treatments for arsenic toxicity

Arsenic is quickly metabolized and distributed throughout the body via the lungs, liver, and kidneys, where it settles into keratin-rich tissues like the hair, nails, and skin. While the half-life of inorganic arsenic in the body is relatively short—the majority of it is excreted within less than a day—chronic, repeated exposure to arsenic is where the real danger lies. Currently, there are no 100 percent cure-alls for mitigating arsenic's carcinogenic effects.

Well-known treatments for arsenic poisoning include chelating the metalloid with several agents including British anti-Lewisite (BAL), sodium 2,3-dimercaptopropane 1-sulfonate (DMPS), and meso 2,3-dimercaptosuccinic acid (DMSA), among others. These chelation agents bind with arsenic and allow it to be flushed out of the body via excretion.[67]

In 1938, it was discovered that arsenic actually protected against selenium poisoning. Shortly after, arsenic began to be used as a tonic by industrial hygienists to cure workers of selenium poisoning.[68] More recent research with animals has shown selenium is effective at countering arsenic toxicity, and studies are eyeing selenium supplementation as a low-cost way to counter chronic arsenic poisoning.[69]

Several studies have linked the use of garlic to decreased effects of arsenic toxicity on cells.[70,71,72]

Natural arsenic binders

My own laboratory research at the Natural News Forensic Food Labs (labs
.naturalnews.com) has identified many substances that have a natural affinity
for binding to arsenic. Throughout 2013, I developed a testing methodology called "Metals Capturing Capacity" (MCC), that is able to determine
how well any given substance naturally binds with and captures free arsenic.
Metals Capturing Capacity is explained in more detail in videos found at labs
.naturalnews.com/videos.html.

After testing more than 1,000 substances for their natural arsenic binding
properties, I found that the substances with the highest arsenic MCC were:

- Powdered fruit seeds
- Sodium alginate
- Dehydrated powders of certain rare seaweeds

After completing the research, I formulated a series of dietary supplements that maximize the binding and capturing of heavy metals, including
arsenic. This resulted in the release of a fruit-based formula with an arsenic
reduction of 14.8 percent, and then a much stronger "Metals Defense" formula with an arsenic reduction of 92.9 percent and an MCC of 6.0, meaning
each gram of the formula binds with 6.0 micrograms of free arsenic. (See
more scientific results at www.HeavyMetalsDefense.com.)

Importantly, this formula only binds with arsenic during digestion,
before it is absorbed into the bloodstream. Once arsenic enters the blood
and latches on to cells and tissues, it is extremely difficult to remove from the
body without using aggressive interventions such as intravenous chelation
agents. Hair, nail, and skin cells (where arsenic eventually settles) fall away on
their own, of course, demonstrating one of the body's elimination pathways.
Ultimately, it is important to avoid ongoing exposure to arsenic (and other
toxic elements) while giving the body time to rid itself of the offending elements through routine processes of growth and regeneration.

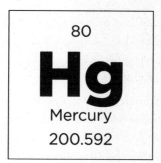

80

Hg

Mercury

200.592

MERCURY (Hg)

ATOMIC NUMBER: 80

GROUP 12: CADMIUM AND ZINC

Shiny, metallic, and intoxicatingly strange in its elemental liquid form, mercury has long been known as a poison, as intriguing as it is deadly. One of the most toxic elements on the planet—especially in organic form—mercury has long been known to be poisonous to humans, animals, and the environment.

With more than thirteen times the density of water, a sea of mercury would be dense enough to theoretically walk on, or break apart most things plunged into it. And that's only the beginning of its unique properties. In the ancient, occult-driven pursuits of alchemy, it was thought to be an element of central importance to achieving transmutation to gold. It was even a key ingredient in a popular elixir-of-life formula, believed to bestow eternal life despite its toxic qualities.

Mercury is a particularly insidious heavy metal that appears in three forms: organic, inorganic, and elemental, the latter of which is familiar to most as the curious liquid metal that responds to air pressure, which has been widely used in thermometers. Like other harmful heavy metals, it is frequently extracted as a by-product alongside other ores, but it has also been mined deliberately for the useful pigment properties exhibited by the reddish cinnabar, a crystalline mercury-based ore formed by volcanic activity or alkaline conditions, such as those seen in hot springs. Crushed cinnabar is burned, separating sulfur from the alluring liquid mercury yielded for industrial production.

Legends say China's first emperor, Qin Shi Huang, died after imbibing a powdered jade and mercury mixture his alchemists told him would bring eternal life. Although the location of his tomb has been discovered, archeologists are still unsure of how to go about excavating the site due to the underground moat of liquid mercury surrounding it and the cloud of toxic vapors that opening the area would surely unleash.[73]

Over the last several hundred years, the study of chemical reactions when mercury is combined with other elements has led to the development of mercurial compounds, believed to be useful in treating numerous diseases. Arabs created quicksilver ointments for the treatment of skin conditions based on knowledge of Greek medicine and other reputed remedies. After the Renaissance era, an understanding of the principles of metal oxidation lent to its use in apothecary drugs and attempts to create antiseptic treatments.[74]

While mercury does have antimicrobial properties, which led some cultures to recognize how it may be beneficial for killing bacteria, fungi, and mold, it's also extremely toxic to nearly all forms of life, making it a less-than-desirable medicine.

Mercury exposure leaves workers "mad as a hatter"

The rise of the industrial age has revealed the dangers and downsides of increasing societal exposure to mercury and its various chemical compounds. A trend of workplace hazards began to emerge during the nineteenth century, bringing into view new diseases that could befall laborers subjected to mercury vapors and direct skin contact.

The most infamous are the so-called Mad Hatters, seen prior to but made famous in Lewis Carroll's 1865 novel *Alice in Wonderland*. Unnerved, edgy, and tormented by a complex of erethism symptoms, these tradesmen actually suffered from mercury poisoning. Industrial hat workers engaged in curing pelts to make felt hats, as well as other related furrier activities, were known to frequently suffer symptoms including mental instability, irritability, and tremors from exposure to mercury.[75] The common thread behind the phrase "mad as a hatter" was the workplace use of mercury nitrate, which left many sickened, debilitated, or simply off kilter. Miners, gilders, and mirror makers in the Renaissance era and Middle Ages were known to suffer similar ailments as well, though it would not be attributed to mercury for centuries to come.[76]

Mercury commonly found in consumer goods

There's a common, but potentially deadly, misconception that mercury has been banned from everyday products. In reality, the relatively rare earth mineral is widely used in the production of many consumer goods. In addition to its use in thermometer bulbs, mercury is also used in batteries, pesticides, and now in large quantities as an element of energy-saving CFL fluorescent light bulbs.

We face exposure through broken thermometers or light bulbs, both of which can emit vaporous mercury that's quickly inhaled. That's why instructions for cleaning up a mercury-containing CFL light bulb include an extensive list of steps to ensure basic safety, despite CFL's touted reputation as "green" technology. A health study found that if a single CFL bulb breaks, mercury gas concentrations released can reach 800 μg/m^3, more than eight times the Occupational Safety and Health Administration (OSHA) limit of 100 μg/m^3 for adults in an eight-hour period.[77] A research team also found that because an electrical current is charging the mercury vapor contained in all CFLs, and the curly shape of the bulb can make it more prone to tiny cracks in the phosphor coating that would otherwise protect people from those rays, the bulbs were giving off cell-damaging UV radiation.[78] They recommended keeping one's distance from these bulbs and encasing them in an extra glass structure just to be safe.

Mercury is even used in vaccines given to children. On October 9, 2015, California Health and Human Services Agency Secretary Diana S. Dooley issued a directive that suspended the ban on mercury in vaccines given to children, allowing those children to be injected with a mercury-containing vaccine preservative known as thimerosal. "I am granting a temporary exemption from California Health and Safety Code Section 124172 for seasonal influenza vaccine with trace levels of thimerosal to be administered to children younger than three years from October 9, 2015, through December 31, 2015, because the current supply of thimerosal-free vaccine for young children is inadequate," wrote Dooley. In doing so, she demonstrated that even when governments recognize the threat of mercury toxicity to children, they will nevertheless allow mercury to be injected into children whenever supply conditions demand it.[79,80]

Flu shots, by the way, typically contain over 50,000 ppb of mercury—about 25,000 times the concentration limit of mercury allowed by the EPA in drinking water.

Coal-burning power plants

The argument made by many CFL proponents for CFL's viability as an environmentally friendly technology despite its dangerous mercury content is that the energy it saves results in a net reduction of mercury emissions from coal-fired power plants. Such power plants eject thousands of pounds of mercury into the air every year, where it eventually settles to the ground, contaminating soil, water, and products for human consumption. According to the National Resources Defense Council, 33 tons of mercury pollution are emitted from power plants each year just in the United States alone.[81]

Limitations on these mercury emissions have only recently been put into place. In an attempt to curb such emissions, the EPA announced the Mercury and Air Toxics Standards (MATS) for power plants in December 2011, limiting the amounts of mercury and other hazards such as arsenic that power plants are legally allowed to emit.[82] However, the rule still allows for 1.2 pounds of mercury per trillion BTUs of energy produced, and because even tiny levels of mercury accumulate in the environment, the cycle of pollution will undoubtedly continue.

Wildfires and mercury pollution

Wildfires are another significant cause of mercury pollution. A 2007 *Global Biogeochemical Cycles* journal article noted that mercury in the atmosphere accumulates on foliage, and when it dies or decomposes, that mercury then enters the soil where it is taken up by roots and incorporated into tree leaves and structures. When a forest fire sweeps the area, mercury is emitted and carried by the rising heat and smoke into the atmosphere. The authors concluded that forest fires comprise a fourth of all mercury emissions in the United States.[83]

Mercury emitted via waste disposal

Mercury is also burnt or disposed of throughout industry, creating a pattern of contamination that has yet to be reined in. Everything from hospitals to dentists' offices, veterinary clinics, laboratories, septic haulers, residential neighborhood waste, batteries, printing, painting, pottery, scrap metal, and industrial laundry contribute to the mercury waste burden. Unburned quantities of waste materials are often dumped back into croplands and waterways via sludge-based fertilizers. Either way, these mercurial compounds reemerge in the environment and continue to pose health risks.[84]

Mercury in pesticide use and residual effects in croplands

Although the majority of agricultural inorganic mercury uses have been banned or discontinued in most countries throughout the world, mercuric chloride, an inorganic mercury-chlorine compound, is still allowed for use in some pesticides in the United States and other countries—while the residue from decades past still impacts background metal exposure.[85] Populations that eat grains sprayed with those pesticides (or meat from animals that ingest those grains) also accumulate toxic mercury.

Production, use, and emission figures for this compound are unknown, but it's estimated to be in the hundreds of metric tons in the United States alone.[86] Avoiding or restricting imports cultivated with the use of this harmful pesticide from places like China, where regulations are lax or difficult to enforce, may prove difficult or impractical. Thus, banned formulations still appear in foods consumed by millions of people.

For example, my own research into certified organic vegan protein products made predominantly from rice protein grown and processed in China found mercury concentrations as high as .036 ppm.[87] Given that some consumers of such products eat over 100 grams of these proteins each day, their mercury intake from this one product can exceed 3.6 micrograms. It's also important to note that nearly all of these rice protein products are certified USDA organic, which most consumers assume means "free of toxic substances." Yet, as I discussed previously, USDA organic standards have no limits whatsoever on mercury or any other heavy metals.

Mercuric chloride has also made headlines for its negative effects when found in industrial waste. Following complaints of a strong chemical smell making villagers ill, news outlet RIA Novosti reported that 200 tons of a banned mercury pesticide was discovered dumped in a Russian village in 2011.[88] Just as with all heavy metals, once mercury is in the environment, it is exceedingly difficult to remove.

The EPA has listed inorganic mercury as a Class C "possible human carcinogen," as the agency's own Office of Research and Development acknowledges it is a developmental toxicant that can cause gastrointestinal erosion and kidney damage in addition to DNA damage and cancer in lab animals.[89]

Mercury in dental fillings

Elemental mercury is still used in amalgam dental fillings, which contain, on average, 50 percent mercury. The FDA issued a final rule in 2009 that reclassified mercury from Class I (least risk) to Class II (more risk) and officially classified encapsulated dental amalgam—a mixture of silver, tin, copper, elemental mercury, and a powdered alloy—as a Class II restorative medical device.[90]

Although the American Dental Association has released a statement claiming that dental amalgam "is considered a safe, affordable, and durable material,"[91] studies specific to dentists and mercury exposure via amalgam have produced worrisome results. In a neurobehavioral study of dentists exposed to elemental mercury at work, researchers found that the dentists did significantly worse on mental acuity and motor skill tests than control subjects; in addition, as years of exposure to elemental mercury in amalgam increased, a dentist's test performance significantly decreased.[92] Female dentists and dental assistants exposed to mercury in another study were also found to have significantly more reproductive failures, including more painful and irregular menstrual disorders as well as more miscarriages and increased congenital malformations in infants.[93]

Dental amalgam has also been shown to leach mercury into the mouth, which can emerge in the form of vapors or be swallowed if pieces of amalgam break off. Studies vary widely on the quantities of mercury people are exposed to in this fashion. Researchers with the Department of Materials Science at University of Virginia's School of Engineering and Applied Science found that both stannous and sodium fluorides, active ingredients in commercial toothpastes and mouthwashes, played a role in increased corrosion rates of mercury fillings.[94] Multiple adverse health effects have been correlated to the presence of dental amalgams, including one study that found that mothers who had six or more dental amalgams during pregnancy and later had a child diagnosed with autism were more than three times more likely for that diagnosis to be severe autism than the autistic children of mothers with five or fewer mercury fillings.[95]

Studies have also revealed that microwave radiation from cell phones and magnetic resonance imaging (MRI) significantly accelerated the leaching of mercury from dental amalgams, giving cause for concern about people who have these fillings coming into everyday contact with electromagnetic fields (EMF), including the ubiquitous Internet Wi-Fi hotspots found in most urban and suburban areas.[96]

Interestingly, even the cremation of human bodies releases enormous quantities of mercury vapor into the atmosphere due to the burning of mercury amalgam dental fillings found in most people. A 1994 study conducted by Japanese researchers found that a single crematorium released approximately 9.4 kg of mercury into the atmosphere each year.[97]

DENTISTRY—HOW DENTISTRY POLLUTES OUR BODIES WITH MERCURY

One source of potential toxin exposure that may not immediately come to mind is dental amalgam (i.e. "fillings").

Over 90 percent of American adults have received one or more dental fillings as a remedy for their cavities. The vast majority of these fillings are "silver" amalgams composed of 50 percent elemental mercury (Hg) and other metals that are less toxic than mercury.

According to the CDC and the National Institute of Dental and Craniofacial Research, Americans between the ages of twenty and sixty-four have an average of 3.28 cavities each.[1,2] Though some 23 percent of these cases go untreated, a staggering number of cavities (in the hundreds of millions) have been treated with fillings containing mercury, a well-known heavy metal toxin and brain-damaging element.

Research has shown that these amalgams pose an ongoing risk, as they continuously release mercury vapor, which is in turn inhaled into the body, where it wreaks havoc on cell integrity. About 80 percent of the elemental mercury vapor is inhaled through the lungs and enters into the bloodstream.[3] From there, significant amounts of mercury can cross the blood–brain barrier where it is transported to the brain via blood. Additionally, small pieces of mercury can also be swallowed if the amalgam breaks or chips.

Studies on the impact of mercury-containing fillings have concluded that amalgams contribute the vast majority of mercury that accumulates in the human body,[4] with the World Health Organization naming amalgams as the most significant source of inorganic mercury in the general population, contributing to half of overall exposure. The WHO further reported that frequent activities among the entire population such as chewing, including both eating and chewing gum, and brushing teeth can increase mercury vapor emissions by more than fivefold.[5] Worse, the active ingredients in commercial toothpastes and mouthwashes—stannous and sodium fluorides—have been found in studies to increase amalgam corrosion rates.[6] Higher rates of mercury uptake among the general population have additionally been documented for frequent gum chewers and those who grind their teeth.[7]

Research has also shown that when people with dental amalgams are exposed to microwave radiation from cell phones and magnetic resonance imaging, mercury release from dental amalgam accelerates.[8] Considering how many urban and suburban areas are bathed in perpetual electromagnetic fields due to a myriad of Wi-Fi hotspots these days, this finding is particularly worrisome and demands further study.

According to the EPA, the adverse health effects of breathing elemental mercury vapor include mood swings, irritability, nervousness, tremors, insomnia, muscle atrophy and twitching, headache, nerve response and sensation changes, cognitive dysfunction, and—at very high levels—kidney and respiratory failure, and even death.[9]

Autopsies used in an Italian research study concluded that subjects with twelve or more fillings had significantly higher levels of mercury in the brain and other tissues than did subjects who had three or fewer fillings.[10] Rat studies confirmed that exposure to amalgam vapors increased concentrated brain mercury by as much as eight times, while accumulation in kidney tissue after exposure was also high.[11]

Several studies have linked the presence of mercury fillings with mental disorders. One found that multiple sclerosis patients with fillings had far higher levels of depression and sudden feelings of anger and irritability than those who had their amalgam fillings removed.[12] A related study found that mental issues were improved or eliminated within about ten months of removing mercury amalgams. A study on women found those with amalgams showed tendencies toward uncontrolled anger, a lack of happiness and satisfaction, and an inability to make decisions as compared with those without amalgams.[13] Even low doses of exposure to dental amalgam mercury have been shown to contribute to adverse behavioral effects in relation to toxicity burden in the body.[14]

After years of complaints of symptoms and worry over the risk of toxicity by the public and researchers, the FDA reclassified mercury from a Class I (least risk) to Class II (more risk) in 2009. Further, the agency officially classified dental amalgams (composed of elemental mercury, silver, tin, copper, and a powdered alloy) as a Class II restorative medical device.[15]

Exposure to mercury is also a serious concern for dentists and dental assistants. Studies have long found suicide rates among workers in the dental industry to be significantly higher than in other occupations.[16] Although the full explanation for this is not clear, chronic exposure to mercury vapors and elemental mercury may play a significant role.[17] Already the available data show that there is reason to associate amalgams with mental health issues and depression. Considerable research has gone into investigating occupational exposure for dental professionals, who consistently have higher rates of mercury in their bodies alongside notable adverse health effects.

A 2001 study published in the *British Dental Journal* found elevated blood mercury levels not only in the dental students working with restorative amalgams, but also in surrounding students and staff who worked in the same environment but had no direct contact with the materials.[18]

Researchers have further concluded that, when compared to control subjects, dentists perform significantly worse on mental acuity and motor skills tests; worse, the longer a dentist had been exposed to elemental mercury, the poorer his or her performance was on the tests.[19] Researchers have also found memory disturbances and kidney disorders among dentists.[20]

Studies have also shown that female dentists and dental assistants are prone to significantly more irregular periods, miscarriages, and giving birth to infants with congenital deformities than women who are not consistently exposed to mercury at work.[21] Moreover, it was found that occupational exposure to mercury vapor lowered fertility rates among female dental assistants.[22] A study examining the relationship between amalgam fillings and mother's breast milk taken shortly after birth found a correlation between the concentration of mercury in milk and the number of fillings in the mother.[23]

Though the damaging effects mercury can exhibit on the reproductive system are known, many dental journals insist that the risk is low if proper mercury hygiene is used and mercury accumulation remains below the established "threshold limit value"—though no true "safe" level for mercury has ever been established.[24,25]

Even waste disposal has been an issue for the mercury used in amalgams by dental practitioners, and amalgams have now been identified as a significant source of environmental pollution.[26] With a focus on the tons of mercury-amalgam waste dumped into sewers or on land in the United Kingdom, the WHO reported that as much as 53 percent of total environmental mercury emissions come from dental, laboratory, and medical waste.[27] An estimated one-third of mercury waste collected in sewage sludge comes from dental discharge.

Methods for separating amalgam and reducing mercury levels in waste have been deployed, while many countries have begun regulating dental disposal practices.[28,29] However, discharged mercury remains a widespread and significant environmental problem, and many dental amalgam separation technologies have been found inadequate at reducing pollution levels.[30]

Once mercury enters the environment, microbes readily convert the elemental mercury into methylmercury, which significantly bioaccumulates and becomes a major issue in the food chain, as with fish (see "Methylmercury in fish" on page 49 for more information).

Mercury from dental amalgams has even been found to be a significant environmental pollutant through its release into air after deceased people who had fillings in their teeth are cremated, as mercury is being released at levels similar to other industrial emissions.[31]

Even with the abundance of studies noting the adverse effects and the FDA's reclassification of amalgam as riskier to health, the American Dental Association (ADA) continues to assert that mercury-containing dental amalgam is "a safe, affordable, and durable material."[32]

Banned in the EU; concerns about exposure in U.S. products

The European Union has banned nearly 1,400 chemicals from being used in the production of cosmetics based on health risk assessment that they may be carcinogenic, mutagenic, or reproductive toxicants.[98] In an attempt to rein in mercury by-products, the European Union enacted a ban, beginning in 2011, on the export of mercuric chloride, cinnabar ore, and many derivatives.[99] Mercury, lead, and arsenic (among others) have all been banned as cosmetic additives in Canada.

By contrast, the U.S. FDA has only banned ten ingredients, and even though mercury is on that short list, up to 65 parts per million of mercury is still allowed in cosmetics applied to the eye area.[100] As of 2007, Minnesota was the first U.S. state to officially outlaw thimerosal, a mercury derivative, in some cosmetics, including mascara, eye liners, and skin-lightening cream—a far stricter rule than the federal standards currently in place. One of the concerns Minnesota officials considered was that fumes from these cosmetics could build up within the containers and users might inhale them upon opening the products. Minnesota Senator John Marty, who sponsored the ban, noted, "Mercury does cause neurological damage to people even in tiny quantities."[101]

Bioaccumulation

The senator is correct: Mercury damages human health. Even low-level mercury exposure can be toxic, and chronic exposure bioaccumulates in the body.

Mercury poisoning can induce reproductive, developmental, systemic, immu-nological, genotoxic, and carcinogenic adverse effects, potentially impacting every single body system.[102] Although science has long established the grave effects that mercury poisoning can have on the brain and nervous system—especially for fetuses and developing children—exposure to even smaller amounts have been linked to cardiovascular disease and neurotoxicity.[103]

Methylmercury in fish

Methylmercury, as found in tuna and other large fish, is the primary source of dietary mercury consumed today. Once ingested, methylmercury is absorbed through the gastrointestinal tract where it is eventually converted to inor-ganic mercury. Five percent of bodily mercury load is found in the blood and another 10 percent is found in the brain. The metabolism rate for mercury is slow, so less than 1 percent of the total mercury in the body is actually excreted in a given day.[104]

The New York Times conducted an investigation on mercury involving twenty Manhattan sushi restaurants and stores in 2007 and found that eat-ing a mere six pieces of sushi a week would actually surpass EPA limits on mercury. Five of the twenty restaurants had mercury levels high enough to warrant FDA action.[105]

To avoid mercury toxicity, considering your fish intake is important.

Mercury and Seafood: Eating Guide

Highest Mercury (Avoid eating)
Mackerel (King), Marlin, Orange Roughy, Shark, Tilefish

High Mercury (Eat only three servings or less per month)
Bluefish, Grouper, Sea Bass (Chilean), Tuna (Yellowfin, Canned Albacore)

Moderate Mercury (Eat six servings or less per month)
Bass (Striped, Black), Carp, Cod (Alaskan), Croaker (White Pacific), Halibut (Atlantic, Pacific), Lobster, Mahi Mahi, Perch (Freshwater), Sablefish, Sea Trout, Snapper, Tuna (Canned Chunk Light, Skipjack)

Least Mercury

Anchovies, Butterfish, Clam, Crab (Domestic), Croaker (Atlantic),
Flounder, Hake, Herring, Mullet, Oyster, Plaice, Pollock, Salmon, Sardine,
Scallop, Shrimp, Sole (Pacific), Tilapia, Trout (Freshwater), Whiting

Source: The Natural Resources Defense Council (based on FDA and EPA data). NRDC.
org. www.nrdc.org/health/effects/mercury/walletcard.PDF.

In my lab testing, I have primarily found mercury in fish—and shellfish-
derived food products, including those harvested from the North Atlantic
region. Beware of high mercury content in pet food treats derived from fish,
where I've spotted some products containing more than 1,000 ppb mercury
(1 ppm).

THE SYSTEMIC, APOCALYPTIC POLLUTION OF THE WORLD'S OCEANS

Approximately ten years ago, Newcastle yachtsman Ivan
Macfadyen decided to sail from Melbourne, Australia,
to Osaka, Japan, then on to San Francisco, California. In
a 2013 interview with the *Newcastle Herald*, Macfadyen
recalled how the ocean was teeming with life: sounds of
sea birds and an abundance of fish to catch with a simple
bait and line.[1]

Expecting a similar journey, Macfadyen recently
decided to redo the trip only to find a very different ocean
waiting for him. For 3,000 nautical miles, Macfadyen said
he saw very few signs of life. He said there were hardly
any fish to catch. A creepy quiet filled the air where the
noise of sea birds should have been. In fact, the only sound
consistently heard amid the lapping ocean waves was that
of garbage hitting the hull of his boat. Macfadyen was sail-
ing through the aftermath of the 9.0 earthquake and subsequent tsunami that hit the Daiichi Nuclear Power Plant
at Fukushima, Japan, in 2011. "The wave came in over the

land, picked up an unbelievable load of stuff and carried it out to sea. And it's still out there, everywhere you look," Macfadyen said. He also noted that something in the water near Japan reacted to his boat's bright yellow paint job, causing the craft to lose its sheen in what he described as a "strange and unprecedented way."

When he finished his voyage, Macfadyen declared it official: "The ocean is broken."

To be fair, even though an estimated 25 tons of debris were said to have been swept out into the Pacific Ocean after the tsunami hit,[2] the sea was already in deep trouble way before the Fukushima earthquake.

Ever heard of the Great Pacific Garbage Patch? Right now as you read this, an island twice the size of America comprised entirely of rubbish—everything from water bottles to used syringes to broken boats and storm-captured houses—all kept together by swirling currents is floating out in the Pacific Ocean.[3] In fact, five garbage patches are perpetually accruing trash out in the subtropical oceans between the continents. An Australian research team investigating the ocean garbage dumps concluded, "humans have put so much plastic into our planet's oceans that even if everyone in the world stopped putting garbage in the ocean today, giant garbage patches would continue to grow for hundreds of years."[4] And that was before the Fukushima earthquake and tsunami hit, with its 25 tons of debris.

Until it was banned by the U.S. Congress in 1988, America used the ocean as a giant toilet—literally. That is, thousands upon thousands of tons of processed municipal sewage were regularly dumped into the ocean for decades. The last 400 tons were dumped by New York City in 1992.[5] Too many oil spills have occurred over the years . . . so many that the well-known 1989 *Exxon-Valdez* spill and the BP oil spill in 2010 do not even make the "top ten worst" list (for the record, according to *Popular Mechanics*, the worst oil spill in history happened during the first Gulf War, when somewhere between 240 and 336 million gallons of oil were purposefully dumped into the Persian Gulf by Iraqi forces attempting to slow American troops as they fled Kuwait).[6]

Before all that, the ocean was used as a testing ground for America's atomic bomb development at the Bikini Atoll islands, where twenty-three surface and subsurface nuclear devices were detonated between 1946 and 1958.[7] In addition, decades of toxic runoff from industrial pollution—everything from agriculture to mining—has allowed all manner of noxious chemicals and heavy metals to seep into the ocean. In the wake of the 2011 Fukushima disaster, the Tokyo Electric Power Company (TEPCO) that owns the crippled Daiichi Nuclear Power Plant has admitted that some 400 tons of irradiated groundwater is continually being dumped into the plant's harbor in the Pacific Ocean every single day.[8] Somehow, though, TEPCO claims the radioactive water is magically confined to the 0.3 square kilometers (0.12 square miles) within the bay in front of the nuclear plant—a claim scientists have outright called "silly."[9]

This puts a whole new perspective on eating so-called "bottom feeders" like shrimp, crabs, and other shellfish that have subsisted off of ocean waste even before the ocean became as filthy and polluted as it is today. Fish in general, especially larger fish that live longer, such as tuna and shark, tend to accumulate toxic heavy metals. The primary pathway to mercury exposure in most humans is through eating these fish. Studies have also shown that bluefin tuna have been able to carry poisonous, radioactive cesium 134, with a half-life of a little over two years, and cesium 137, with a half-life of a little over thirty years, all the way from Japan to the United States—cesium that is traceable to Fukushima.[10]

It has long been common knowledge that seaweed is an efficient metal ion absorber as well; in fact, European researchers in 2005 demonstrated the use of seaweed as a way to decontaminate heavy metals such as cadmium and zinc from toxic water runoff continuing to drain from old metal mines.[11] This fact hasn't stopped the commercialization of several types of seaweed for human consumption, promoted as a "healthy" snack food option before any real testing was done on the heavy metals accumulated in them. For example, a 2009 analysis of six different edible seaweed products from Spain showed that all contained levels of toxic cadmium exceeding French regulations and one type contained particularly high levels of total and inorganic arsenic.[12]

Studies have also shown that marine life experiences stress from continued pollution exposure, exhibiting physiological symptoms such as thinned stomach linings and ulcers, high blood glucose levels, decreased hormone levels, and weight loss.[13] Just imagine what it does to humans who consume those stressed, sickened creatures.

In short, the sea has been used as a gigantic garbage can for hundreds of years, and now, nearly everything in the ocean is polluted. Simply put, whatever goes into the ocean goes into the food chain there, where it will ultimately wind up in some form or fashion on someone's dinner plate.

Dietary defense against mercury in sushi, fish, and other foods

Although mercury is present in alarmingly high concentrations in sushi and fish, my research into the Metals Capturing Capacity of foods and dietary supplements has revealed a surprisingly positive finding: Many foods naturally bind with and "capture" dietary mercury during digestion, surviving the "acid bath" of the stomach and likely preventing the mercury from being absorbed through intestinal walls.

In fact, mercury is the easiest of all heavy metals to capture in this fashion, and seaweeds tend to have very high efficiency in capturing free mercury during digestion. Even the nori seaweed often used in sushi is able to capture around 85 percent of dietary mercury, according to my lab tests. Other seaweeds are more effective, however. One brand of dulse seaweed, for example, showed an ability to capture 99 percent of dietary mercury.

In the lab, mercury is well known as a "sticky" element that sticks to everything, including sample tubing on laboratory equipment. This stickiness makes mercury easy to capture in the gastrointestinal tract using natural foods that contain insoluble fibers, such as fruits and vegetables.

Nearly all whole foods containing natural fibers show some affinity for capturing elemental mercury, including cereals and fruits. Strawberries and camu camu were the most effective fruits for this purpose, and nearly all grass powders (such as alfalfa grass powder) and chlorella superfood supplements showed high affinity for mercury.

The "Metals Defense" formula I developed at the lab captures nearly 100 percent of elemental mercury, leaving almost no mercury available for absorption during digestion. (See the full laboratory details on this formula at www.heavymetalsdefense.com.)

Methyl- versus ethylmercury

Both ethyl- and methylmercury are organic mercury. Organic mercury readily builds up in the environment. While some mercury apologists claim that ethylmercury is not harmful (they ridiculously compare it to ethyl alcohol), ethylmercury is actually far more harmful than methylmercury *once it enters your body's cells.* As stated in the abstract of a published study entitled "Toxicity of ethylmercury (and Thimerosal): a comparison with methylmercury":

> EtHg's [ethylmercury] toxicity profile is different from that of meHg [methylmercury], leading to different exposure and toxicity risks. Therefore, in real-life scenarios, a simultaneous exposure to both etHg and meHg might result in enhanced neurotoxic effects in developing mammals. However, our knowledge on this subject is still incomplete, and studies are required to address the predictability of the additive or synergic toxicological effects of etHg and meHg (or other neurotoxicants). [106]

Another study entitled "Thimerosal-Derived Ethylmercury Is a Mitochondrial Toxin in Human Astrocytes: Possible Role of Fenton Chemistry in the Oxidation and Breakage of mtDNA" explains how ethylmercury damages mitochondria:

> We find that ethylmercury not only inhibits mitochondrial respiration leading to a drop in the steady state membrane potential, but also concurrent with these phenomena increases the formation of superoxide, hydrogen peroxide, and Fenton/Haber-Weiss generated hydroxyl radical. These oxidants increase the levels of cellular aldehyde/ketones. Additionally, we find a five-fold increase in the levels of oxidant damaged mitochondrial DNA bases and increases in the levels of mtDNA nicks and blunt-ended breaks.[107]

Because the oceans are polluted with it, methylmercury is typically found in fish and shellfish. The larger the fish and the longer the lifespan, the more mercury is accumulated; the most contaminated include tuna, swordfish, king mackerel, and shark. The EPA warns that nearly all fish are tainted with at least trace amounts of mercury. Some of the more health-conscious grocery stores even include warnings on store shelves about methylmercury in tuna, and many recommendations caution people from eating tuna more than once a week (pregnant women are cautioned to eat it sparingly, if at all).

MINAMATA DISEASE: MERCURY POISONING VIA INDUSTRIAL POLLUTION IN JAPAN

The most significant mass acute mercury poisoning in recent history was seen in cases of Minamata disease in Japan, officially attributed to industrial contamination. Wastewater dumped into Minamata Bay containing high levels of inorganic mercury was converted to methylmercury through biological processes, bioaccumulated up the food chain, and ingested in large quantities by local residents.

In the short term, about one hundred people were killed by intense industrial-based mercury poisoning. Decades later, thousands of people from the region had been officially diagnosed with Minamata disease, while over one thousand of those diagnosed have died from the effects of mercury poisoning since the 1950s.[1]

Mercury's debilitating effects include sensory damage, muscle weakness, paralysis, coma, and possible death. The Chisso Corporation, responsible for the pollution, has paid out more than $80 million in damages to tens of thousands of affected people and has been ordered to clean up the sources of waste.[2]

Meanwhile, numerous other sources of toxic mercury pollution remain barely noticed and under-regulated.

Thimerosal in vaccines

The ethylmercury preservative thimerosal is found in personal care products like lotions, cosmetics, and contact lens solution; over-the-counter medications including some nasal and throat sprays; and in some vaccines including many widely available flu shots officially recommended to pregnant women. Because a vaccination is injected directly into the bloodstream, all of the ingredients are allowed to bypass the digestive tract where many of the body's natural defenses are located.

According to the state of Wisconsin's Department of Natural Resources, "Vaccines with 1:10,000 or 0.01 percent thimerosal have about 50 mg/L mercury, which exceeds the 0.2 mg/L hazardous waste toxicity characteristic regulatory level for mercury." This means that discarded vaccines containing the preservative may need to be officially handled as a hazardous waste per state and federal standards.[108] The Environmental Working Group (EWG) has listed thimerosal as a 10, or high hazard, on the organization's health hazard scale—the highest ranking an ingredient can receive.[109] The CDC continues to assert that thimerosal in vaccines is safe and denies links to adverse health effects including autism on the insistence that ethylmercury is much less dangerous than methylmercury.

Shorter half-life for ethylmercury

Part of this argument rests on the observation that ethylmercury has a shorter half-life in the blood, but some researchers have advised caution in making ethylmercury safety determinations based on this criterion. A comparative ethyl- to methylmercury toxicology study found little difference between the neurotoxicities of either compound, and detected concentrations of inorganic mercury in treated rats was higher after an ethylmercury dose.[110] Further research corroborated these findings.

A 2005 study assessing human ethylmercury risk noted that much higher levels of inorganic mercury were found in the brain than with methylmercury, where it remains much longer than organic mercury at a half-life of more than a year. The author cautioned that neurotoxic potential in developing brains exposed to inorganic mercury "are unknown" and that thimerosal risk assessments based on blood mercury measurements alone may be invalid and require further research.[111] Researchers reviewing medical literature in combination with U.S. government data have concluded that thimerosal induces autism and related its symptoms in some children who suffer the effects of mercury poisoning due to the preservative.[112]

Mercury in high-fructose corn syrup (HFCS)

High-fructose corn syrup (HFCS) is a highly processed sweetener made primarily from corn and found in a plethora of food and beverages on grocery store shelves. The U.S. Department of Agriculture's Economic Research Service estimated in 2011 that the average consumer per capita consumes nearly 42 pounds of high-fructose corn syrup per year.[113] Not one, but two studies in 2009 found that HFCS commercially produced in America and American-bought HFCS products were tainted with mercury.

The first study published in the peer-reviewed journal *Environmental Health* found that, of twenty samples collected and analyzed from three different manufacturers, nine, or 45 percent, came back tainted with mercury.[114] The second study by watchdog group Institute for Agriculture and Trade Policy (IATP) purchased fifty-five food items from popular brands off grocery store shelves in the fall of 2008—items in which HFCS was the first or second principal ingredient—and detected mercury in nearly a third of them.[115]

The contamination may have been due to the fact that mercury cells are still used in the production of caustic soda, an ingredient used to make HFCS.

The HFCS-mercury plot thickens, however. Online news outlet Grist reported that the lead researcher in the *Environmental Health* study, Renee Dufault, previously worked as an FDA researcher. Dufault had apparently turned over the information contained in her HFCS-mercury study to the agency back in 2005, but the FDA reportedly sat on it and did nothing, so Dufault went public with it after she retired in 2008.[116]

A breakthrough in converting dextrose to fructose with the use of a microbial enzyme in 1957 set the stage for a commercially viable process to produce what became known as high-fructose corn syrup. The development was pursued by the Clinton Corn Processing Company, which was later acquired by Archer Daniels Midland in 1982.

The Clinton Corn Processing Company's work in the mid-1960s with the Japanese Agency of Industrial Science and Technology led to the discovery of HFCS in 1966 by Dr. Yoshiyuki Takasaki, who was granted a patent on the substance in 1971 alongside development of the sugar substitute's commercialization.[117,118]

HFCS is created through a complex process in which cornstarch undergoes acid hydrolysis and becomes dextrose,[119] the glucose sugar produced from corn. A secondary process uses the enzyme glucose isomerase to convert glucose into fructose.

Clinton Corn created different formulas of HFCS, including a 42 percent fructose concoction that contains 58 percent glucose, which is frequently used to sweeten solid foods, as well as a purified 90 percent fructose formula (with only 10 percent glucose) that is rarely used directly. Instead, the 42 percent and 90 percent fructose formulas are blended to create a high-fructose corn syrup that is 55 percent fructose and 42 percent glucose (or alternately 45 percent glucose and 52 percent fructose).[120] This liquid corn-derived 55 percent fructose variety is the most widely consumed, typically used to sweeten sodas, fruit drinks, and more. By comparison, sucrose, or table sugar, has 50 percent fructose and 50 percent glucose.

How this chemical corn derivative became a staple of the American diet is rather interesting.

Initial attempts to get corn syrup widely dispersed into the U.S. food supply in the 1970s didn't really take off because sugar was so cheap and abundant at the time. However, this changed, as U.S.-imposed tariffs decreased sugar imports throughout the 1970s and early 1980s, making sugar significantly more expensive in America than in other parts of the world.[121,122,123]

The surface explanation for these tariffs was to protect American sugar farmers; behind the scenes, however, Big Agra interests had lobbied for the policy to promote what would become a new source of sugar—derived from corn—which soon emerged as a popular commodity that was sold at a price significantly cheaper than cane sugar or beet sugar.[124]

Archer Daniels Midland opened the first large-scale plant in 1978 (before they acquired the Clinton Corn Processing Company) to produce 90 percent HFCS and 55 percent HFCS. By January 1980, Coca-Cola began allowing high-fructose corn syrup to be used as a sweetener at 50 percent levels with regular sugar; Pepsi Cola followed suit by 1983.[125] By November 1984, both major soft drink brands had approved full sweetening with HFCS, and HFCS quickly captured 42 percent of the sweetener market. The rising dominance of HFCS allowed it to maintain commercial prices similar to sugar until the 1990s.[126]

For the past several decades, the U.S. government has paid subsidies to American farmers to grow tons of corn (much of which—nearly 90 percent—is genetically modified) and shifted domestic agricultural policy to maximize corn crops. This made high-fructose corn syrup and other corn-derived processed ingredients much cheaper for industrial food manufacturers to use.

Today, HFCS is nearly ubiquitous on American grocery store shelves. It can be found in a wide range of items, including candy, ice cream, bread, chips, snacks, soups, soft drinks, fruit drinks and other beverages, condiments, jellies, deli meats, and much, much more.

HFCS is not just a cheaper sweetener than sugar, but also useful in stabilizing and extending the shelf lives of many products.[127] Moreover, it was not only used to replace sugar, but also infused in new recipes. It became so pervasive, often lurking in unexpected foods, that the TIME writer Lisa McLaughlin commented in 2008, "unless you're making a concerted effort to avoid it, it's pretty difficult to consume high-fructose corn syrup in moderation."[128]

The average American consumes 12 teaspoons of HFCS per day, but for many (and especially children and teenagers who crave sweets), consumption can frequent 80 percent above this average amount.[129] By 2004, about 8 percent of total calories consumed by the average American came from high-fructose corn syrup.[130] Overall, Americans consume about fifty to sixty pounds of high-fructose corn syrup per capita—an insane amount.

HFCS has been linked in scientific research to obesity, diabetes, heart disease, fatty liver, and other contributors of bad health and early death.

A 2004 study linking high-fructose corn syrup to the rising obesity epidemic shocked the market and national consciousness. It asserted that the increased consumption of HFCS since 1970, which increased more than 1,000 percent by 1990, mirrored the rapid increase of obesity in America. The study argued that HFCS's abundant fructose sugar promotes new fats, and its interaction with insulin and leptin prevents appetite regulation and encourages the consumption of more and more calories.[131]

In experimental conditions, another study also found that consumption of the sugar alternative damaged metabolism, contributing to disease, even when weight gain did not take place, while it also contributed toward hypertension and cardiovascular disease.[132]

As the biggest dietary source of fructose, HFCS also promotes insulin resistance and increasing uric acid levels, which contribute to metabolic dysfunction and type 2 diabetes.[133,134]

Further, researchers in 2008 found a correlation between high fructose consumption and liver scarring in nonalcoholic fatty liver disease (NAFLD), which is present in nearly a third of American adults.[135,136]

Of course, sucrose (table sugar) is also very detrimental to health.[137,138] While excess intake of both can contribute toward weight gain, studies found that rats fed fewer calories of HFCS (at 8 percent) gained more weight than those eating sucrose (at 10 percent).[139]

In both cases—in ordinary sugar and high-fructose corn syrup—it is the fructose rather than the glucose that is spiking insulin and damaging the body.[140] Though glucose theoretically counterbalances fructose, studies have found that both HFCS (55) and sucrose, which have both glucose and fructose in close-to-equal proportions, act on the body almost exactly like pure fructose, which is rarely used in food production.[141,142]

The body's response to highly refined liquid sugars fails to satiate appetites and contributes toward eating more.[143] But the relative inexpensiveness of high-fructose corn syrup, in contrast to the other two, allowed food manufacturers to indiscriminately increase package sizes and amounts of calories. Cane sugar was relatively expensive and statistically less likely to become an overindulgence.

As consumers added high-fructose corn syrup to their diet for the first time, they increased total sweet calories on top of increasingly already high added-sugar intake.[144] Bottom line, eating more sweet calories and more calories overall went hand in hand with the age of cheap and overabundant high-fructose corn syrup. The rise in HFCS intake outpaced that of any other food during this period.[145]

Of course, it's worth keeping in mind that high-fructose corn syrup is not naturally occurring, nor is it easily made. It requires sophisticated industrial-scale processing with multiple transformations of the base corn raw material.

Technically, it is possible to create this concoction at home, but it requires unique and expensive ingredients. Preparing HFCS takes significantly more effort than your average cookbook recipe.

Just boil water, add a drop of sulfuric acid, heat to 140 degrees Fahrenheit, reduce, and add the corn to soak overnight. The next day, add a teaspoon of alpha-amylase, stir until viscous and thin, cool to room temperature, add a teaspoon of glucose-amylase, and pour the mixture into a cheesecloth-lined bowl. Sprinkle on a teaspoon of xylose and strain the resultant slurry through the cloth. Reheat back to 140 degrees, add some lab-created glucose isomerase (genetically modified from the streptomyces rubiginosus bacterium) and boil; then cool and enjoy![146]

Beyond the impact that high-fructose corn syrup has on American waistlines, Western fructose consumption, and the food market, this bittersweet foodstuff is adding very harmful and very hidden food additives as well.

HFCS is everywhere, but most people who eat it never even consider that it could be contaminated with toxic mercury.

Chlor-alkali plants produce chlorine and caustic soda using something called mercury cell technology. Even though it has been well-known for hundreds of years that mercury is a poison, and more energy-efficient, mercury-free technologies exist, approximately fifty mercury cell chlor-alkali plants are still in operation worldwide.[147] As of 2009, eight such plants operated in the United States. Each plant's cells can contain as much as 448,000 pounds of mercury, and unaccounted-for mercury losses get reported to the EPA every year.

Aside from all that toxic mercury poisoning the air, water, and soil, it also directly contaminates the food supply in so many of the products containing HFCS. How? Caustic soda is a main ingredient in the corn conversion process used to turn corn into HFCS. Four of the big plants that manufacture HFCS in the United States still use mercury cell technology to do it.

Two studies came out in 2009 exposing mercury-tainted products containing HFCS. First, the Institute for Agriculture and Trade Policy published "Not So Sweet: Missing Mercury and High-Fructose Corn Syrup" following an investigation of mercury content in fifty-five foods and beverages from popular brands including Kraft, Hershey's, Smucker's, and Quaker. The sampled products included HFCS as the first or second most predominant ingredient. All told, mercury was detected in nearly one-third of the fifty-five products tested.[148]

That same year, Dufault et al. (2009) published a paper in *Environmental Health* in which twenty samples of HFCS from three different U.S. plants were tested for the presence of mercury. Of the twenty samples, nine were contaminated with detectable levels of mercury (≥ 0.005 µg/g), ranging from 0.012 to 0.570 µg/g HFCS.[149]

As consumption of this relatively new sweetener remained historically high, and with the presence of mercury at concerning levels in a wide array of foods containing HFCS, the regular consumption of these foods by children and adolescents grew in significance.[150]

The Institute for Agriculture and Trade Policy and Center for Science in the Public Interest highlighted the high consumption of sodas and other drinks containing HFCS by these vulnerable and developing members of society.[151] It found that some 20 percent of children one to two years old drank sodas, while half of children ages six to eleven consumed an average of 15 ounces of soda per day. Teenagers who drink soda tossed back three or more high-fructose beverages per day on average.[152]

Hopefully, this is beginning to change.

HFCS consumption climbed steadily from the early 1980s through 2000, but sales slumped a significant 11 percent from 2003 to 2008 as concerns about its contribution to obesity and other issues reverberated in the media, even as sugar consumption surged about 7 percent over the same period. The term "high-fructose corn syrup" gained a definite negative connotation.[153]

On top of lobbying efforts, the Corn Refiners Association, an industry organization of which Archer Daniels Midland is a key member, launched the website SweetSurprise.com as a media relations ploy to debunk "myths" about HFCS and clarify "the facts about high-fructose corn syrup."[154]

It also ran well-funded TV advertising starting in 2008 sticking up for the industry's favorite sweetener and asserting that "sugar is sugar," which prompted a lawsuit by sugar producers claiming false advertising in 2011.[155] The FDA also demanded the corn industry stop using the term "corn sugar" without approval.[156]

In 2012, the FDA rejected a petition filed by the Corn Refiners Association in 2010 to change the name of high-fructose corn syrup to "corn sugar" for the purposes of food labeling and advertising. The Corn Refiners Association claims that it wanted the name change to "educate consumers," the majority of whom are "confused about HFCS."[157]

What seems perfectly clear is that most consumers in Western culture, and increasingly many people in the developing world, have adjusted to drinking and eating far too much fructose—both from high-fructose corn syrup and from ordinary table sugar. Americans in particular rode a wave of cheap corn, subsidized by the taxpayer, which was added to foods across the spectrum. While it sweetened the deal on fast and easy calories, tasty snacks, and sugary drinks, that wave has brought with it a severe backlash of obesity, with more people than ever being overweight and unwittingly following at-risk lifestyles.

Strategies for chelation and removal of mercury

Avoiding or limiting fish intake, particularly of those higher up the food chain and more inclined to accumulate harmful mercury, is one way to limit exposure to this toxin, but the extensive presence of it in the environment due to modern industrial practices means that no one can reasonably avoid it altogether.

Pregnant mothers and young children should not eat tuna or other large fish, and adults should eat no more than a few servings per month. Moreover, health-conscious individuals should minimize or eliminate their intake of many common processed ingredients, including high-fructose corn syrup.

For those who choose to consume fish on a regular basis, a defensive strategy against mercury is crucial for self-protection.

Fortunately, several essential nutrients, which can be obtained from foods or by vitamin supplementation, play an important role in defusing the effects of mercury. This means you can, to some extent, eat your way to natural mercury elimination. The original research I have conducted at the Natural News

Forensic Food Lab shows that fresh, raw strawberries, when eaten in conjunction with mercury-tainted meals, bind with and capture over 90 percent of dietary mercury during digestion, effectively "locking up" the mercury in fibers that pass through the body undigested.[158]

The use of detoxifying foods, nutrients, and activities that support the elimination of heavy metals may be necessary for those concerned about the buildup of significantly high levels. Regular ongoing, long-term detox efforts to encourage the elimination of toxins through sweat and excretion may be among the safest and most effective methods. Chelation has proven to be effective as well but should only be pursued under the direction of qualified, licensed chelation practitioners with significant experience in the art.

Outside the body, several well-known compounds demonstrate strong affinity for mercury, including activated carbon charcoal, sulfur, and selenium. Activated charcoal—based around oxygen-treated carbon—is used widely to effectively remove toxins (and other materials) in a vast array of potential bodily infiltrators due to its sizable surface area.[159] Air and water filters, as well as oral consumption, are used to administer carbon as a purifier element. Of course, charcoal has long been used to intervene in cases of poisonings and drug overdoses of all kinds.

Many important sulfur compounds have a particularly strong affinity for binding with mercury. These include sulfhydryl-containing thiols, which attract many heavy metal ions—including mercury, cadmium, lead, chromium, zinc, and arsenite—and allow chelation from the body through metabolic pathways.[160] Thiol solutions have also been used successfully to remove mercury from scrubber tanks in coal-fired power plants.[161] Important sulfhydryl compounds in various bodily processes involving antioxidant protection and DNA transcription include the sulfur-containing amino acids cystine, cysteine, methionine, and taurine.[162,163]

Mercury also binds to glutathione, perhaps the most important form of cysteine in the body, which some doctors have referred to as "the mother of all antioxidants," allowing for significant heavy metal removal.[164] Glutathione, which regenerates other oxidized antioxidants such as vitamins C and E, is critical to a fully functioning, healthy immune system. Whey protein has been identified as an important dietary source of glutamine, the primary precursor to glutathione.[165]

Both cysteine and glutathione are effective at detoxifying heavy metals but are also depleted by heavy metals' presence and may require supplementation.

Although normally recycled in the body, glutathione becomes depleted when toxic loads become too great, rendering a person unable to rid their body of toxins and opening them up to free radical damage, illness, infections, and cancer. Selenium, by the way, is a necessary dietary mineral for glutathione production. Thus, maintaining proper levels of selenium through a well-balanced diet remains imperative to maintaining proper health, as well as in reducing heavy metal toxicity.

Selenium and mercury: a highly specific and significant relationship

Mercury's binding properties with selenium are also highly significant, as this essential nutrient can block heavy metal bioavailability and reduce toxicity. However, in turn, mercury can deplete selenium, making it insoluble and reducing its protective abilities as an antioxidant, opening the body up to free radical attack.

Mercury's ability to cross the blood–brain and placental barriers allows it to deplete important stores of selenium components located there, which are essential to critical bodily functions. Mercury's powerful affinity for this element disrupts the metabolic processes of selenocysteine, and by binding into mercury selenides, it makes them unavailable for protein synthesis.[166]

Selenium is naturally absorbed by most foods when present in soils. While many people's diets are deficient in selenium, too much, if repeatedly ingested at sustained high levels, can also be damaging. Selenium is available in vitamin supplements as selenium methionine and is a significant nutrient in several dietary sources, which have known antagonistic effects with mercury.[167]

Brazil nuts are known for yielding the highest serving of food-based selenium, with more than 767 percent the established daily value (DV) in just a single one-ounce serving. The actual selenium content in the harvested nuts, of course, depends on the availability of selenium in the soil. So concentrations in such nuts may vary widely. However, many warn against eating more than one or two of these nuts per day on a regular basis, due to concerns about the possibilities of selenium toxicity (though high levels of selenium must accumulate over time before any adverse effects could occur). Numerous seeds including sunflower, chia, and others contain significant levels of selenium, as do many commonly consumed meats, though none of them reach the concentrations of Brazil nuts.[168]

Ironically, tuna fish and oysters, known for their high mercury content, are the next largest food-based sources of selenium, with some researchers demonstrating the ability of the selenium inside these seafoods to bind with mercury and make both unavailable for bioabsorption, though the ratio between the two elements is highly relevant to the risk of mercury exposure.[169] Whether the mercury is organic or inorganic is relevant as well; methylmercury irreversibly blocks selenium-related enzymes from functioning correctly.

Registered pharmacist and nutritionist Barbara Mendez notes that even low-level mercury poisoning can cause a number of symptoms that might easily be mistaken for other health issues, including rashes, inflamed gums, mood disturbances, insomnia, anxiety, and depression.[170] Mendez recommends a diet that helps optimize liver function, including garlic, cilantro, Brazil nuts, pumpkin seeds, and ground flaxseed. In studies, garlic has been effective against methylmercury-induced cytotoxic (toxic to living cells) effects.[171]

The therapeutic compounds BAL, DMPS, and DMSA have all been shown to chelate mercury. Researchers at the University of Lisbon's Research Institute for Medicines and Pharmaceutical Sciences found selenite helped detoxify cells and make these chelators more effective.[172]

Researchers who exposed mice to mercuric chloride pesticide were able to ward off oxidative stress and liver cell damage using propolis, the resinous botanical mixture honey bees mix with their beeswax to glue their hives together. A treatment for inflammatory disease and infections, propolis was found to protect antioxidant defenses against mercury poisoning in the mice.[173]

There are possibilities for mitigating the harm imposed by mercury-based pesticides as well as the environmental pollution imposed by industrial contamination, though concerned individuals should focus on personal strategies to limit their exposure.

| 82 |
| **Pb** |
| Lead |
| 207.2 |

LEAD (Pb)

ATOMIC NUMBER: 82

GROUP 14: CARBON, SILICON, GERMANIUM, AND TIN

Lead is a shiny, bluish-white heavy metal that dulls to gray when it comes into contact with air. Its legendary usage is closely tied to the rise and fall of civilization, used in water-carrying pipes, glazed pottery, cooking utensils, and even the preservation of wines by the ancient Romans, who produced some 40 percent of the world's lead alongside their abundant quantities of silver and other precious metals.

Plumbing itself draws its name from *plumbum*, the Latin word for lead (abbreviated as Pb), a luxury afforded only to the patrician class in what was once the world's greatest empire. The poisonous effects of lead were known to antiquity, and lead was noted among some thinkers of the time for its effect on shipbuilders. High levels of lead have been found in the bones of patrician gravesites, leading historians to believe it played a role as a regular part in the decadent lifestyle of the upper class, who suffered stillbirths, lower fertility, brain damage, and deformities as Rome's glory faded.[174]

But these lessons, misplaced in the dark ages, had to be relearned again in the modern industrial age. The "epidemic" effects of lead exposure were starkly noticed alongside spikes of disease in the nineteenth century during the production and manufacture of spreading industrialization.

Industrial assault of lead

In considering the modern-day hazards of lead exposure, tainted paint chips and the environmental disaster that was leaded gasoline might immediately come to mind. Although a small amount of lead is naturally occurring, the industrial revolution that began in the latter half of the eighteenth century created the conditions for widespread contamination all over the world. Today, everything from agricultural pesticide use to cosmetics, bullets, batteries, and pipes, to industrial practices such as mining and smelting continue to contribute to overall environmental lead contamination.

Unsafe at any level

No safety threshold for lead has ever been established—a multitude of studies have proven time and again lead is downright dangerous to health at any level. Government organizations such as the EPA have admitted there is no safe allowable level of lead intake.[175] Even in small amounts, this cumulative toxin competes with calcium, iron, and zinc, blocking absorption of these necessary nutrients and wreaking havoc.

Unlike other metals, which play a role in biochemical reactions, lead is just a pollutant. It has no known essential function within the human body, and science has long acknowledged that lead is poisonous to every bodily system. Once lead enters the air, it can travel far distances before it falls to the ground where it readily contaminates water and soil. This cycle inevitably leads to lead-tainted crops that are cooked into lead-tainted dinners. Exposure to cigarette smoke, even secondhand, can also mean exposure to dangerous amounts of lead.

As such, the EPA regulates it under seven different acts: the Toxic Substances Control Act (TSCA), the Residential Lead-Based Paint Hazard Reduction Act of 1992 (Title X), the Clean Air Act (CAA), the Safe Drinking Water Act (SDWA), the Clean Water Act (CWA), the Resource Conservation and Recovery Act (RCRA), and the Comprehensive Environmental Response, Compensation, and Liability Act (CERCLA).[176]

Unrestricted and pervasive exposure in foods

Although we are exposed to lead in the air we breathe and the water we drink, many organizations, including the European Food Safety Authority, have concluded that the majority of human lead exposure actually comes from the food we eat.[177]

However, even with all the bad press about lead—as with toys from China and cosmetics—and despite the fact that we know lead inflicts neurological damage on the brain and contributes to cancer, there is no fundamental framework for limiting exposure to food-based lead contamination. Despite the effort to control the impact of this dangerous substance on the environment under the guise of the EPA, there are no limits on the concentration of lead allowed in food sold in the United States, apart from a few specific products, such as candy and food additives, where the FDA has set maximum allowable thresholds.

Because lead pollution taints the water and soil that is used to grow crops for human consumption, all foods may contain some trace amount of the heavy metal (parts per trillion). Shoddy industrial food-processing practices have led to even more contamination from heavy metals. Trace levels of lead exist in many important foods we consume every day, but, generally speaking, regulators consider trace levels of lead contamination to be safe only because they are focused on short-term, acute effects that are almost never linked directly to the consumption of a food or dietary supplement. Lead is accepted in food at varying trace levels in part because lead is so difficult to avoid, and furthermore, because harmful effects from lead bioaccumulation typically show up years down the road with little solid connection to any specific foods, nutritional supplements, or protein powders.

When I appealed to Whole Foods to pull the lead-contaminated protein powders from their store shelves, the retailer did nothing to halt their sales of such products. Whole Foods continues to sell vegan, organic protein powders in its stores that show alarming concentrations of lead contamination because the raw materials are sourced from China. Instead of responding in a responsible way to my appeal, Whole Foods managers and employees began spreading rumors that claimed my laboratory didn't exist and that none of their protein powders contained lead. Whole Foods is another example of a corporation that misleads health-conscious consumers into thinking they're buying "clean" foods when, in reality, many of those foods have been contaminated with significant concentrations of lead.[178]

Children most affected by lead

According to the EFSA's *Scientific Opinion on Lead in Food*, cereals were found to contribute most to a person's daily dietary lead intake.[179] This is particularly troubling in light of the fact that many cereals are marketed to the most vulnerable members of our society—kids. Children have been found to be at greater risk for lead poisoning and toxicity than adults because their systems are still developing and they usually absorb more lead than adults do. Fetuses and nursing infants are way less equipped to metabolize harsh toxins, so amounts that are harmful to an adult can be downright deadly to a baby.

Lead accumulation, as well as that of other heavy metals such as cadmium, in the roots and shoots of wheat and other grains remains a major concern, as a direct result of heavy metal contamination in soils around the globe.[180] A study conducted by the University of Valencia in Spain compared the lead and cadmium content of twenty-nine different infant cereals commercially available on the market, and found consistent contamination levels of both cadmium and lead in the milk-free varieties, and even higher levels of lead in milk-containing infant cereals, foods that comprise a major part of an infant's diet starting between four and six months old.[181]

Another dietary staple for infants who are not breast-fed is infant formula. As infant formulas require reconstitution before they are prepared, if lead is already in the drinking water (especially in a home more than twenty years old that may have old pipes) and the water is heated during the preparation process, infants can be exposed to dangerously high levels of lead through a diet completely reliant on formula as the main source of nutrition. While the FDA assumes that manufacturers will adhere to rules when creating a new formula product, the agency warns on its website that infant formulas may be marketed without prior FDA approval.[182]

Lead confirmed in more than 80 percent of food samples

Too many times, foodstuffs purchased at grocery stores across the country have later been found to be tainted with troubling levels of lead, though not enough testing is done to prevent lead from entering into the food supply. These high levels of lead are not limited to conventional or imported foods but also appear in foods raised organically. The Environmental Law Foundation commissioned a study in 2010 that sampled nearly 150 popular children's food products, including fruit juices, fruit cocktail mixes, and even processed baby food. Foods tested were chosen from both conventional and organic sources. Using an EPA-certified lab in Berkeley, California, to test the nearly 400 samples taken, it was determined that an astounding 125 out of 146 foods contained disconcerting amounts of lead.[183] The results were so damning that the FDA was compelled to respond, although the organization only tested thirteen samples of similar foods for comparison. The agency claimed that, while lead was found in the items, it was in parts per billion, and thus, was less than the 0.1 parts per million, or 100 ppb, the agency had set for candy in children.

My own testing of foods for lead has found alarming results, including:

- 500 ppb lead in cacao superfoods
- Over 500 ppb lead in certified organic rice protein
- Over 11 ppm (11,000 ppb) lead in organic mangosteen powder
- Over 300 ppb lead in turmeric supplements
- Over 400 ppb lead in green superfood powders
- Over 800 ppb in sea vegetable superfoods
- Over 150 ppb in healthy breakfast cereals
- Over 300 ppb in cilantro powder
- Over 1,000 ppb in chopped clams
- Over 300 ppb in maca root powder
- Over 100 ppb in some spirulina powders (from India)
- Over 500 ppb in common cooking spices
- Over 1,800 ppb in popular pet treats (made in China)
- Over 8,000 ppb in calcium supplements
- Over 600 ppb in some trace mineral supplements
- Over 7,000 ppb in citrus tree fertilizers
- Over 900 ppb in chlorella supplements grown in China (other samples of chlorella were far cleaner)

These are extraordinary results the FDA seems to pretend do not exist. Yet these results were derived from off-the-shelf purchases of foods and supplements consumed by people every single day.

During the government shutdown of October 2013, the FDA held off on sending out food-safety recall e-mails, so all of the releases between October 1 and October 17 were batch e-mailed on October 17. Among those were several warnings from different distributors that PRAN brand turmeric powder was contaminated with dangerously high levels of lead. Some batches reportedly contained 48 and 53 ppm, as the powder delivers a concentrated form of the root vegetable's background exposure.[184]

Although the FDA began considering a limit on lead exposure from foods back in the 1930s due to lead-containing pesticides and the lead-based solder on food cans, even to this day, the agency has yet to establish a regulatory limit for lead levels in *all* foods across the board. Instead, the FDA has set limits on specific items in response to pressure by consumer advocate groups, such as bottled water (5 parts per billion[185]), children's candy (0.1 parts per million[186]), and food additives (varies widely by additive). For example, the lead content in candy wasn't even on the regulatory radar until 1994, when authorities in California discovered inks used in printed candy wrappers were seeping into the candy, causing the FDA to react in 1995 with the new standard.

In 1994, the FDA set a tentative daily limit for lead intake, which it termed the *provisional tolerable total intake level* (PTTIL), and included both food and nonfood sources. The bar was set so that the resulting daily lead limits in blood would be 75 µg/dL for adults, 25 µg/dL for pregnant women, 15 µg/dL for children over seven years, and 6 µg/dL for infants and children under six years.[187] (It was not explained how the threshold of adulthood suddenly makes a person safely eligible for much higher levels of daily lead exposure.) Within the same document that proposed these limits, the FDA admitted studies have shown lead presence in the blood as low as 25 to 30 µg/dL could trigger high blood pressure and eventually cause cardiovascular disease.

Federal regulators have repeatedly claimed that actual daily food exposure to lead and other metals is much lower, but without tests on specific foods and production lots, how would the average individual gauge the threat from this toxin, particularly as data have shown significant harm from the bioaccumulation of this heavy metal over time?

Rare in nature, most prevalent in its inorganic form

Though it comes in organic and inorganic forms, lead is rare in nature and it is mostly the inorganic form of lead that continues to proliferate, contaminating everything with which it comes in contact. The use of leaded gasoline began being phased out in 1973, but it was not entirely banned for sale until the U.S. Clean Air Act went into full force in 1996. Although no longer allowed in formulations after 1978, lead dust from lead-based paint decay continues to be a danger in older homes, especially to growing infants and toddlers who like to crawl around on the floor. The U.S. Consumer Product Safety Commission lists over 3,400 documents regarding the regulations and recalls of lead-contaminated items, including toys, electronics, clothing, and medicines.[188]

Lead was not officially banned in toys in America until 2008, when the U.S. Congress passed the Consumer Product Safety Improvement Act (CPSIA) following a series of high-publicity recall cases on toys and baby products found to contain dangerously high amounts of the heavy metal. Mattel Inc., the parent company of Fisher-Price, was forced to recall nearly 20 million toys worldwide including 9.5 million in the United States back in 2007. Of the two recalls announced in a two-week period, the first involved 1.5 million toys marketed to preschool-aged children. Manufactured in China, where roughly 80 percent of the world's toys are made, Mattel's toys were pulled from store shelves due in part to the discovery of lead-based paints that could flake off and cause harm. Considering that preschoolers like to put things in their mouths, the potential for lead poisoning went well beyond a toddler merely touching a leaded item.

Still, even today, researchers and watchdog groups continue to find lead well above regulatory levels lurking in consumer goods. For other products that come into contact with a consumer's skin and mucous membranes, the FDA may or may not have set an allowable lead limit. For example, the agency has not set a limit on lead in cosmetics,[189] but has established a maximum threshold for the color dyes used in cosmetics, which the FDA regulates to 20 parts per million lead.

Studies continue to show that, even though not directly ingested in large amounts, repeated daily application of lead-containing cosmetics can add up to significant and cumulatively dangerous exposure.[190] The FDA commissioned a study of lead in popular U.S. cosmetics in 2010 and found that, of over 400 lipsticks, including samples of the most popular brands purchased at retail stores, every single one contained lead—all of them.[191]

Developmental impairment as neurotoxin

Science has already confirmed a link between neuropsychological damage during development in early childhood and the intake of low levels of lead through chronic exposure. Brain and nervous system damage caused by lead poisoning can disrupt a child's actual learning ability, and it can cause behavioral issues such as hyperactivity and aggression. Lead crosses the placenta, harming the brains of unborn children in the womb, and it has been directly linked with lowered IQ scores throughout life.

Researchers at Australia's Commonwealth Scientific and Industrial Research Organisation conducted a survey of seven year olds and lead exposure, which showed an average decrease of four to five IQ points when raised blood lead measurements were recorded between fifteen months and four years of age.[192] A 2013 study found that children who sustained lead poisoning and presented with blood lead levels of greater than 10 but less than 20 µg/dL before age three did significantly worse on end-of-grade elementary school exams than children with lower blood lead levels.[193] Researchers have also concluded that lead exposure both before and after birth can cause significant memory impairment.[194]

Lead has, of course, been found to be a neurotoxin in adults as well. Elevated bone lead levels in the elderly have been associated with dementia and other negative mental health issues.[195] In a fifteen-year study on the effects of lead in the adult brain, researchers concluded that past exposure to lead may actually be attributable to a significant amount of what the medical profession generally considers "normal" age-related cognitive decline.[196]

Contributions to cardiovascular disease, the world's leading killer, and more

According to the World Health Organization, cardiovascular diseases are the leading cause of death globally.[197] While there are many important factors at play, lead and other dangerous heavy metals have been linked to heart disease and related effects. Lead toxicity may play a particularly significant role in cardiovascular diseases, as high blood lead levels have been positively correlated with high blood pressure.[198] The exact mechanisms involved are unknown, but researchers believe these metals impair the metabolism of antioxidants,

resulting in years of sustained oxidative stress driven by chronic low-level exposure.[199] Anemia can result from the disruption of hemoglobin synthesis.[200]

Lead also damages kidneys, as do most toxins that inflict major consequences on total health. Renal damage, along with impairment issues in
the liver, are hallmark signs of dysfunction by chronic poisons, disabling the
body's ability to break down and excrete toxins. Moreover, the long-term
storage of harmful substances such as lead in the liver and kidney are enough
to shut down normal detoxification processes in the body, and allow accumulated heavy metals to be re-released into the bloodstream, causing further
damage by inhibiting essential enzymatic actions and white blood cell immunity response.[201]

Further, lead has been connected to reproductive issues. A 2008 study
found decreased sperm motility in what should otherwise be healthy, fertile
metal workers exposed to high amounts of lead,[202] while another study in
2010 concluded that increased lead concentrations in blood and semen were
directly linked to decreased sperm counts.[203] In women, lead has been linked
to miscarriage and infertility, while inflicting additional damage to unborn
babies.[204]

Research has even linked toxic lead exposure to dental diseases, including
gingivitis, periodontitis, significant decay, and more missing teeth.[205]

Finally, lead exposure has been associated with increased mortality rates.
In the final analysis, the heavy metal, in all of its aspects of exposure, wages a
war against the body's ability to function. It's killer stuff.

Long-term accumulation of lead in bones

The power of this toxin to tear people up comes from its ability to remain in
the body for a long period of time. Over 80 percent of the lead we take in
settles into our soft tissues before being stored in the bones, and some studies
even show that as much as 90 percent of ingested lead may persist in our tissues for a prolonged period.[206] While lead's half-life in the blood and soft tissues averages about a month to a month-and-a-half, lead can remain in bones
for up to thirty years.[207] During certain times throughout life, the bones will
dump lead back into the blood where it can further contaminate tissues and
organs, such as during pregnancy and breast-feeding, when a bone gets broken, or even as someone naturally grows older and bone mass deteriorates.

As with all chemical sensitivity, a myriad of biological and environmental factors is at play, leaving some people simply more prone to lead's toxic effects than others. Again, because lead accumulates in the body, even ingesting repeated small amounts of it over time can eventually add up to enough to cause significant harm.

Eliminating lead quantities through chelation and natural health remedies

The good news is that research has shown several foods, vitamins and nutrients, and chelating techniques can help rid the body of toxic lead deposits.

Vitamins B1 (thiamine) and B6 (pyridoxine) have both been shown to aid in reducing lead toxicity, with B1[208] reportedly shown to lower lead levels in the liver and kidneys, and B6[209] acting as an antioxidant and moderate chelator of the heavy metal. Vitamin C not only works to help chelate lead, but it can also prevent oxidative cells. Researchers at the Wuhan University Department of Toxicology found that when vitamin C was supplemented in combination with vitamin B1, it actually reverted some of the oxidative stress as well as DNA damage to the liver caused by lead, undoing overall damage.[210] The antioxidant vitamin E has also been shown in research to detoxify by scavenging free radicals, thus undoing lead-related cell damage.[211] In that same study, garlic oil was found to have similar effects on counteracting lead damage.

Several bioflavonoids, the polyphenic compounds synthesized by plants, also help to undo lead damage and chelate lead from the body. Quercetin, found in grapefruit, onions, apples, and red wine, has not only been shown to stabilize free radicals to prevent lead damage,[212] but it is also a lead chelation agent.[213] Alpha-lipoic acid, or thioctic acid, is an organosulfur compound found in vegetables such as beets, carrots, and spinach. This antioxidant has the power to restore other antioxidants, such as vitamins C and E, both of which help fight lead damage. In short, alpha-lipoic acid is an oxidative stress-fighting powerhouse.

Some foods have also been shown in studies to reduce lead in the body. Across several studies, the principal substance in turmeric, curcumin, has been found to chelate lead from the body and significantly reduce lead burden in many organs including the brain.[214,215] (Be careful with consuming turmeric to acquire more curcumin, however. Nearly all common sources of turmeric are contaminated with lead and sourced from India. In my lab testing, it is rare to find a turmeric raw material that isn't significantly contaminated with lead. The only way to find low-lead turmeric is to ask the manufacturer or retailer for scientific lead test results on that particular batch. Sadly, virtually no one in the industry tests their turmeric for lead.) Garlic oil has also been proven to reduce lead within the soft tissues.[216,217] Researchers at the School of Veterinary Medicine at Shahrekord University found that the administration of both fresh garlic and garlic tablets led to a significant lead-burden decrease in blood, kidneys, liver, and bones, with little difference in powerful health benefits between each type (fresh or tablet). Sesame seed oil, which contains the natural antioxidant sesamol, has been found to work as a chelator against liver and kidney lead poisoning without adverse effects.[218]

A lot of research has focused on the binding of lead for removal through dimercaptosuccinic acid (DMSA) chelation therapy. DMSA has been shown to be a potent lead chelator in both animal and human studies, especially in combination with other vitamins and minerals, such as vitamin C and calcium.[219] A 2012 study even revealed that metals chelated and removed with DMSA reduced behavioral effects in autistic children.[220] The research on DMSA is so promising, scientists at the University of Birmingham City Hospital declared it to be an effective antidote for lead poisoning in 2009.[221]

In terms of lead binding during digestion, my laboratory research allowed me to develop an ion-exchange lead binder made from dehydrated seaweed and seawater extract that shows a near-100 percent efficacy at binding with free lead.

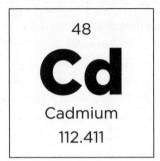

48

Cd

Cadmium

112.411

CADMIUM (Cd)

ATOMIC NUMBER: 48

GROUP 12: MERCURY AND ZINC

Cadmium is a heavy metal that can damage all cells in the body. It poses a significant threat to human health through continued and sustained exposure via tainted foods, water, soil, polluted air—and especially through smoking cigarettes.

As with other heavy metal pollution, the total environmental exposure to cadmium has been drastically increased as a result of industrial mining practices. Cadmium is a by-product of zinc, copper, and lead ores, which outweighs naturally occurring sources of the toxic element by several factors.

Today, cadmium is most widely used in creating rechargeable nickel-cadmium (Ni-Cd) batteries, metal plating, and pigments, and as a stabilizer for PVC products.

Ni-Cd batteries are major polluters when disposed of improperly, as they are often thrown out indiscriminately with household trash. Burning municipal waste containing cadmium releases toxic air particulate, as does burning fossil fuels.

Found in phosphate rock, cadmium is used to create phosphate fertilizers and, as a consequence, shows up in significant quantities as its application to soil is almost universal. Cadmium concentrations in soil are also increased by other agricultural input components such as sewage sludge and manure from large concentrated animal feeding operations where drugs, steroids, and/or antibiotics are applied. Sewage sludge and manure also amplify the presence of lead, arsenic, and other harmful metals, particularly when concentrated in the topsoil strata where plant nutrients and contaminants are absorbed.[222]

Both waste products are used for crop fertilization, when they then leach cadmium into soils and plants. While sewer sludge, concentrated from urban populations, has been known to contaminate foods with cadmium, it is the widely used manure sourced from cattle and other livestock that is most dangerous. Approximately 90 percent of the cadmium consumed by these animals passes through to the manure and the manure helps mobilize plant absorption of other cadmium compounds in the surrounding soils.

Some soil treatments, including liming and zinc fertilizers, have been formulated to reduce cadmium uptake in crops by increasing pH levels and immobilizing metals in the soil, thus making them less bioavailable and less concentrated in the resulting food crops.

Exposure to cadmium through contaminated crops

Crops planted for human consumption in metal-tainted soils have shown significant increases in the uptake of cadmium, affecting a wide range of fruits, vegetables, nuts, and even certified "organic" produce.

With a remarkably long half-life, cadmium can remain in the body for up to twenty to thirty years and is stored in soft tissues where it specifically targets the liver, kidneys, and vascular system. There is substantial research proving cadmium's ability to adversely affect all body systems; however, possibly the most troubling is its ability to interfere with mechanisms responsible for DNA repair, according to some scientists.

Cadmium exposure also disrupts mitochondrial activity. Found in the cells of animals, plants, and many other living things, mitochondria are responsible for producing energy from nutrients, inducing important cell activities and mediating both a cell's growth and a cell's death. They're essentially a cell's "powerhouse." Studies show that cadmium actually alters mitochondria, interfering with and inhibiting core biological functions.

Tobacco and rice are most affected by cadmium due to their tendency to absorb more nutrients—and toxins—than other plants. This proclivity goes hand in hand with the warnings over poisons in cigarette smoke as well as in rice in connection with recent food scares over inorganic arsenic.

As a result of large-scale lawsuits against the tobacco industry and continuing public relations campaigns, the majority of people are aware of the dangers that smoking poses to health. But few realize that among these risks are tobacco's propensity to accumulate metals and other toxins from the soil, concentrating them until their release into the lungs during inhalation. Though lead, mercury, arsenic, formaldehyde, and other toxins are significant ingredients in tobacco, documentation has shown cadmium exposure to be particularly significant as well, and tobacco constitutes the element's most dangerous pathway for toxicity.

Rice, one of the world's most important staple crops, absorbs far greater amounts of water than other crops because the fields where it is grown are flooded during cultivation, drawing in with it greater levels of trace toxins. Despite the fact that heavy metals are spread out in soil and water at levels of only a few parts per million, the superabsorptive crops accumulate and often distribute toxic elements to the edible parts of the plant, passing them along for dietary intake.

Leafy green vegetables, including lettuce, cabbage, and spinach, and root vegetables, including potatoes, carrots, radishes, and beets, are also heavy to moderate accumulators of cadmium given the tendency of their leaves, stems, and roots to absorb contaminants and chemicals.[223]

The surrounding sources of pollution, the condition of the soil, and the use of inputs such as pesticides and fertilizers all weigh heavily in assessing heavy metal exposure for a given crop. Without testing, it is difficult to know which crops, including organically grown produce, may contain dangerously high levels of heavy metals such as cadmium.

The extent of soil and water contamination, which amounts to almost ubiquitous exposure to cadmium and other metals through food and water, has made total avoidance almost impossible. Scientists have recognized several "irreversible agronomic problems" due to the toxic elements in fertilizers and pesticides since at least the 1920s, yet no serious regulations were set until the 1970s after the formation of the EPA, and it would be decades more before states such as California began subjecting soils to risk assessment for heavy metals and other contaminants, establishing limits on arsenic, cadmium, and lead in phosphate fertilizers, sewage waste products used in fertilizers, and other micronutrient materials.[224]

The EPA and other regulatory agencies have claimed these fertilizers add no significant load to the background levels of metals present in the soils, even as bans on dumping human waste into the oceans has driven industrial companies to market toxic sludge to farmers as cheap or even free fertilizer. Even if current regulations limit adding new heavy metals to soils, many soils are already contaminated, perhaps permanently, from decades of inundation with fertilizers and pesticides that relied on elemental metals as significant constituents. Perhaps more importantly, few risk assessments take into account the total cumulative exposure to toxins across the spectrum of air, food, water, and skin intake, which may overlook the combined effects with other significant metal exposures in the human life cycle.

Soils contaminated with city sewage waste, cadmium, and other heavy metals

The accumulation of toxic constituents in soils from waste-derived fertilizer is a large contributor to ground pollution, which eventually also affects our air and water. These pernicious fertilizers include "biosolids," which consist of city sewer waste full of drugs and hormones excreted or disposed of as medical waste, as well as chemicals and other by-products that are improperly disposed of by reckless industries. The rise of pathogens, antibiotic-resistant superbugs, and disease-promoting conditions in humans is recycled back into soils, affecting food production and the delicate balance of ecosystems essential for healthy life on Earth. As another example of a persistent contamination crisis caused by industry, consider how soils and streams are routinely "carpet bombed" with antibiotic chemicals via livestock feedlot operations.

Dr. Arjun Srinivasan, with the U.S. Centers for Disease Control and Prevention, announced in a 2013 interview on PBS's *Frontline* that we've reached "the end of antibiotics, period."[225] Dr. Srinivasan was referring to the rise of superbugs, bacteria that has grown resistant to antibiotics due to their overuse. The FDA estimates a whopping 80 percent of antibiotics sold on the American market are intended for use in food-producing livestock.[226] Antibiotics are common livestock- and poultry-feed additives in the

United States—the majority of which are fed preemptively to healthy animals because of the unsavory conditions found in most factory farms where animals are crammed in to relatively small spaces, making the fast spread of disease nearly inevitable. For example, more than 10 million pounds of the antibiotic tetracycline was sold specifically for use in livestock in 2009, more than all other antibiotics sold for human use that year combined.[227]

Studies have shown that tetracycline, in particular, has played a significant role in contaminating groundwater with antibiotic-resistant bacteria, which may alter the accumulation and bioavailability of metals in soils.[228] In 2010, environmental science researchers at China's Nankai University concluded that the presence of tetracycline also increases cadmium's ability to accumulate in soil.[229]

In turn, other pollutants amplify the soil's exposure to heavy metals contamination, which is especially dangerous to humans as crops absorb these heavy metals, resulting in these impurities being passed on to humans through the food chain. This cycle of environmental pollution can be blamed on modern trends in industrial production, the medical industry, big agriculture, and urban living, as well as many other factors.

Like other industries, modern agriculture operates on the basis of short-term profitability: favoring larger livestock raised on the cheapest feed and most bountiful crop yields controlled by the greatest effective technology to give shareholders the best return on their investment with minimal concern for long-term environmental impact or the chronic accumulation of toxins by humans, even when it is potentially life threatening. Big Agra has a social mandate to feed the world at any cost, and rules for regulators at the federal and international level are often heavily influenced by lobbyists for dominant corporate players in food production, with potential conflicts of interests watering down protections for consumers and compromising public safety.

Agencies in Washington, D.C., such as the FDA, the USDA, and the EPA, have often delayed issuing warnings, bans, and recalls on products containing harmful substances. Frustrating delays in protecting public health are also frequently observed, with regulatory delays at the U.N. World Health Organization and Food and Agriculture Organization, as well as most nations around the world using modern agricultural techniques.

Compared with the United States, regulatory authorities in the European Union have been much more proficient at setting limits on heavy metal content in food and water. Europe has also acted more swiftly in banning products like bisphenol A (BPA) and certain pesticides and herbicides that disrupt endocrine systems or contribute to cancer, which remain legal in the United States. Showing a willingness to regulate genetically modified organisms (GMOs), Europe, Japan, and other nations have also required labels for foods containing genetically modified ingredients. The global differentiation of restrictions on certain foods, including GMOs, leaves a large portion of the world exposed to harmful ingredients, including those in the United States, who often fall victim to false assurances that these questionable foods are safe.

Cadmium's adverse health effects, including its role as a reproductive toxin

Even at low concentrations, cadmium is a reproductive toxin. Environmentally relevant levels affect reproductive system development with lifelong consequences, damaging even the most basic procreation processes[230] through interferences such as prematurely killing off egg cells or significantly slowing the ability to produce sperm.[231]

Cadmium consumption leads to demineralization, specifically because it competes with calcium, magnesium, iron, zinc, and copper. It also inhibits vitamin D3 and adversely affects our body's nutritional levels of that crucial anticancer nutrient. Researchers have discovered a link between cadmium and poor bone mineral density, as studies have revealed the higher the cadmium exposure, the more prone someone will be to fractures, especially in the elderly, whose risk for osteoporosis rises with age.[232]

These effects were best illustrated during the first half of the 1900s when many Japanese suffered from itai-itai ("ouch-ouch") disease following the most infamous case of wide-scale cadmium poisoning to date.[233] Due to the Russo-Japanese War and World Wars I and II, cadmium production was greatly increased as a by-product of ore mining. The Mitsui Mining and Smelting Company had been dumping cadmium into the Jinzū River for decades, tainting drinking water and irrigation for rice paddies in the Toyama and surrounding prefectures. A 1961 investigation led to the Japanese government officially admitting cadmium was the culprit of itai-itai disease in 1968, making it the first disease to officially be blamed on environmental pollution.

There is some evidence that certain nutrient-rich superfoods may help protect the body from the most harmful effects of cadmium-induced cellular damage. High doses of antioxidant-rich spirulina, for example—the protein-rich, blue-green, freshwater algae—were found to dramatically decrease fetal abnormalities caused by dosing pregnant mice with cadmium.[234] The mechanism of spirulina's apparent protection against cadmium is not known and merits further study.

Natural News Forensic Food Lab test results on cadmium-containing food products

As illustrated, cadmium can have dangerous health effects in humans, highlighting the need to avoid exposure to this heavy metal as much as possible. Below are the results of some testing I conducted on off-the-shelf products at the Natural News Forensic Food Lab that may help you make informed decisions about your food:

- Over 2,000 ppb in sea vegetables
- Over 70 ppb in healthy breakfast cereals
- Over 200 ppb in acai superfruit powders
- Over 200 ppb in cinnamon spice powder
- Over 180 ppb in barley grass powder
- Over 200 ppb in lower-cost forms of chlorella
- Over 1,800 ppb in certified organic rice protein powders
- Nearly 600 ppb in maca root powder
- Over 2,300 ppb in organic cacao powder
- Over 400 ppb in ginkgo biloba herb powders
- Over 200 ppb in prenatal vitamins
- Over 750 ppb in fish treats for cats
- Over 1,300 ppb in sunflower seeds
- Over 1,700 ppb in a popular mineral supplement
- Over 700 ppb in a popular baking powder
- Over 600 ppb in popular calcium supplements
- 200 ppb in fast-food French fries
- Over 1,400 ppb in "natural" cigarettes
- Over 1,000 ppb in white willow bark powder herb

Our lab testing has also consistently found cadmium in coffee products. At the time of this writing, we had not yet completed an exhaustive analysis of cadmium in various coffee products, but we hope to complete that work (and publish the results) in early 2017.

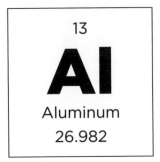

13
Al
Aluminum
26.982

ALUMINUM (Al)

ATOMIC NUMBER: 13

GROUP 13: BORON AND GALLIUM

Aluminum is the third most abundant element on the planet, after oxygen and silicon. Nearly 32 million tons of aluminum is produced globally each year[235] for use in a plethora of consumer goods, including airplane parts, building materials, baseball bats, soda cans, kitchenware, foil, cosmetics, antiperspirants, antacids, vaccines, and many more. Despite its ubiquitous nature, the lightweight silvery white metal is not an essential nutritional element, serving no known biological function for any plants or animals.

Though aluminum is seemingly everywhere, the majority of it is not bioavailable to humans and animals—or at least it wasn't, until certain types of coal began being burned for energy over a century ago. Since then, over a hundred years of acid rains from unchecked industry emissions and air pollution have significantly altered minerals in the Earth's soils.

Industry emissions have released aluminum into the soil, contaminating water sources, decreasing pH levels, and creating a toxic environment for fish.[236] Aluminum is taken in by plants and then passed to both animals and humans who feed on the plants. Humans can also be exposed to dangerous levels of aluminum by consuming animals that have eaten aluminum-contaminated vegetation. Thanks to humans, aluminum toxicity can no longer be avoided, prompting some to refer to the problem as "a postmodern rite of passage."[237] While aluminum is everywhere, and the metal can be inhaled, absorbed through the skin, and imbibed in contaminated liquids, food is the largest pathway of human exposure to aluminum.[238] A European Food Safety Authority panel concluded that while the majority of unprocessed foods already contain small amounts of aluminum, some foods such as flours, breads, cakes, and pastries; vegetables such as spinach, radishes, and lettuce; dairy products; sausages; and shellfish contain higher-than-average levels of aluminum. The highest levels were found in cocoa, tea leaves, and herbs and spices.[239] In addition, using aluminum pans, foil, and trays could increase aluminum concentrations specifically in tomato and apple purees, rhubarb, salted herring, and certain types of pickles and vinegar (because they are more acidic than other foods and therefore cause more aluminum to be released from pans and trays).

Certain infant formulas also appear to contribute to high levels of dietary aluminum. In a 2013 U.K. study titled "The aluminum content of infant formulas remains too high," researchers found that all thirty formulas they tested were contaminated with aluminum. Soy-based formulas in particular had the highest aluminum content overall, leaving some infants who primarily rely on that type of formula more at risk as they consumed more than 700 µg/L per day (in terms of micrograms consumed compared to their body blood volume).[240]

This contamination is in addition to all of the food additives made of aluminum, which are still widely used throughout the United States. Two of the most common aluminum-containing compounds are potassium aluminum sulfate and aluminum oxide.[241] Although aluminum has been approved by the FDA as a food additive, the word *alum* itself can refer to any form of aluminum sulfate, including toxic versions.[242] Ingesting a single ounce (28.34 grams) of alum could be lethal for the average adult.[243] Potassium alum is an inorganic salt added to many over-the-counter drugs and used as a pickling agent. Aluminum chlorohydrate is typically used in industrial wastewater treatment and often found in many popular deodorant products. Sodium aluminum phosphate is used as a leavening agent in many baking powders as well as in some cheeses.

These additives were banned by the European Parliament in 2008 for their carcinogenicity and ability to damage DNA,[244] but they are still considered generally recognized as safe (GRAS) by the U.S. FDA. While the EFSA has noted that aluminum bioavailability from water is less than 0.3 percent and from food about 0.1 percent, the agency also recognizes that, depending on what combination of chemicals are present in the food and water, aluminum's bioavailability can increase up to tenfold.[245] The agency also contends that certain dietary molecules have been shown to increase aluminum-ion bioavailability. One culprit is fluoride—a common additive in many municipal water supplies in the United States and throughout many other parts of the world.

All told, the Health Sciences Institute estimates the average person absorbs 10 to 100 mg of aluminum each day through all these various sources.[246] Once aluminum is absorbed by the body, it settles in all tissues and accumulates in the bones. Aluminum is able to cross the blood–brain barrier and deposit in the brain; the metal can also cross the placenta and affect the fetus. Scientific studies are continuing in regard to aluminum's destructive effects in humans, but research dating back to the 1920s demonstrates the toxic effects of aluminum in animals. Mice that were fed bread leavened with aluminum phosphate not only had fewer babies than their counterparts, but researchers discovered that lesions had developed on their ovaries.[247]

In people whose kidneys do not function at optimal levels, aluminum is much more dangerous, as it is more easily absorbed through the gut and eventually deposits in the brain, where it has been found to accumulate and contribute to dialysis encephalopathy syndrome, a form of dementia that can afflict dialysis patients.[248] Although researchers went back and forth for many decades over whether aluminum could contribute to Alzheimer's disease, a recent study published in the *Journal of Alzheimer's Disease* concluded, "The hypothesis that Al significantly contributes to AD is built upon very solid experimental evidence and should not be dismissed. Immediate steps should be taken to lessen human exposure to Al, which may be the single most aggravating and avoidable factor related to AD."[249]

People who lived in Camelford, a town and civil parish in north Cornwall, England, were exposed to toxic levels of aluminum when 20 tons of aluminum sulfate were dumped into the water supply there in 1988. A post-mortem study of the brain of Camelford resident Carole Cross, a woman who died at the relatively young age of fifty-eight in 2004, revealed she had a rare form of dementia induced by the high levels of aluminum that had accumulated in her brain over the years.[250]

If small amounts of aluminum are allowed to accumulate in the body over time, it can have toxic effects once tissue concentrations exceed a certain threshold. In another example, scientists have linked aluminum chloride salts used as the main active ingredient in many antiperspirants to breast cancer, as they have been found to induce proliferation stress, DNA double-strand breaks, and the speeding up of the aging process over time in otherwise normal mammary epithelial cells.[251]

Aluminum in vaccines

Many vaccines contain aluminum adjuvants (chemical additives). An American infant administered every vaccine on the schedule could receive 4.225 milligrams of aluminum shot directly into his or her bloodstream in the first year of life.[252] The hepatitis B vaccine typically given at birth contains 0.25 milligrams of aluminum in just that one dose. An infant has the potential to be exposed to much more aluminum if fed formulas contaminated with the heavy metal, especially soy-based formulas as mentioned earlier.

Due to increasing scientific confirmation and the growing public concern that aluminum can be a dangerous neurotoxin and the adjuvants in vaccines might pose a risk to infants, FDA toxicologists completed a study in 2011 and concluded the risk from combined aluminum in childhood vaccines is extremely low and the benefits of these shots outweigh the risks.[253] The study was published in the journal *Vaccine*. The editor-in-chief of *Vaccine* happens to be Dr. Gregory Poland,[254] founder of the Mayo Clinic Vaccine Research Group and regular consultant for vaccine production companies. Poland has held the position of chairman on a safety-evaluation committee for investigating vaccine trials at Merck Research Laboratories, a mega pharmaceutical corporation that makes billions of dollars each year off its vaccines.[255]

In fact, as the FDA hailed its latest aluminum-adjuvant safety study, two other studies were published in the same year warning of the dangers of this type of adjuvant being injected into large swaths of people and calling for further independent studies and a reevaluation of current vaccination policies in specific regard to aluminum. Canadian researchers published a study in the journal *Current Medical Chemistry* noting that aluminum adjuvants put everyone, including infants, at risk for long-term brain inflammation and other neurological complications in addition to autoimmune disorders. They concluded, "In our opinion, the possibility that vaccine benefits may have been overrated and the risk of potential adverse effects underestimated, has not been rigorously evaluated in the medical and scientific community."[256]

The same researchers published another study in the same year in the *Journal of Inorganic Biochemistry* in which they investigated the correlations between exposure to aluminum adjuvants and the rise of autism in seven Western countries, including the United States, over the past two decades. Children in these countries are often administered multiple doses of up to eighteen aluminum adjuvant–containing vaccinations. The researchers concluded that the aluminum in vaccines may in fact share a causal relationship with the rise in autism.[257]

In January 2016, the American College of Pediatricians (ACP) issued a warning statement about the toxic side effects of the Gardasil vaccine, citing aluminum and polysorbate 80 as two vaccine adjuvants of concern. "It has recently come to the attention of the College that one of the recommended vaccines could possibly be associated with the very rare but serious condition of premature ovarian failure (POF), also known as premature menopause," said the ACP.[258]

The ACP also explained that aluminum toxicity in Gardasil vaccines may have been concealed by the design of Gardasil clinical trials, which deliberately added aluminum and polysorbate 80 to the placebo group injections. As the ACP stated, "Pre-licensure safety trials for Gardasil used placebo that contained polysorbate 80 as well as aluminum adjuvant . . . Therefore, if such ingredients could cause ovarian dysfunction, an increase in amenorrhea probably would not have been detected in the placebo-controlled trials."

As the FDA has stated, aluminum food additives are GRAS; there is no limit to their use, although the agency has set an upper limit on certain pharmaceuticals and bottled water. The EPA's Secondary Maximum Contaminant Level (MCL) limit of aluminum in drinking water is 0.05–0.2 mg/L, but it is not based on what levels will affect humans or animals; rather, it is based on taste, smell, or color.[259] The European Union's EFSA panel established a tolerable weekly intake (TWI) for aluminum of 1 mg/kg of body weight per week, but found that the limit was likely exceeded in many parts of Europe with findings of up to 2.3 mg/kg in highly exposed consumers.

Just like arsenic, aluminum competes with phosphorus in the body.[260] The aluminum hydroxide in antacids has specifically been shown to cause phosphate depletion syndrome.[261]

Reducing aluminum in the body

Chelation can help eliminate aluminum from the body's tissues, but aluminum can be one of the more difficult metals to chelate. Professionals warn that chelation therapy should only be attempted with expert guidance after a hair strand test or similar test for aluminum content has first determined someone is carrying dangerously high aluminum levels. Naturopathic doctor Marty Milner of the Health Sciences Institute studied the effects of malic acid on chelating aluminum in his fibromyalgia patients. Using hair strand tests, Milner was able to determine that malic acid dramatically lowered aluminum levels in their tissues.[262]

The cultivation of green algae chlorella has also been very effective in scientific studies at safely removing concentrations of several heavy metals from wastewater, including aluminum.[263] Below are some of the more spectacular findings the Natural News Forensic Food Lab has uncovered on aluminum in foods and supplements. Note the use of ppm rather than ppb in this list, as aluminum levels are often 1,000 times higher than cadmium or lead:

- Over 700 ppm in seaweed superfood granules
- Over 40 ppm in organic, healthy breakfast cereals
- Over 500 ppm in some traditional Chinese medicine herbs
- Over 135 ppm in cilantro spice
- Over 1,100 ppm in a popular "detox" liquid supplement
- Over 300 ppm in a popular wheatgrass powder
- Over 200 ppm in sugary rolls sold at the grocery store
- Over 75 ppm in popular children's drink mixes
- Over 400 ppm in popular green superfood powders
- Over 1,000 ppm in popular ginkgo herb supplements
- Over 2,500 ppm in popular children's multivitamins based on TV cartoon characters
- Over 1,700 ppm in calcium supplements
- Over 26,000 ppm in baking powder

The bottom line is that while aluminum in food is virtually impossible to avoid, much of that aluminum cannot be readily absorbed by the body. The most important sources of aluminum to avoid are vaccine injections, aluminum-containing medicines, and any liquids or beverages that may contain aluminum.

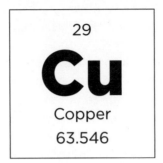

29

Cu

Copper

63.546

COPPER (Cu)

ATOMIC NUMBER: 29

GROUP 11: SILVER AND GOLD

Copper, known for its reddish-orange color and electrical conductivity, is an essential chemical (in trace quantities) to all living things and a valuable metal commodity.

Copper's antimicrobial properties have been known for thousands of years, dating back to an ancient Egyptian text that discussed using the mineral as a sterilization agent for injuries and purifying water.[264] The father of modern medicine, Hippocrates, even mentioned copper as a treatment for leg ulcers circa 400 B.C. Ancient copper treatments span the gamut, as they were used as a cure for ear infections, for purging the stomach of intestinal parasites, and for controlling lung diseases.

Copper is essential for maintaining good health, unlike many other metals that play no beneficial role, but its quantities must be in balance with zinc, iron, and other essential minerals; otherwise, it instead blocks necessary enzymatic activities and may contribute directly to detrimental health effects.

Foods high in copper include red meat, eggs, whole grain breads, shellfish, nuts, seeds, dried legumes, and dark green, leafy vegetables. While it is possible to be overexposed to copper via whole foods, copper toxicity typically occurs primarily through multivitamins that contain high levels of this heavy metal. This is documented via Natural News Food Forensic Lab results, which are made available toward the end of this section. While copper deficiency can be dangerous—a severe lack of the metal can lead to irreversible neurological damage, for instance—excess copper in the body can be equally dangerous (and is more common). Copper is required for healthy biological functions but only in small amounts within a strict intake range. Too much copper not only impairs the thyroid, but it can decrease liver and kidney function, and, ultimately, copper poisoning can cause brain damage and lead to death. Copper is particularly important to the glandular, reproductive, and nervous systems. Graves' disease, an autoimmune disorder that leads to hyperthyroidism, can be caused by a copper imbalance. Studies have found that copper and estrogen metabolism are inextricably linked. According to Dr. Lawrence Wilson, "[Copper] imbalance can cause every conceivable female organ–related difficulty such as premenstrual syndrome, ovarian cysts, infertility, miscarriages, sexual dysfunctions, and more."[265] Copper imbalance is also closely associated with mental deficiency, neurological dysfunction, and psychological disorders.

Copper's adverse health effects

In the industrial world, copper is the choice metal for plumbing, carrying much of the drinking water to faucets all over the world. Scientists recently found that overexposure to copper in drinking water could lead to oxidation and a buildup of amyloid beta protein, a hallmark plaque found in those suffering from Alzheimer's disease.[266] Researchers have studied the senile plaques in Alzheimer's patients and found significantly elevated levels of both copper and zinc present.[267]

Copper's ability to damage the body comes from its ability to throw a person's nutrient balance out of whack. Zinc and copper compete with each other for absorption, so having both in the proper balance is essential. Participants with anxiety disorder in a 2011 study were found to have significantly higher levels of copper and significantly lower levels of zinc in their blood than their control counterparts.[268] The same researcher was able to show a similar correlation between depression symptoms, decreased zinc, and increased copper plasma levels—concluding that the higher the level of copper found in the person's blood, the more severe their depression symptoms were.[269] The same principle also proved to be true for autism symptoms.[270] Accumulation of copper nanoparticles over time has also been found to cause death in animal studies.[271]

As new research in this area is currently emerging, the full implications on public health when it comes to heavy metals and other additives in food have not been fully evaluated. For example, ligands are ions that bind to metal atoms to form a new complex. Aspartame has been studied as a ligand for copper, leading scientists to only recently discover in the fall of 2013 that the two interact in the body and the resultant complex binds to DNA much more strongly than aspartame alone.[272]

The trace element molybdenum and copper are antagonistic, meaning the former prevents molecules in the blood from binding to the latter, and thus, more copper is excreted from the body. Molybdenum has been used as an effective copper-chelating agent in the form of tetrathiomolybdate.[273]

University of Kentucky researchers from the Department of Pharmaceutical Sciences found that copper levels were significantly high in various types of cancer tumors, including reproductive and digestive system malignancies, lung tumors, and leukemia. Using the copper chelator D-penicillamine (D-pen), the researchers were able to effectively reduce cytotoxicity in both breast and leukemia cancers by substantially lowering copper concentrations in those tissues.[274]

Sulfur binding blocks copper retention in the body, underlining yet another good reason to make sure your diet includes a healthy dose of food-based sulfur compounds like those found in garlic or whole eggs.

At the Natural News Forensic Food Lab, I found very high levels of copper in many multivitamins, including "natural" vitamins for children. Copper is deliberately added in high concentrations to these vitamins in order for the vitamin manufacturer to achieve a "100% RDA" rating on the product label. Often, children's vitamins are formulated to deliver 2,000 micrograms of copper per day. For example, Flintstones Children's Complete Multivitamin Chewable Tablets deliver exactly that amount (2,000 micrograms, or 2 mg), yet according to the Dietary Reference Intakes Tables and Application from Institute of Medicine of the National Academy of Sciences, a four-year-old child should only be consuming 440 micrograms per day, with an upper limit of 3,000 micrograms per day.[275]

Flintstones multivitamins, in other words, are delivering over 450 percent of the recommended daily allowance of copper for a four-year-old child.

Given the potential toxicity of copper in children, this seems very concerning. (I have begun urging parents to avoid feeding their children multivitamins that deliver over 500 micrograms of copper.)

Considering all the other sources of copper intake (including contamination from copper pipes), 2,000 micrograms from a multivitamin alone may be hazardously excessive, especially when that multivitamin is taken daily alongside other dietary sources of copper.

Here are some of the highlights of our findings on copper:

- Over 75 ppm in a sea vegetable supplement for pets
- Over 1,200 ppm in children's multivitamins based on cartoon TV characters
- Over 3,000 ppm in an organic line of "raw" multivitamins
- Over 5,000 ppm in a popular mineral supplement

Because the daily intake range of efficacy versus toxicity is so narrow, copper intake sources need to be meticulously monitored to avoid a copper overdose. Given the widespread use of copper pipes in residential and commercial construction, it is reasonable to conclude that many people may be taking in a very high dose of copper from water pipes alone, and that when multivitamin or mineral supplements are added to that intake, they are exceeding safe copper limits, potentially unleashing disastrous health effects.

50

Sn

Tin

118.711

TIN (Sn)

ATOMIC NUMBER: 50

GROUP 14: CARBON, SILICON, GERMANIUM, AND LEAD

The use of tin has been dated as far back as 3,000 B.C., culminating in some 5,000 years of interaction with human society. Trademark to its best-known properties, when a bar of tin is bent, the silvery-white crystalline metal makes a crackling sound known as the "tin cry."

Dating back to the 1880s, workers exposed to tin vapors complained of symptoms such as headaches, nausea, and general fatigue. A medical treatment for skin infections using triethyltin was known to poison more than 200 patients in the 1950s in France, ultimately killing more than one hundred of them via cardiac arrest, coma, or serious complications from convulsions, while those affected but not killed frequently suffered years of ongoing headaches and visual complications.[276] However, health risks from tin exposure are not nearly as acute as risks from more toxic metals such as lead and mercury.

While tin is familiar in its many industrial uses, its organic form, organotin, poses the most risk to humans due to its toxic effects. These forms include trimethyltin and triethyltin, both of which exhibit neurotoxicological effects.[277]

Initial studies on tin found relatively low toxicity in its inorganic compounds, which were poorly absorbed, limiting accumulation to low levels.[278] The same data show that organic tin compounds synthetically produced starting in the 1960s posed potential problems for the growth, survival, and continuity of animal species, while observed disruptions in exposed animals' behavior suggest neurological development issues. Animal studies involving tin(II) fluoride and tin(II) chloride have been shown to reduce or inhibit the natural functions of the liver.[279,280,281] Tin(II) tartrate was also found to cause a decrease in the antioxidant glutathione, ultimately leading to liver damage in animal studies.[282]

The public's biggest exposure to tin comes from foods containing trimethyltin and triethyltin organic compounds, with a focus on seafoods and canned foods. Like other metals, tin has been found to accumulate in shellfish and other seafood. According to the Food Safety Authority of Ireland (FSAI), fish is the number one tin-accumulating culprit, food-wise.[283] In environmental impact studies, trialkyltins have been found to be the most toxic on algae species.[284]

Because tin is relatively noncorrosive, it is almost ubiquitously used in food cans. Tin may leach into the food stored inside those cans, though the amount varies widely based on a multitude of factors including the type of food, pH of the food, food additives present, and so on. Canned foods—some 90 percent of which use tin compounds—with high pH levels have been found to contain between 100 and 500 ppm of tin.[285]

Although trace amounts of tin can be found naturally in water, inorganic tin-based pesticides and industrial waste are substantially increasing the tin contamination of waterways.[286]

Agri Tin®, registered trademark of the Nufarm brand, is a fungicide/pesticide composed of triphenyltin hydroxide and used on potatoes, sugar beets, pecans, and other crops. Its product label warns that, like other toxic substances, it must be handled by a certified applicator due to the danger of "affecting fetal development" and its general carcinogenicity at high doses.[287]

These issues may be amplified by the fact that tin has a strong affinity for soils, persisting in soils for long periods of time while presenting the possibility of bioabsorption and accumulation through human consumption.[288]

"Swallowing large amounts of inorganic tin compounds may cause stomachache, anemia, and liver and kidney problems," says the Tin Compounds fact sheet for the Agency for Toxic Substances and Disease Registry. "Humans exposed for a short period of time to some organic tin compounds have experienced skin and eye irritation and neurological problems; exposure to very high amounts may be lethal."[289]

Tin is commonly referred to in industrial uses by the term *stannous*, reflecting the Latin base for tin, *stannum*, and its abbreviation on the periodic table, Sn.

Stannous fluoride—composed of tin(II) and fluoride—is one of the most common types of fluoride applied in dentistry to prevent cavities (administered orally) and put in many toothpastes to prevent tooth decay, often under the trade name "Fluoristan," a generally more expensive variation of fluoride that has been used in formulas for name brands such as Crest and Oral B. The fluoride salt is also added to some municipal water supplies, though the sodium fluoride and fluorosilicic acid species are more common. Stannous fluoride carries the risks of other fluoride compounds, including osteosarcoma, osteoporosis, and fluorosis.[290]

Fluoristan (containing tin and fluoride) has caused death in at least a few acute cases of poisoning. In January 1979, the parents of a three-year-old boy in New York were awarded $750,000 after the child ingested a lethal dose of stannous fluoride gel that was spread on his teeth to prevent decay.[291] The hygienist had reportedly failed to give proper instruction to spit out the solution, instead neglecting to prevent the young child from swallowing some 45 cubic centimeters of 2 percent stannous fluoride solution, estimated by the Nassau County toxicologist to be three times a deadly amount for the boy's size and weight.

Studies involving humans have confirmed that tin competes with zinc, so too much dietary tin can decrease the amount of zinc a person can absorb.

One way to remove heavy levels of tin in the body is through the chelation properties of quercetin, a flavonoid found in many fruits and vegetables, according to a study that revealed it was effective for removing stannous versions of tin.[292]

Overall, though tin poses some dangerous neurotoxic and carcinogenic effects, its potential for cancer and disease via chronic, low-level exposure throughout the food chain is typically lower than many other common heavy metals. It is not as absorbable and often less accumulated than more problematic peers such as mercury, lead, cadmium, and arsenic.

At the Natural News Forensic Food Lab, we don't currently test for tin. It requires a unique methodology, so it's not something we currently track in foods.

CHEMICAL CONTAMINANTS

The sophistication of science in developing new chemicals for application in industry, businesses, homes, schools, agriculture, and food preparation has ultimately led to the widespread contamination of the food we eat.

What's worse, while everyone is focused on fat, calories, carbohydrates, and, in some cases, food preservatives, few are really paying attention to chemicals that may be unlabeled in their groceries or even in the packaging that their food comes in.

The body assembles, binds, absorbs, or passes various nutrients through the elements and compounds we ingest or inhale, depending on the chemical structure. But many complex lab creations aren't recognized by the body so they cannot be properly processed. As a result, they often become hazardous to our health.

BISPHENOL A (BPA)

Scientific literature is awash with studies on the numerous and compounding health issues surrounding the chemical compound bisphenol A or BPA. Since early on in the twenty-first century, BPA has also become a worrisome consumer issue, with widespread campaigns to end its use and to avoid products that contain it.

Though papers on its synthesis were published as far back as 1905, commercial production of BPA did not start until 1953, when both Bayer and General Electric independently began the development of polycarbonate materials with BPA as a key component.[293] BPA's infinite uses in modern convenience goods was indispensable, as it produced clear and virtually unbreakable plastic that was readily molded into practically any shape and size, and today it is used for every application you might imagine: in household appliances, business utility, construction, electronics, medical equipment, dental sealants, eyeglasses, and, especially, food containers.

Some tin-containing food cans are often lacquered with a bisphenol A lining. BPA leaches from these cans into food, particularly with acidic fruits and vegetables such as tomatoes and tomato-based foods. This realization has led to widespread concerns, as BPA has been tied to hormonal disruption and reproductive dysfunction. The amount of risk posed by this chemical is unclear; it builds up over time and accumulates in lipids, contributing to potentially long-term imbalances in sex function.

Today, several billion pounds of bisphenol A are produced annually (to say nothing of other plastics components), compared with just 16 million pounds in 1991[294]—and a sizable portion of that BPA is made in the United States.

BPA is a synthetic carbon-based chemical created for use in plastics and epoxy resins. Bisphenols are characterized by their typical diphenylmethane structure, meaning that two phenyl groups are joined with methane and hydrogen. Type "A" is not the only type by far. Other bisphenols make up a long list of laboratory derivatives and variations, including bisphenols: -AP, -AF, -B, -BP, -C, -E, -F, -G, -M, -S, -P, -PH, -TMC, and -Z. Bisphenols -A and -S are most widely publicized for, and commonly associated with, causing endocrine disruption in living things both human and animal.[295]

By disrupting hormones in the endocrine system, distorting gene signals, mimicking estrogen, and even amplifying the effects of estrogen in the body, bisphenol A contributes to reproductive problems, including infertility and birth defects, developmental issues, autism, diabetes, obesity, cancer, irregular heartbeat, and a host of other health detriments. Both a fetus and a developing child are particularly vulnerable to BPA health hazards, as the xenoestrogen imposes itself in estrogen receptors both in genes and within cells in the delicately developing brain, immune system, and body at large.[296] Like other toxins, BPA poses the biggest threat through long-term exposure.

Bisphenol's estrogen-mimicking ability should not be a surprise to anybody who knows the chemical's history; although it was first synthesized in 1891, it became more widely known in the 1930s when a London chemist named Edward Charles Dodds studied it in his attempt to develop an estrogen-replacement therapy.[297]

Despite industry claims that bisphenol A is quickly excreted from the body, studies have found that it is stored in body fat.[298] At least ten studies have found BPA in human fetal tissues, including umbilical cord blood.[299] Even trace amounts of bisphenol A may cause harm; ultimately, BPA has been demonstrated to be a possible carcinogen, triggering prostate[300] and breast[301] cancers in animal studies.

Recent research led by Dr. Ruth Lathi, a Stanford University reproductive endocrinologist, found that high levels of BPA could also raise the chances for miscarriage in women who have had difficulty getting pregnant.[302] Of her study, Lathi told the Associated Press, "It's far from reassuring that BPA is safe."[303]

Bisphenol A is everywhere: Exposure to bisphenol A in humans is so widespread that urine tests find it in nearly everyone in Western society, with well over 90 percent of the population registering at least trace levels of the chemical. This contamination comes almost exclusively from food and beverages, with minimal environmental exposure through air and water.

BPA is most widely found in food cans, as it is frequently used in the interior lining, where studies have found it leaches into foods and beverages, especially when cans are exposed to heat. It is additionally found in hard, clear plastics like those with recycling numbers 3, 6, and 7—namely PVC, polystyrene, and polycarbonates. Most thermal printer receipts contain BPA, including those used in grocery stores, gas stations, and retail outlets, and BPA also shows up in paper currencies, where it is absorbed through the skin via handling.[304]

Perhaps even more alarming, until its recent ban by the FDA in 2012, BPA was found in nearly every baby bottle and sippy cup sold in consumer markets, suggesting a footprint of chemical exposure that may have affected a majority of the vulnerable and developing youth for the past several decades throughout the developing world.

Studies have also shown the chemical can actually cause subtle genetic alterations that can be passed from generation to generation. So the long-term implications of widespread, continual BPA exposure in the most defenseless members of the human race are not inconsequential.

Shockingly, despite the fact that the FDA banned BPA in baby products and despite a wealth of information on the dangerous nature of the chemical, the FDA's position on BPA posted on the agency's website currently states that, because BPA was approved for food contact use over forty years ago, the FDA's regulatory structure limits their ability for oversight.[305] After its approval, hundreds of formulations for BPA epoxy linings were created, and none of the companies that created them were required to inform the FDA. As such, the agency claims that, "if FDA were to decide to revoke one or more approved uses, FDA would need to undertake what could be a lengthy process of rulemaking to accomplish this goal." So, just because the agency might have rushed to judgment in declaring BPA safe, and because it would be such a "lengthy process" to revoke government-approved uses, does that really mean it shouldn't be done if it is in the best interests of humanity and the environment?

While further research may be required to determine the exact dosage, mechanisms, and circumstances in which real world harm can occur, there are many studies that show BPA is a serious issue and that exposure needs to be reined in. However, regulatory agencies have avoided fully admitting these risks.

A 2008 National Toxicology Report expressed only "some concern" for BPA's effects on the brain, behavior, and the prostate gland for prenatal and early-age exposure based on current levels of exposure, and even less concern for its possible links with early onset of puberty, birth defects, low birth weight, or infant mortality, or for that of reproductive effects in adults including those who suffer workplace exposure.[306] Again, this is despite a plethora of independent studies confirming the detrimental health effects of the chemical.

One of the most pervasive aspects of BPA's impact, and that of industrial chemicals in general, is the cost trade-offs between rising food costs and the ability of the food industry to better preserve foods with a longer shelf life, a strategy that aims to make food more affordable to the consumer. This effort to make food more affordable results in a need for long-lasting ingredients, processed blends, meals, mixes, and preservatives both in the food and in the packaging, as well as elements that help create a desirable appearance, color, and smell. For many foods, this results in a seemingly endless list of ingredients and additives that few could navigate without significant knowledge of food science, leaving undereducated sectors of society and children particularly vulnerable.[307,308]

Both critics and academics have suggested that lower-income families may be exposed to more BPA and other toxins. Diabetes, linked with BPA, disproportionately affects African-American[309] and Hispanic[310] individuals, those living at or near poverty lines,[311] and those living in inner-city urban areas.[312,313] There is ample evidence to suggest that the affordability of these foods may very well play a significant factor in determining the quality and level of food safety in someone's diet.[314,315,316,317,318]

Adam Drewnowski, Ph.D, a professor in epidemiology at the University of Washington in Seattle and the director of the Nutritional Sciences Program, has argued that this creates a difficult dilemma in lowering the risk of poverty-stricken families who are often faced with "nutrient-poor" but "energy-dense" and "good taste, high convenience" low-cost food options with a prevalence of red-flagged ingredients that contribute to or compound obesity, diabetes, and other health risks.[319] This not only goes for the process used to create the food itself, but also the packaging the food comes in.

Not only is BPA still currently approved for general food-contact use in the United States (aside from baby bottles and sippy cups), but similar rules apply in the European Union and Japan as well as many other countries.[320] Despite the lack of regulation forcing BPA's total removal in these countries, others have moved to rid their food supplies of the harmful chemical. In

2012, the French parliament voted to ban the use of BPA in *all* food containers by 2015, making France the first country to do so.[321] Sweden has also proposed a total BPA ban.[322] Despite the chemical industry's strenuous objections, the Canadian government formally declared BPA to be a "toxic substance" in 2010, paving the way for a future full ban.[323]

As knowledge regarding BPA's negative health effects becomes more common, more and more health-conscious companies are voluntarily regulating BPA out of their own products, with the number of goods encased in materials deliberately made without BPA baring "BPA-free" labels. With the de facto beginning of a voluntary ban on the use of these plastics, this necessary action, in time, will likely be recognized by the FDA, other American regulators, and other nations around the world.

As the chemical is ubiquitous and the potential health hazards too great to ignore, eliminating as many sources of BPA from one's environment as possible is the best strategy. However, there are numerous health and nutritional strategies for eliminating BPA buildup in the body and ways to flush the chemical out of bodily stores where it poses continual harm.

Minimizing body fat is the number one best way to keep down BPA. The chemical loves to latch on to lipids, and that is primarily where it gets stored in the body. So along with its many life-extending benefits, slimming down if you are overweight will leave BPA, as well as other harmful chemicals, fewer places to be stored.

The lower you are on the body-mass index (BMI), and the closer to your ideal weight you fall, the less likely you are to have as much BPA stored up. The good news is that improving your diet to avoid sources of dangerous chemicals will also go hand in hand with eating healthy and supporting weight loss and proper nutritional management.

BPA is notorious, as noted previously, for its propensity to leach out of plastics and food cans into the things we eat and drink, especially when those containers are holding warm or acidic foods. So start with seeking out labels that declare containers to be "BPA-free." Such labels have become much more common than in the past, appearing at many consumer outlets from low- to high-end, and particularly at health and wellness stores. Do not consume hot foods in plastic containers or put such containers in the microwave to heat foods, and choose wooden or bamboo kitchen utensils over plastic options. Also, try to limit canned goods as much as possible, opting for fresh or frozen foods instead when possible.

BPA isn't the only endocrine disruptor in this chemical family on the market by a long shot. Warnings have already been sounded over bisphenol S (BPS),[324] while many antibiotics, pharmaceuticals, and Big Agra foods contain other culprits for destabilizing hormones (some of which may not even have been recognized as a problem).

Questions over the safety of BPA-free alternative plastics on the consumer market have already been raised as well. Studies found that brand names like "Tritan" and "EcoCare" did indeed keep bisphenol A from leaching into liquids under recommended usages.[325] However, other researchers have raised questions about just how much is known about the alternative plastics currently replacing BPA in the consumer market. George Bittner, a professor of neurobiology at the University of Texas, and his team conducted a study on numerous plastics advertised on the market as alternatives to BPA and subjected them to stress tests. The results of that study found that the vast majority of these BPA-free containers also leached estrogen-imitating hormones that interfered with the endocrine system into food.[326,327]

While some of these results have been challenged, including in a prominent lawsuit by Eastman Plastics (the makers of Tritan), important questions remain about the safety of these products that are being used by millions around the world every day. BPA may well be best identified as the tip of the iceberg of many chemical derivatives of plastic production that pose potential health risks.

Phthalates, various types of polymer materials, and other chemical contaminates in plastics have also been suspected of causing negative health effects.[328]

Perhaps when it comes to these plastics, the precautionary principle is the most appropriate approach.

As consumers, we would all be wise to exercise some common sense and think twice about we're eating and drinking from, particularly when it involves hot and acidic foods. The convenience and flexibility of plastics gives several obvious advantages, perhaps in particular with children. However, glass, stainless steel, ceramics, and other containers are much safer bets and should be used when possible or appropriate.

We can reduce our time exposure to BPA by using foods and safe detox agents to promote the excretion of offending chemicals from the body. Leafy green vegetables, healing and edible herbs, clean-sourced animal livers, numerous fortified foods, and especially probiotics and fermented foods provide ample folate (or B-9 folic acid) to the body and help eliminate BPA, all while boosting body function and immunity.

Many of the nutritional supplements and foods—including maca root, royal jelly, black tea, and beets to name a few—that promote sex drive also help regulate the body's hormone levels and will promote the elimination of BPA and other endocrine disruptors.[329]

HEXANE

Hexane is a volatile, flammable petrochemical solvent. There are five iso-mers known as hexanes. N-hexane is the unbranched, basic version, while the other four are methylated derivatives of both butane—a flammable gas used in fuel blending—and pentane—an agent most widely known for its use in creating polystyrene foam. Obtained primarily through the crude oil refining process, hexane has a wide variety of industrial uses, including the formulation of glues and the manufacturing of many textile goods. Hexane is a primary ingredient found in many types of gasoline including jet fuel. Some 85 percent of the jet fuel used by the U.S. military for example is called JP-4, which is made up of 22 percent n-hexane.[330]

Hexane is listed as a hazardous air pollutant and its dangerous neuro-toxic effects are noted on the EPA's Technology Transfer Network. Long-term hexane exposure via inhalation causes polyneuropathy, manifesting in weak-ened muscles, blurred vision, fatigue, headaches, nausea, and numbness in the arms and legs.[331] Chronic hexane exposure can also cause dermatitis, con-fusion, jaundice, and coma.[332] Mice exposed to hexane in laboratory studies were shown to have epithelial lesions in their nasal cavities,[333] while rats in further inhalation studies showed severe flaccid paralysis and signs of axo-nopathy (a neurological disease of the axons). Pregnant rats exhibited devel-opmental toxicity including cell death, abnormal cell growth, and genetic alterations.[334,335,336] Pulmonary lesions also formed in both rabbits and mice in further studies.[337]

The last thing this description of hexane would conjure up in most peo-ple's imaginations is food. Hexane, however, serves another industrial pur-pose: The food industry uses hexane to extract proteins from soybeans and oils from other grains such as canola and corn. Many soy food additives are derived through a process that uses hexane. Soy lecithin, an emulsifier, is commonly found in a vast array of products on grocery store shelves including everything from chocolate to margarine to bread and beyond. Soy protein isolate is routinely found in everything from breakfast cereals to veggie burgers to soups and sauces; it's also added to many "health food" products such as protein bars and meal-replacement shakes. Both of these are commonly extracted using hexane. Sadly, even foods labeled as "all natural" may contain soy by-products and other ingredients that were derived using the hexane extraction process.

In addition, cornmeal and soybean meal extracted during this process are given to all grain-fed livestock in the United States, including cows, poultry, hogs, and even some farmed fish that are being raised on completely unnatural grain diets. In other words, when people eat those meats on top of their regular soy- and corn-based diets, they may be consuming an extra dose of hexane residue.

The EPA, while listing hexane as a dangerous air pollutant with neurotoxic effects, claims that hexane is not officially classifiable in regards to human carcinogenicity because there is not enough information available on hexane's carcinogenic effects.[338] As the EPA's Integrated Risk Information System notes, "No epidemiology or case report studies examining health effects in humans or chronic laboratory studies evaluating potential health effects in animals *following oral exposure* to n-hexane are available."[339]

The one study the EPA cites in its hexane-toxicity evaluation, regarding hexane and cancer in humans, was seemingly not a substantial enough study on which to base a judgment call. In the study, published in the *Journal of Environmental and Occupational Medicine*, Beall et al. (2001) determined no relationship between hexane and cranial tumors in employees at petrochemical research facilities who self-reported both cranial tumors and n-hexane exposure. However, only a small fraction—12 out of the 2,595 workers surveyed—even self-reported such tumors, and petrochemical workers are obviously exposed to substantial concentrations of chemicals on any given day, so singling out hexane as a cause for brain tumors on such a small number of cases would be next to impossible in that study design.[340]

The CDC's Agency for Toxic Substances and Disease Registry public health statement on hexane mentions a study where female mice exposed to commercial hexane for two years had an increase in liver cancer; however, the ATSDR goes on to say, "Commercial hexane is a mixture, and we do not know what parts of the mixture caused the cancer in the female mice," and that n-hexane is not characterized as a carcinogen.[341]

Most of the research on hexane's toxicity is based on it being inhaled instead of it being ingested, due to the fact that hexane would normally never be ingested under natural circumstances; inhalation is typically regarded as the primary route of hexane exposure. Now that hexane is so widely used in modern food production, however, there is no telling how much hexane is being ingested by someone subsisting on the average diet in the developed world.

Because it is added to so many foods by its numerous by-products, soy is by far the biggest potential dietary source of hexane. According to the report, "Behind the Bean: The Heroes and Charlatans of the Natural and Organic Soy Foods Industry," by watchdog group The Cornucopia Institute, "The effects on consumers of hexane residues in soy foods have not yet been thoroughly studied and are not regulated by the U.S. Food and Drug Administration. Test results obtained by The Cornucopia Institute indicate that residues—ten times higher than what is considered normal by the FDA—do appear in common soy ingredients."[342]

In 1995, the EPA released a report on the emission factors for vegetable oil processing. The report described how there are two main processes for extracting oil from soybeans, and the traditional method of using a screw press is not widely used because the efficiency is much lower than using a solvent.[343] The common approach to extracting oil from soybeans and other grains is to literally wash the grains with a solvent, and hexane is the food industry solvent of choice.

Not only are the soybeans washed with hexane/oil mixtures during this process, but they are also eventually washed with pure hexane, sometimes referred to in the industry as a "hexane bath." To desolventize the oil, the oil/solvent mixture is exposed to steam, pumped through heaters and film evaporators, and run through a stripping column, theoretically separating hexane out to get reused again and again. But not all the hexane gets removed. Ultimately, some hexane residues wind up in the foods created through this process. Even the EPA admits that small quantities of hexane are left behind after the solvent extraction is complete. The EPA has no data on when the hexane volatilizes, but the agency says it will "probably" happen during cooking, as if that is any kind of reassurance to anyone eating this stuff.[344]

At these processing plants, hexane emissions are released into the atmosphere through vents and from the transfer and on-site storage of hexane at these plants. In addition, the chemical also knowingly winds up in the plants' wastewater (though wastewater emission data has not been collected). Millions of pounds of n-hexane are released into the environment from U.S. facilities alone every single year.[345] In fact, of all the industrial uses for hexane that exist—from shoe glue to paint thinner to tire manufacturing to roofing materials to jet fuel—over two-thirds of all hexane emissions in the United States are actually released by food processing plants.[346]

One of the scariest issues with hexane-tainted foods concerns baby formulas. In another investigation, The Cornucopia Institute filed a Freedom of Information Act (FOIA) request and found nearly one hundred adverse reactions of infants fed formula with hexane-extracted DHA/ARA oils added.[347] These oils are marketed as an additive that makes formula more like breast milk and aids in brain development. Adverse effects reported include infants suffering painful gastrointestinal problems, including vomiting and diarrhea. Some parents even reported seizures. These symptoms only stopped once the formula was switched to one that did not have those added hexane-extracted oils. The problem is only compounded when the baby is given a soy formula created using hexane. Infants who rely on soy-based formulas as their main nutrition source for the first six months of their lives are relying on a food composed of a main ingredient that is soaked in a neurotoxic petrochemical bath before it gets to their bottles.

As for safety regulations, the U.S. Occupational Safety & Health Administration has set a permissible exposure limit on n-hexane at 500 ppm in workplace air. OSHA attempted to set a 50 ppm limit, but it was remanded by the U.S. Court of Appeals, so that limit is not enforced.[348]

The bottom line is that the U.S. government does not require food companies—even those that produce infant formulas—to test any of their products for hexane residues before they are shipped to supermarkets. Just because something touts the "organic" label does not automatically mean it is hexane-free. In 2009, the U.S. National Organic Standards Board (NOSB) voted to allow a de-oiled, nonorganic formulation of soy lecithin prepared with both hexane *and* acetone to remain on its approved ingredient list for inclusion in foods labeled as "USDA certified organic."[349] In addition, food companies can mislead consumers with foods that proudly bear the phrase "all-natural" on their packaging, even though there's nothing even remotely

EVERYTHING YOU NEED TO KNOW...   111

natural about washing some of the ingredients in a volatile, toxic petrochemical solvent bath.

The Cornucopia Institute has been working to get hexane in food regulated by the U.S. government, even filing legal complaints against supposed organic food manufacturers who add hexane-derived ingredients to their products, but so far, nothing significant has been done on the regulatory front to protect consumers from hexane in their food.

PESTICIDES

Large-scale, modern agricultural practices have changed the way that billions of people eat, and the heavy use of pesticides has become a substantial, even integral, part of that food production system. While there is a significant movement to buy foods such as organic produce that are not cultivated with the use of pesticides, the simple fact is that most people are regularly consuming foods grown with pesticides and often consuming a multitude of pesticide chemicals.

The widespread use of pesticides has become a worldwide trend, with farmers in virtually every part of the world adopting Western agribusiness models that include pesticide and herbicide treatment as a standard and significant input for nearly every conceivable crop. Global pesticide use has continued to increase since the second half of the twentieth century, with more than 5.2 billion pounds of herbicides, insecticides, and fungicides in use as of 2007,[350] and global sales headed toward an estimated $57 billion by 2016.[351] According to Food & Water Watch, herbicide use has increased by 26 percent in the United States just since 2001.[352]

About 89 percent of this lucrative multibillion-dollar agrichemical market is dominated by the top ten firms, and the vast majority of the money flows to the six largest companies—Bayer, Syngenta, BASF, Dow AgroSciences, Monsanto, and DuPont.

The huge role these companies play in the production of the world's food supply is compounded by the fact that they don't just produce pesticides; they are also involved in specialized seed production, including genetically engineered varieties. Bayer is the largest agrochemical producer, and also the seventh largest seed company, while Syngenta is the second largest in agrochemicals and third in seeds. Monsanto weighs in as the fifth largest in agrochemicals but the premiere giant in seeds, while DuPont follows Monsanto in both categories.[353]

In many of these cases, the proliferation of pesticides is tied to the use of crops that are genetically modified by these companies to resist specific pesticides, as with Monsanto's widely used Roundup Ready soy, corn, and other genetically modified seed varieties. But various types of herbicides, insecticides, fungicides, and other toxins are used during one or more stages of plant growth in most modern agricultural practices, regardless of whether the crop is genetically engineered or not.

At least 60 percent of the herbicides used in global agriculture (by poundage) and many types of insecticides, fungicide, rodenticides, and other pesticides have been demonstrated to interfere with the endocrine system and reproduction.[354] Glyphosate, 2,4-D, and atrazine have estrogen-mimicking properties and constitute a nearly ubiquitous presence in global agriculture, with millions of tons sprayed annually on crop acreage.

Manmade endocrine-disrupting chemicals have been connected with infertility and reduced fertility, deformities, low birth weight, reproductive and developmental issues, early onset of puberty, endometriosis, breast cancer, and other cancers.[355] Many pesticides also target the neurological and nervous systems.

As a result of the 1996 Food Quality Protection Act (FQPA) in the United States, the EPA created the Endocrine Disruptor Screening Program (EDSP) to monitor pesticides and other chemicals that exhibit activity similar to estrogen (female hormones) or androgen (male hormones), or affect the thyroid system or fish and wildlife.[356,357] The EPA identified priority chemicals for testing and monitoring, and began evaluating dozens of pesticides for endocrine disruption and possible restrictions back in 2009.[358]

Dozens of first-priority chemicals that are found in pesticides are being officially studied and evaluated for their toxic effects. Atrazine, 2,4-D, benfluralin, chlorthal-dimethyl, norflurazon, fenbutatin oxide, propargite, acephate, chlorpyrifos, diazinon, dimethoate, disulfoton, ethoprop, malathion, methamidophos, methidathion, methyl parathion, phosmet, tetrachlorvinphos, carbaryl, carbofuran, methomyl, oxamyl, carbamothioic acid, bifenthrin, cyfluthrin, cypermethrin, esfenvalerate, permethrin, piperonyl butoxide, dicofol, endosulfan, propachlor, metolachlor, flutolanil, chlorothalonil, linuron, metalaxyl, simazine, propiconazole, tebuconazole, triadimefon, myclobutanil, trifluralin, glyphosate, abamectin, toluene, isophorone, phthalates, and methyl ethyl ketone are all under review, though it is unclear how long it will take the EPA to make its final decisions, what thresholds will be considered "safe," and what actions they will take to restrict the use of dangerous compounds in pesticides.[359]

Glyphosate

Glyphosate, a derivative of glycine, known more formally as glycine phosphonate or N-(phosphonomethyl) glycine, is used as a broad spectrum herbicide. Discovered in 1971 and subsequently patented by Monsanto chemist John E. Franz,[360] today glyphosate is the most widely used herbicide in agricultural production, with more than 15 million tons used annually. Roundup, Monsanto's best-selling hallmark product, is composed of 41 percent glyphosate, which makes it the primary active ingredient.[361]

Glyphostate usage drastically increased as crops genetically modified to be glyphosate-resistant came to the market. Patented Roundup Ready GMO seeds were developed for use in conjunction with Roundup pesticides, allowing farmers to essentially douse entire fields and kill weeds while preserving the glyphosate-resistant crops. Pesticide-resistant genetically modified varieties of corn, soybean, canola, cotton, alfalfa, and sugar beets are all widely used today and are sold sterile, preventing their reproduction and requiring farmers to buy new patented seeds each year rather than saving them, thus increasing profits for the issuing company.

However, since the introduction of Roundup Ready soybeans in 1996 and Roundup Ready corn in 1998, the heavy use of pesticides required to cultivate these genetically engineered seeds has triggered new agricultural problems that have only increased the prevalence of pesticides. Today, pesticide-resistant "superweeds" have become a worsening scourge on agriculture, with even the American Chemical Society, one of the largest industry organizations, recognizing that pesticide use has doubled and even tripled, increasing costs for farmers while simultaneously giving rise to unwanted growth that is increasingly unchecked.[362] Overall, there has been a threefold increase in herbicide-resistant weeds found in farm fields from 2001 to 2011. Sounds like a real money maker for the agrichemical business.[363]

Indeed, scientists have recognized chemical resistance to be inevitable, and chemical firms have taken notice of the problem. The dwindling effectiveness of glyphosate, used so universally on the world's biggest cash crops, has forced farmers to find a new approach to protect crops and harvest earnings.[364] New strategies are being aggressively rolled out to prevent the "useful lifetime" of herbicide-resistant genetically modified seeds and herbicides like glyphosate from being "cut short."[365]

In 2013, the EPA raised glyphosate residue tolerance limits based on a petition the agency received directly from the company that produces the most glyphosate herbicide in the world, Monsanto.[366] The Big Agra giant's request was quietly fulfilled, in some cases doubling tolerances.

Allowable amounts of glyphosate in oilseed crops such as soybeans and flax were officially increased from 20 ppm to 40 ppm. In sweet potatoes and carrots, the limit was raised from 0.2 ppm to 3 ppm and 5 ppm, respectively. Glyphosate limits on some food crops such as potatoes were raised from 200 ppm to an astonishing 6,000 ppm. Consumer protection group GMWatch began sounding the alarm, warning that genetically modified foods have shown a pattern of continual increase in the amount of herbicides necessary over the years since they have been introduced, and citing research showing that simply raising the oilseed levels from 20 to 40 ppm elevates them to over 100,000 times the concentration necessary to cause human breast cancer cells to grow in a lab.[367,368]

In the Federal Register listing the FDA's decision, under the question, "Does this action apply to me?" the agency wrote, "You may be potentially affected by this action if you are an agricultural producer, food manufacturer, or pesticide manufacturer." What about if you are simply someone who buys and eats food in America?

At the same time the FDA was upping the allowable limits for glyphosate residues in food, the nation of El Salvador actually outright banned the herbicide (along with fifty-two other chemicals) altogether.[369]

As a result of increasing tolerance to herbicides by weeds, many industry voices are advocating a basket application of glyphosate in addition to several other herbicides used in rotation or combination to deter superweeds not effectively dealt with by the one-size-fits-all approach that had been used on many large-scale monoculture farms. Biotech corporations are now advocating that farmers approach crop control with a varied technique: include not only a variety of herbicides but a rotation of both crops and herbicide-tolerant traits.[370,371] Bayer CropScience is expanding the platform for its Liberty-brand herbicide by promoting the use of LibertyLink-brand genetically engineered seeds, which are resistant to Liberty's active ingredient herbicide, glufosinate-ammonium.[372] Likewise, Monsanto has rolled out its Roundup Ready Xtend Crop System line[373] to manage issues of glyphosate-resistant weeds by introducing stacked genetically modified–resistant traits, such as both glyphosate and dicamba or glufosinate.[374]

The chlorophenoxy herbicide 2,4-dicholorophenoxyacetic acid (2,4-D) that was infamously used as the active ingredient in Agent Orange[375]—a defoliant sprayed during the Vietnam War that harmed millions of both Vietnamese people and American troops—is now being promoted as a popular alternative and complement to a glyphosate approach. It is one of the most widely used, highly toxic pesticides in existence.

All of this leads to greater and greater volumes of glyphosate and other pesticides dispensed during crop production, increasing the potential exposure for consumers, in ground and surface water, and in the environment in general. Scientists have found the potential for gene flow between genetically engineered crops given trait resistance to weed species, causing a decrease in the effectiveness of herbicides in controlling competing growth.[376]

Despite the widespread use of glyphosate in commercial agriculture, neither the FDA nor the USDA test for glyphosate residue on food in either the FDA's Pesticide Residue Monitoring Program (PRMP) or the Department of Agriculture's Pesticide Data Program (PDP).[377]

A study on the negative health impacts of human exposure to glyphosate concluded that, "Negative impact on the body is insidious and manifests slowly over time as inflammation damages cellular systems throughout the body."[378] If this study holds true, it indicates that glyphosate could be inflicting long-term, virtually untraceable damage to the health of millions of individuals.

Researchers from MIT and former government environmental contractors found that glyphosate "enhances the damaging effects of other food-borne chemical residues and environmental toxins" by interfering with certain enzymes and healthy gut bacteria levels. By magnifying unhealthy toxins and contributing to chronic inflammation, glyphosate contributes to a wide range of ailments including gastrointestinal disorders, obesity, diabetes, heart disease, autism, Alzheimer's disease, infertility, and cancer.[379,380]

The organization Earth Open Source compiled existing data and scientific papers to demonstrate that glyphosate exposure is linked to a variety of birth defects but that industry and government entities have done little to warn the public or stop potentially harmful effects from occurring.[381,382]

In February 2016, the FDA reluctantly announced it would begin testing foods for glyphosate, but the agency failed to declare what laboratory methodology it would use, leaving open the possibility that they would rely on a method that could be deliberately chosen to demonstrate very low "recoverability" rates (effectively downplaying the actual glyphosate contamination of foods). [383]

Triazines

Triazines are organic chemicals made up of three carbons, three hydrogens, and three nitrogens in a heterocyclic ring, with several herbicides and the toxic substance melamine as key derivatives. Atrazine, simazine, and propazine are all well-known and widely used chlorinated triazine pesticides.

Atrazine is the active ingredient in the best-selling herbicides sold by agrichemical and seed giant Syngenta. Their products have been some of the most widely used pesticides in the United States for decades, with more than 75 million pounds applied each year to corn, sorghum, sugarcane, and other crops and golf courses, lawns, rights-of-way, and other locations to control broadleaf and grassy weeds.[384] Altrazine is also one of the older pesticides still in use, having been first registered back in 1958.

This class of herbicides has been closely tied to neuroendocrine developmental and reproductive effects in laboratory studies, and there is ample reason to believe this pattern of toxicity could impact both human health and environmental systems.[385]

Along with both neurological- and endocrine-disrupting toxic effects, atrazine has been shown to synergistically amplify the harmful attributes of other pesticides, such as organophosphates, through an oxidative enzymatic process for a greater total toxicity.[386,387] However, this is not the case in every combination with every chemical.

Banned in the European Union for its known harmful effects since 2004, atrazine has been identified as a major water supply contaminate inside the United States, with the EPA and other regulators acknowledging widespread incidents of the herbicide above the set Maximum Contaminant Level of 3 ppb.[388] The World Health Organization has set a guideline value for atrazine in drinking water of 2 µg/L (ppb).[389]

More than 2,000 water districts in the U.S. Midwest filed a class action lawsuit against Syngenta over high levels of groundwater contamination from cropland runoff of atrazine, which affected the drinking water of a reported 52 million Americans.[390] The case was settled for $105 million, with Syngenta agreeing to pay filtration costs while admitting no liability and maintaining that "no one ever has been or ever could be exposed to enough atrazine in water to affect their health."[391]

Atrazine is the most commonly detected pesticide in U.S. waters,[392] and it has been linked to altering the sex of amphibians, including the development of female sex organs and eggs, and hermaphroditism in male frogs at levels in the water of 0.1 ppb, contributing to a decline in the species.[393] A study on atrazine's effect on plankton found that it inhibited photosynthesis and slowed phytoplankton growth in ponds at a relatively low concentration of 1 to 5 µg/L.[394]

Another study found that poplar trees were able to uptake and metabolize atrazine into less harmful derivatives, suggesting that trees and other vegetation could be used to remediate tainted soils.[395]

Simazine and propazine pose similar issues of runoff and toxic environmental contamination where they are used as herbicides.[396,397]

Organochlorines

Composed of chlorine, carbon, and hydrogen, organochlorines are one of the most widely used classes of agricultural chemicals, used particularly as an insecticide. Many of its compounds are of serious concern as "persistent organic pollutants" readily biomagnify in animals and the environment, posing risks throughout the spectrum of life. Like other organic pesticides composed with hydrocarbons, organochlorines are known for their neurotoxic effects.[398]

"Reproduction and endocrine dysfunction, immunosuppression, and cancer" are just some of the known impacts on human health associated with these compounds. The United Nations Environment Programme (UNEP), under advisement from the International Programme on Chemical Safety (IPCS), classified many of the known and likely carcinogens that fall under organochlorines, including many types of pesticides.[399]

Though some of these are still in use, many have been banned or heavily restricted in nations throughout the world in light of evidence about their potential and actual harm. Among these are dioxins, DDT, heptachlor, pentachlorophenol, 1,1-dichloro-2,2-bis-(p-chlorophenyl)ethylene (DDE), mirex, aldrin (which converts to dieldrin), chlordane, dieldrin, endrin, hexachlorobenzene, polychlorinated biphenyls (PCBs), polychlorinated dibenzo-p-dioxins, polychlorinated dibenzofurans, and toxaphene. According to the UNEP, researchers, and many other authorities, "convincing substantive evidence exists for the actual and potential toxic impact of these substances."

The United Nation's Stockholm Convention on Persistent Organic Pollutants established a treaty in 2001 in an attempt to limit organochlorine use.[400] However, studies have found dozens of these compounds throughout the developed and developing world, in spite of bans on their use in many countries.[401]

The environmental persistence of these toxins and their ability to transport across the food chain after being produced under industrial conditions poses an ongoing threat. For instance, samples show significant levels of organochlorine pesticides still accumulate in many soils, even decades after their use is discontinued. These compounds don't break down easily and are attracted to fat tissue, where they are readily stored.

People are primarily exposed to organochlorines through their diet. Organochlorines are most prevalent in animal fats, meat, and dairy,[402] but have been known to accumulate significantly in other sectors of the food supply as well.[403]

DDT (dichloro-diphenyl-trichloroethane) was one of the most widely used insecticides during the middle of the twentieth century, playing a significant and celebrated role in combating malaria around the world. Later, it became infamous as an environmental super-toxin in Rachael Carson's 1962 *Silent Spring*, which sparked media coverage, global concern, and bans on DDT, as well as the efforts of environmentalists and U.S. Fish and Wildlife Service employees that led to wider reforms and played a role in the creation of the U.S. Environmental Protection Agency. DDT, however, continues to have its defenders, including expert opinions and industry players that have disputed the evidence for its toxicity and carcinogenic effects,[404,405,406,407,408] particularly in the wake of the book's release.[409,410,411] Regardless, studies continue to produce ample reason for concern, and the precautionary principle should apply for those making an effort to reduce their exposure to toxins.

Other studies have shown that DDT exposure has been connected to breast cancer,[412] testicular cancer,[413,414] and endocrine disruption, while animal studies have found it to be tumor promoting.[415] DDT, along with many other organochlorines including DDE, dieldrin, methoxychlor, dicofol, and many types of PCBs, is an estrogen mimicker, and therefore can affect the endocrine system, fertility, prenatal development, and sexual function. Ultimately, many organochlorines are suspected human carcinogens and linked to cancers in animal studies, including aldrin, dieldrin, chlordane, DDT, HCH, heptachlor, toxaphene, and dicofol.[416,417]

A study lasting more than a decade found that exposure to persistent organochlorines increases the risk of diabetes, even at relatively low levels of contact.[418] Studies in New York linked DDE to a significant increase in breast cancer, suggesting that other organochlorines may also play a role.[419] Meanwhile, some studies have considered the impact of combinations of organochlorines in estrogen-like effects on the human body, including their contribution to reproductive issues,[420] breast cancer,[421] and testicular cancer.[422]

With the lingering presence of organochlorines in the environment and their propensity to bioaccumulate up the food chain while resisting degradation and metabolizing, it is worth noting that certain species of fungi have been found to break down persistent organic pollutants (POP) and other pesticide pollutants. The white rot fungi species Phlebia was found in studies to remove between 71 percent and 90 percent of the pervasive toxin heptachlor, giving promise to strategies to reduce levels of these harmful synthetic chemicals.[423]

Organophosphates

Organophosphate pesticides, another of the most widely used types of insecticides, are recognized by the EPA for their toxic effects on life, with data showing adverse effects on humans and wildlife, especially to bees, which can suffer acute deadly poisoning.[424,425] They have known effects on the nervous system in acute as well as chronic doses, and also count among their class several types of nerve agents, including Sarin and VX gases.

Malathion and parathion, among other organophosphates, became popular substitutes for organochlorines, as the highly toxic phosphates are more readily broken down under environmental conditions than the "persistent" chlorines and thus are assumed to be safer. Parathion was first synthesized in 1944 by IG Farben chemist Gerhard Schrader, who also discovered Sarin and

Tabun nerve agents for the first time in the late 1930s under Nazi scientific research efforts.[426] Schrader's pioneering research into organophosphates was carried over into Bayer AG, which broke off from IG Farben after World War II, while global chemical companies used his research to develop a commercially successful parathion pesticide.

Parathion is one of the most toxic substances used in agriculture. It is listed as a possible human carcinogen under EPA guidance[427] and further as a known endocrine disruptor, according to several authorities.[428] It has impacted the reproduction, development, and behavior of amphibians, fish, and aquatic life, and is highly toxic to honeybees.[429] Ethyl parathion is acutely toxic in high doses, and poses an immediate risk to farm workers and others.[430] Hundreds of workers have been killed or severely harmed by this pesticide. Its application and usage is severely restricted, and it has now been banned in the United States; all registered use ceased in 2003.[431]

The EPA banned the use of organophosphates for residential applications in 2001.[432,433] However, it remains a mainstay of commercial agriculture, though there are signs that many varieties will eventually be phased out. Studies have shown numerous connections between organophosphorus pesticides and poor neurobehavioral development, particularly with regard to exposure to fetuses and young children.

A May 2010 National Health and Nutrition Examination Survey confirmed a connection between organophosphate pesticide levels in the urine of children and the prevalence of attention-deficit/hyperactivity disorder (ADHD).[434] This prompted widespread media coverage on CNN and other outlets warning consumers that pesticide exposure could be harmful and recommending organic and local produce, both of which have significantly lower levels of pesticide residue, as a route to minimize risk.[435]

New research continues to show that the organophosphates already understood to be harmful to the nervous system and brain can also affect short-term memory, reaction time, and other neurological and developmental issues, even from chronic low-level doses,[436] particularly during the vulnerable periods in the womb and first few years of life.[437,438,439] Occupational exposure through farming is also a significant source for neurotoxicity[440] and nerve-function decline.[441] Berkeley researchers found that exposure in the womb to organophosphates, including among pregnant farm workers, correlated with a five-point drop in IQ measured in children at seven years of age.[442]

Among the most common organophosphate pesticides and other compounds that have been or are still in use are parathion, methidathion, malathion, chlorpyrifos, diazinon, trichlorfon, dichlorvos, monocrotophos, dimethoate, dicrotophos, oxydemeton-methyl, disulfoton, mevinphos, methamidophos, acephate, stirofos, profenophos, sulprofos, isofenphos, azinphosmethyl, phosmet, and dialifor herbicides, including phenoxy herbicides.

Fenitrothion, another organophosphorus pesticide, has exhibited anti-androgen effects in studies on insect species,[443] demonstrating further endocrine-disrupting results of estrogen and androgen mimics among chemical pollutants.

Pyrethroids

Pyrethroids are another class of organic pesticides that offer both powerful effects as insecticides and the potential for neurotoxicity as, essentially, another type of nerve agent.[444] They have been used worldwide to control mosquitoes and other flying insects, as well as in agricultural and household applications.[445] Mosquito-control departments across the world have sprayed pyrethroid compounds into their municipalities to undermine mosquito populations.

Pyrethroid compounds are commercially stabilized from naturally occurring pyrethrins, which are extracted from chrysanthemum flowers grown in and near Kenya that hold natural botanical insecticidal properties. However, these pyrethrins are not very persistent and typically break down in sunlight and water,[446] so scientists in the early twentieth century synthesized pyrethrins with agents such as MGK-264 and piperonyl butoxide to preserve the pyrethrins and allow them to resist breaking down in the environment. While this makes them more potent insecticides, it also enhances their risk of bioaccumulating in mammal fat tissues and in the food chain.[447]

Allethrin, bifenthrin, cyfluthrin, cypermethrin, cyphenothrin, deltamethrin, esfenvalerate, fenpropathrin, fenvalerate, flucythrinate, flumethrin, imiprothrin (Raid-brand products), lambda-cyhalothrin, metofluthrin, permethrin (Biomist-brand products), prallethrin, resmethrin (Scourge-brand products), sumithrin (Anvil-brand products), tau-fluvalinate, tefluthrin, tetramethrin, tralomethrin, and zeta-cypermethrin are all pyrethroids.[448,449]

Pyrethroids pose a danger to bees and other beneficial flying insects, and they have known neurotoxicological impacts on people and the environment. They excite neurological activities and can overstimulate the nervous system; exposure to these compounds has been connected to repetitive sensory organ signals and activity, delayed reaction time, motor nerve and skeletal muscle fiber issues, neurotransmitter-release enhancement, certain negative cardio-vascular effects, respiratory irritation, paresthesia, and sensory irritations.[450,451]

Pyrethroid compounds have been found to pollute surface waters, impacting the populations of aquatic invertebrates that fish and other wild-life depend on. A study of California's Sacramento–San Joaquin River Delta exposed major contamination from runoff and waste disposal, at levels concentrated enough to pose acute toxicity to amphibians.[452] The common use of pyrethroids suggests that such waterway pollution may be widespread.[453]

Additionally, thousands of textile factory workers in China have been exposed to unusually high levels of pyrethroid insecticides that are used to treat cotton, wool, and other textile materials, highlighting another route of potential exposure that may be occurring on a larger scale in the workplace.[454]

Carbamates

Carbamates are a class of organic pesticides composed of carbamic acid. Like organophosphates, they toxify both insects and mammals by blocking an enzyme called acetylcholinesterase, which is important for regulating neuro-transmissions in the nervous system.[455] An inability to break down acetylcholine after signals are passed across the synaptic gap leads to exhaustion and nervous shock in the organism. Carbamate toxicity in humans also triggers depressed levels of red blood cell cholinesterase.[456] Key carbamate insecti-cides include aldicarb, carbofuran, ethienocarb, fenobucarb, oxamyl, and methomyl.

Fenoxycarb, a relatively new and effective mosquito control[457] and insecti-cide agent, stands apart in the carbamate class as a juvenile hormone mimicker, preventing the transformation of egg to larva, or larva to pupa, or other stages of insect metamorphosis.[458] Unlike other carbamates, it is not a neurotoxin and is generally less harmful to vertebrates, though still toxic.[459] Its hormonal activity makes it an endocrine disruptor, and, like many other pesticides and industrial chemicals, it can interfere with the life cycles of crustaceans.[460]

Propoxur (insecticide), carbaryl (nematicide), and chlorpropham and propham (plant growth regulators) are all listed as suspected human carcinogens, though they are not classified as such by the International Agency for Research on Cancer (IARC).[461] Propoxur is highly toxic to honeybees.[462]

Arsenal pesticides

As previously discussed in the arsenic section, many prominent pesticides of the late nineteenth century and much of the twentieth century were composed of toxic arsenic compounds.

"Paris Green" [copper(II) acetate triarsenite],[463] lead arsenate, calcium arsenate, and numerous other types of arsenic-based pesticides were widely used in agriculture and other applications until other pesticides, such as DDT, replaced them.[464] Later, most of these arsenate pesticides were banned or discontinued based on the risk of heavy metal toxicity. Chromated copper arsenate (CCA) was almost universally used in nearly all treated lumber for several decades until it was voluntarily discontinued from most uses toward the beginning of the twenty-first century.[465]

A few arsenic-based pesticides are still used in the developing world, but most usage has now ceased. However, decades of heavy use have led to persistent contamination in a significant percentage of soils where arsenic was used as a pesticide in orchards, row crops, and other types of agriculture,[466] as well as areas near treated lumber.[467] This practice led to arsenic accumulation in foods, as well as other toxic ingredients used in these pesticides, such as lead and chromium VI. See the arsenic section (beginning on page 14) for more information on this issue.

Neonicotinoids

Neonicotinoids are among the newest class of pesticides, synthesized by industry only since the early 1980s. Similar in composition to nicotine, they stimulate acetylcholine areas in nicotine receptors, creating excitotoxicity.[468] These compounds target neurological behavior and the nervous system, as do many other pesticides, but were implemented to replace alternatives like organophosphate and carbamate pesticides that are considered more toxic to mammals.

The neonicotinoid imidacloprid has become the world's most popular insecticide,[469,470] while other neonicotinoids, including acetamiprid, clothian-idin, sulfoxaflor, nitenpyram, nithiazine, thiacloprid, and thiamethoxam, have also been rolled out as insecticides and pesticides.

Studies on the full impact of neonicotinoids on human health are still emerging, but so far, just as with every other chemical pesticide, the results do not look very promising. The immunosuppressive and cytotoxic effects following twenty-eight days of oral imidacloprid exposure in mice, for example, recently caused researchers to determine that "long-term exposure could be detrimental to the immune system."[471]

Clothianidin, imidacloprid, and thiamethoxam have recently come under fire and since been restricted for their connection to honeybee deaths. An investigation by Italian researchers showed that these compounds were undermining bee immune systems and promoting replication of a viral pathogen attacking their health.[472] (See the section that follows for more information.) These neonicotinoids have also severely impacted birds, fish, amphibians, and other wildlife, according to research.[473]

In December 2013, the European Food Safety Authority's panel on Plant Protection Products and their Residues (PPR) announced a potential link between two neonicotinoids—acetamiprid and imidacloprid—and developmental neurotoxicity. According to an EFSA press release, "The PPR Panel found that acetamiprid and imidacloprid may adversely affect the development of neurons and brain structures associated with functions such as learning and memory."[474]

The EFSA recommended further study and a reduction in current guidance limits.

Effect of pesticides on bees

As previously mentioned, honeybees are adversely affected by pesticides, which is largely problematic as bees are responsible for pollinating over one hundred different food crops worldwide. It is widely cited that every third bite of food consumed was produced in part with a bee's help. In short, bees equal food. In the past few years, however, stories of mass bee deaths and colony collapse disorder (CCD) have emerged as a "red alert" issue for many countries around the world. Some reports have even suggested that up to

40 percent of hives have disappeared in the United States.[475] The culprit? Independent scientific consensus says pesticides are to blame.

A recent analysis sampling bee pollen from twenty-three U.S. states and one Canadian province found ninety-eight different pesticides and metabolites, up to 214 ppm in a single sample. Each sample averaged at least six different pesticide residues, with one sample containing thirty-nine different kinds.[476]

As you might naturally suspect, research has verified that repeated pesticide and insecticide exposure exhibits deadly effects on bees.[477] Studies have also concluded that decades of widespread pesticide use is ultimately to blame for mass bee deaths and CCD.[478] One research-supported theory on how this works shows that pollen contaminated with high levels of fungicides weaken the bees' immune systems, allowing them to be much more susceptible to pathogenic bacteria and viruses.[479,480]

Scientists have discovered that treating bees with commonly used neonicotinoid systemic pesticides meant significantly reduced growth rates and a full 85 percent reduction in new queen bees; they concluded, "Given the scale of use of neonicotinoids, we suggest that they may be having a considerable negative impact on wild bumblebee populations across the developed world."[481] Further research on thiamethoxam, another neonicotinoid, found that even nonlethal exposure caused high bee-mortality rates due to homing failure so pervasive, it could collapse an entire colony.[482] When tens of millions of bees died in Ontario, Canada, in 2013, the beekeeper who owned them told CBC Radio it was neonicotinoid exposure.[483]

With the scientific finger pointed at pesticides as the mass murderer of millions of bees, the European Union decided to ban several different types of pesticides in 2013—including neonicotinoids clothianidin, imidacloprid, and thiametoxam—produced by biotech firms Syngenta and Bayer.[484] Clothianidin, you'll recall from earlier, was the industry replacement for organophosphates after they found them to be dangerous to human health and the environment. Syngenta responded by announcing it would take the European Commission to court for what the company felt was a wrongful linkage of Syngenta's thiamethoxam to bee deaths.[485]

The EU ban—backed by research data—did not stop the EPA from boldly concluding that clothianidin posed no imminent hazard to bees, and, as such, the agency denied a petition by beekeepers and environmental watchdog groups to suspend its use.[486] The EPA and the USDA released a joint report in 2012 asserting that further research is needed to find out what is killing the honeybees,[487] but seeing as how the USDA essentially subsidizes the genetically modified food industry (promoting biotechnology exports is officially listed as one of the USDA's four main strategic goals), there is little reason to believe the U.S. government will reasonably rein in agrichemical use in America.

Pesticide food and water residues

What is the end of the line destination of pesticides? Are they ingested into human bodies? Are their known toxic effects contributing toward a sharp decline in the health among Americans, Westerners, and, increasingly, people from developing nations around the world?

Dietary exposure to pesticides is indeed widespread, as the variety of synthetic chemical pesticides used by conventional agriculture has been shown to wind up in residual form in the daily diets of most consumers.

Independent laboratory tests conducted by the USDA's Pesticide Data Program, the Department of Pesticide Regulation's California State Residue Monitoring Program, and the Consumers Union have all demonstrated that pesticide residues found in grocery stores are consistently higher in conventionally grown foods and those grown with integrated pest management (IPM) techniques than in certified organic produce.[488] Lab tests found three times as much potentially harmful residue in conventional produce (73 percent of tested samples) than in organic produce (23 percent of tested samples), while food grown using integrated pest management and classified by the EPA as No Detectable Residues (NDR) had nearly twice the residue levels of organic produce (in 47 percent of tested samples).[489]

Among these residues, banned organochlorine pesticides, no longer used in agriculture, were found as a common contaminant among all three categories, as many soils remain contaminated with organochlorine constituents. About 40 percent of the pesticide residues found in organic foods were hits from this source, demonstrating how the past use of discontinued and banned harmful chemicals can continue to impact food safety and pose potential health hazards.

Worse, many of the conventional and IPM-treated produce contained multiple pesticide residues, with some crops like spinach and green pepper containing traces from as many as fourteen pesticides. Organic products only rarely (6–12 percent of lab tests) showed multiple pesticides.

The organic label has grown in popularity in part as a means of avoiding exposure to pesticides. However, some organic farms do employ some types of organic-approved pesticides, though they are banned from using most of the synthetic chemicals that are widely used in conventional agriculture. USDA-certified organic foods[490] are allowed to be cultivated with certain chemical additives but are legally required to use chemicals that are classified as not harming the environment or human health.[491] While many organic farms may make an earnest effort to produce the cleanest and best foods possible, there is room for concern that some organic-certified producers may in reality be cutting corners and taking advantage of legal loopholes in a way that most conscious consumers would find worrisome and in violation of their reasons for choosing organic options.

For example, some organic produce has been grown using rotenone-pyrethrin, a naturally occurring insecticide and piscicide (fish killer) derived from plant seeds, which is allowed under USDA organic standards but has nevertheless been connected with Parkinson's disease in rat studies.[492,493] Spinosad is another naturally derived insecticide, produced from fermented bacteria, that was given approval for use in organic farming by the USDA National Organic Program (NOP)[494] but has been found to produce toxic effects in rats in both chronic and sub-chronic conditions.[495]

Nevertheless, certified organic produce is a much safer choice than conventional produce—which regularly uses a broad spectrum of pesticides with potential health effects—for anyone applying the precautionary principle to limit their exposure to these chemicals even before their full toxicity has been demonstrated.

Furthermore, food is not the only way we are exposed. Municipal drinking water as well as groundwater-sourced well water both pose frequent and potential sources of exposure to pesticides, as contamination is widespread and under regulated. Thus, the use of a thorough home filtration system for all drinking water, as well as for showers and sinks, is advisable. There are now many options on the market capable of filtering out atrazine, glyphosate, 2,4-D, and others in the spectrum of pesticide contaminates.

FOOD INGREDIENTS AS CONTAMINANTS

The 1958 Delaney Clause, an amendment to the Federal Food, Drug, and Cosmetic Act of 1938, originally said that the FDA could not approve any additives known to cause cancer in lab animals or in humans and that no carcinogenic agents could be allowed in food whatsoever. This all changed in 1988 when Michael R. Taylor, a former Monsanto vice president for public policy and current FDA deputy commissioner for foods, wrote his de minimis interpretation of the clause published in the *International Journal of Toxicology*, stating that if the risk of the carcinogen was "de minimis," or too minor to warrant consideration, then the food should be able to be sold anyway.[496]

Allowing for de minimis amounts of carcinogens only takes into account acute poisoning and does not consider the chronic, long-term effects of small amounts of cancer-causing agents here and there over time. And it throws the door wide open for additives.

As Dr. Jacqueline Verrett, former FDA member–turned–whistleblower who oversaw the approval of aspartame, wrote in her book *Eating May Be Hazardous to Your Health*, "Under the guise of basic research the FDA is using your tax money—quite a bit of it—to try to prove a pet theory that carcinogens can be used safely in food, and to subvert the Delaney clause. The experiments will be used, then, to decide not which chemicals are carcinogens and unsafe for you to eat, but *how much* of a carcinogen you should be allowed to eat."[497]

Until the FDA changes its regulatory decisions, foods purchased every day by American consumers will continue to be formulated with small quantities of carcinogenic chemicals the FDA insists are "safe" in the quantities consumed.

ASPARTAME

If you have chewed a piece of gum purchased from your average grocery store in the past few decades, there's a good chance it contained aspartame. Aspartame—known by its brand names Equal®, NutraSweet®, and AminoSweet®—is one of the most widely used artificial sweeteners on the market today. Many people do not know the history of aspartame or even what it is made out of, just that it's common on supermarket shelves. Aspartame can be found in a wide variety of foods including candy, yogurt, desserts, flavored waters, sports and energy drinks, coffee drinks, instant breakfast shakes, diet beverages (especially diet sodas), vitamins, over-the-counter medicines, and so much more.

Two hundred times sweeter than sugar, aspartame is being consumed by two-thirds of the population in over six thousand products in one hundred countries worldwide.[498]

It's also one of the most addictive neurotoxins still used in the food supply.

What is aspartame?

Aspartame is composed of 40 percent aspartic acid, 50 percent phenylalanine, and 10 percent methanol, and is excreted by genetically modified E. coli bacteria.

Aspartic acid is a nonessential amino acid, meaning the body can produce what's needed on its own. Aspartic acid also functions as a neurotransmitter.

Phenylalanine, another amino acid, helps the brain create active nerve chemicals that affect mood, like epinephrine and dopamine. Too much phenylalanine can lead to chemical imbalances, such as a decrease in the amount of serotonin the body produces over time. Perhaps best known for managing moods, serotonin also regulates sleep, appetite, and muscle contraction, and it even affects memory and learning abilities. Serotonin also keeps us from craving carbohydrates and helps us limit overconsumption; in other words, people with aspartame-induced serotonin inhibition may be driven to eat more.

Methanol—or methyl alcohol—is literally known as wood alcohol. It is an industrial solvent typically found in antifreeze, paint, copy machine fluids, windshield wiper fluid, varnish, and fuel additives. The U.S. National Library of Medicine's MedlinePlus website (with the tagline "trusted health information for you") says methanol is considered a "nondrinking type" of alcohol, and overdose can cause all kinds of awful symptoms, including headaches, blindness, difficulty breathing, convulsions, seizures, low blood pressure, coma, liver dysfunction, nausea, vomiting, abdominal pain, leg cramps, weakness, and even bluish-colored lips and fingernails.[499]

Aspartame is an excitotoxin; that is, ingesting too much aspartame can actually stimulate overexcited neurons to the point of cellular death, which is particularly dangerous for people with weakened immune systems or young children who do not have fully developed blood–brain barriers. In fact, the EPA officially listed aspartame as a "chemical with substantial evidence of developmental neurotoxicity" on its database of developmental neurotoxicants.[500]

There are at least ninety-two side effects of aspartame ingestion that have been reported to the FDA, including headaches, nausea, vomiting, abdominal pain, diarrhea, memory loss, fatigue, dizziness, vision changes, rashes, muscle weakness, insomnia, hives, numbness, tingling, menstrual changes, difficulty breathing, and seizures, just to name a few.[501]

Although it is in so many different foods and drinks, aspartame is most well known for being added to beverages, especially diet sodas. Unfortunately, aspartame has demonstrably limited stability in liquid, and studies have shown that over time, the additive breaks down into formaldehyde and diketopiperazine (DKP), a brain tumor agent.[502] This breakdown happens even if the product containing aspartame is kept chilled in the refrigerator, but it seems to accelerate when the item is left at room temperature or—worse—heated.[503,504] Trocho et al (1998) determined, "aspartame consumption may constitute a hazard because of its contribution to the formation of formaldehyde adducts."[505] This independent research showed that the form-aldehyde from aspartame accumulates in the brain, liver, kidneys, and other organs.

Another study from 2006 demonstrated aspartame's multipotential car-cinogenic effects when rats were given aspartame in their food during an eight-week study. Findings included an increase in kidney tumors in females, an increase in peripheral nerve tumors in males, and an overall increase in leu-kemia and lymphoma in both sexes. Based on the study results, the scientists concluded that a change in the "use and consumption of APM [aspartame] is urgent and cannot be delayed."[506] In a follow-up study in 2007, scientists found that rats fed low doses of aspartame before birth and throughout their lifetimes developed significantly more leukemias and lymphomas, with an additional significant, dose-related increase in breast cancer in females.[507] Five years later, Schernhammer et al (2012) published findings in the *American Journal of Clinical Nutrition* that showed that ingesting aspartame-containing sodas increased the risk of certain cancers, including lymphoma and multiple myeloma, in men.[508]

Even though aspartame is used as a sugar substitute in diet foods and beverages and is promoted to people with diabetes and those who want to lose weight or keep weight off, in one independent study, diet soft drink consumers ended up with a 70 percent greater increase in waist circumference as a group than nondrinkers, and, in another, heavy aspartame exposure was shown to directly contribute to increased blood glucose levels and a higher risk of diabetes in mice.[509]

Due to all the horrid side effects, Dr. Hyman Roberts spent two decades researching the additive, which culminated in a 1,038-page book titled *Aspartame Disease: An Ignored Epidemic* that coined the phrase "aspartame dis-ease" in 2001. Roberts published numerous aspartame studies and responses to studies throughout his career, connecting aspartame to a multitude of neg-ative health effects and diseases, including headaches, high blood pressure, increased pressure inside the skull, brain tumors, low blood platelet count, and allergic reactions/anaphylaxis.[510,511,512,513,514,515]

In a letter to the editor of the *Texas Heart Institute Journal* tying aspar-tame to Graves' disease and pulmonary hypertension, Roberts wrote, "I have written about aspartame disease for more than 2 decades, because of the profound adverse neurologic, cardiopulmonary, endocrine, and allergic effects of aspartame products . . . My own database exceeds 1,300 victims of aspartame-related illnesses, with a 3:1 preponderance of women."[516] In a study of 505 people who reported negative reactions to drinking aspartame, Roberts found that two-thirds of respondents felt their symptoms improve within just two days of not ingesting any aspartame.[517]

If it's so horrible for us, how did aspartame get approved?

Even though the FDA lists aspartame as generally recognized as safe, it didn't start out that way; in fact, aspartame failed to win FDA approval for nearly two decades. When it finally did get approved, it was only under a cloud of controversy.

The facts have been meticulously laid out in evidence file #7 of FDA docket # 02P-0317.[518] While James Schlatter was developing a new ulcer drug for chemical company G.D. Searle (bought out by Big Agra giant Monsanto in 1985) in 1965, he accidentally discovered aspartame. Searle contracted biochemist Dr. Harry Waisman, then director of the University of Wisconsin's Joseph P. Kennedy Jr. Memorial Laboratory of Mental Retardation Research in 1970. As a pediatrics professor and biochemist, Waisman was a respected expert in phenylalanine toxicity, and Searle needed respected experts to perform studies on aspartame to get FDA approval. Of the seven infant monkeys Dr. Waisman fed aspartame-laced milk to, five suffered grand mal seizures and one died. Waisman himself also died unexpectedly at age fifty-eight the following year in 1971, preventing him from any further study on the chemical.

Also in 1970, Dr. John Olney informed Searle of findings from his independent work in which he found that dosing mice orally with glutamate and aspartate in free form (unbound to proteins) caused brain damage.[519] In 1973, G.D. Searle submitted more than one hundred studies to support their position that aspartame was safe, 80 percent of which were completed by Searle or its contractor Hazleton Laboratories, in an attempt to get aspartame approved. An FDA doctor from the agency's Division of Metabolic and Endocrine Drug Products declared the information submitted was scientifically lacking in numerous areas, including missing data on absorption, excretion, metabolism, half-life, and bioavailability. The FDA ruled that it is impossible to scientifically evaluate the clinical safety of aspartame based on the information provided, but it still approved limited use of aspartame. After hearing this, Dr. Olney filed a formal objection against aspartame's use, citing the potential for brain damage, especially in children, that he had told Searle of three years prior.

By July 1975, the breadth of evidence that aspartame posed a health risk caused then-FDA Commissioner Dr. Alexander Schmidt to appoint a special task force to evaluate key studies. Later that same year, the FDA put a hold on aspartame's approval due to the task force's preliminary findings. When the task force's findings came out, one of the lead investigators concluded that the agency had no basis upon which to rely on G.D. Searle's integrity and that G.D. Searle filtered the information presented to the FDA, providing irrelevant animal research that was "poorly conceived, carelessly executed, or inaccurately analyzed or reported."[520] Studies included missing fetuses from experimental animals; undocumented lab method switches during the study; one animal that was reported alive, then dead for several weeks, then alive, then dead again; and one study where 98 of the 196 animals died but were not autopsied for up to a year later, making analysis difficult.

The task force went on to say that even poorly controlled experiments showed some levels of toxicity, providing reasonable basis to assume that a well-designed study would show aspartame's true toxic potential. By then, G.D. Searle had already invested tens of millions of dollars in building new aspartame production facilities. They weren't going to let all that money, time, and effort—and millions in potential future aspartame profits—go to waste.

Following a continued wave of controversial incidents—including G.D. Searle giving the director who oversaw their aspartame research a three-year sabbatical with a $15,000 bonus during the inquiries[521]—G.D. Searle hired Donald Rumsfeld, former U.S. Congress member and chief of staff under Gerald Ford, as company president. Consumer lawyer James Turner, who spent years petitioning to have aspartame banned, alleged that G.D. Searle hired Rumsfeld to deal with the aspartame approval situation as "a legal problem rather than a scientific problem."[522]

The FDA eventually established a public board of inquiry. Based on all available evidence, the board ruled in September 1980 that it could not approve aspartame because it had "not been presented with proof of reasonable certainty that aspartame is safe for use as a food additive under its intended conditions of use."[523] A few months later, when Ronald Reagan was sworn in as the new U.S. president, G.D. Searle CEO Rumsfeld was named as part of Reagan's transition team—the team who just so happened to name a brand new FDA commissioner, Dr. Arthur Hull Hayes, Jr. One of his very first acts as commissioner was to overturn the public inquiry board and approve aspartame as safe for use in dry goods.

The following year, G.D. Searle petitioned to get aspartame approved in beverages; this time, the National Soft Drink Association (NSDA) wrote the FDA attempting to get aspartame's approval delayed due to health and safety concerns.[524] The first carbonated beverages containing aspartame were sold in 1983. The battle to get aspartame stopped continued on, however. Later in U.S. Congressional testimony over the issue in 1985, Massachusetts Institute of Technology's Dr. Richard Wurtman—who studied eighty individuals who suffered seizures after aspartame consumption[525]—testified that the phenylalanine in aspartame is an isolate; that is, it is not bound as it would be in proteins that contain it naturally. As an isolate, nothing blocks that phenylalanine from entering the bloodstream and passing into the brain. Wurtman told Congress, "To my knowledge, no other food that mankind has ever eaten causes the changes in brain chemistry that are provided by aspartame."[526]

By law, food additives are required to be inert; inert ingredients should not break down and cause ninety-two reported side effects, including seizure and death.[527]

While independent studies have shown the horrors aspartame can wreak on health, the food additive industry has funded quite a few studies of its own claiming aspartame is not only safe but in some cases healthy for people. In 2007, following the study that demonstrated the growth of cancers in rats specifically given aspartame-containing feed, a new study, quickly hailed in the media as "the most comprehensive review ever conducted," was released naming aspartame as totally safe across the board.[528] However, closer review of the fine print revealed that the study was conducted by a consulting firm hired by Monsanto and Ajinomoto—two of the world's largest aspartame producers. In addition, all of the study's authors were found to have multiple conflicts of interest. One was a chairman of NutraSweet Co.–funded American Health Foundation (AHF); another was a chairman of a Monsanto- and Ajinomoto-funded chemical and food company research association called the International Life Sciences Institute (ILSI).[529]

Dr. Ralph Walton of the Center for Behavioral Medicine at Northeastern Ohio University College of Medicine performed a meta-analysis of 166 aspartame studies concerning human health and considered the funding sources. Seventy-four of the 166 were reportedly funded by the aspartame industry and the other 92 studies had independent funding sources. Amazingly, Walton found that 100 percent of the 74 industry-funded studies claimed that aspartame was completely safe, while 92 percent of the independent studies not funded in any way by the food additive industry found health issues with aspartame.[530]

Although the conflict between the industry and independent researchers rages on, the FDA continues to assert that aspartame is safe for consumption. The agency has set an acceptable daily intake for aspartame at 50 milligrams per kilogram of body weight. That is equal to about twenty 12-ounce cans of soda or ninety-seven sweetener packets for a 150-pound person. In December 2013, the European Food Safety Authority published a reassessment of aspartame's safety, concluding once again that it and its breakdown products are safe. The EFSA set an acceptable daily intake of 40 mg/kg of body weight.[531]

Acesulfame-K

Aspartame's sweetener cousin acesulfame-K (or acesulfame potassium) was discovered by scientists at the German chemical company Hoechst AG in a similar accident to that of aspartame's discovery. Acesulfame-K is a potassium salt derived from acetoacetic acid and fluorosulfonyl isocyanate and contains the known-carcinogen methylene chloride. Methylene chloride, also called dichloromethane, is a volatile gas that smells like chloroform, and is used in paint stripping, polyurethane foam manufacturing, and metal degreasing.[532] Acesulfame-K is usually mixed with other artificial sweeteners such as aspartame to mask its bitter aftertaste. Of all the artificial sweeteners on the market, the least amount of research has been performed on acesulfame-K. One 2013 study found that long-term acesulfame-K use altered neurometabolic functions in mice, and that chronic use could impair cognitive function.[533]

Avoiding aspartame

Luckily, unlike some food additives such as monosodium glutamate (MSG, discussed on page 138) that continually change names and can be found hiding in foods under multiple monickers, aspartame can be largely avoided at least in the United States, United Kingdom, and Canada because of government requirements that one of its breakdown components, phenylala-nine, be clearly labeled on food and beverage packaging.

However, that may change in America, considering that in early 2013, the International Dairy Foods Association (IDFA) and the National Milk Producers Federation (NMPF) filed a petition with the FDA asking to alter the definition of "milk" to allow chemical sweeteners such as aspartame to officially be considered as optional characterizing flavoring ingredients of milk (along with seventeen other products such as yogurt, egg nog, and whip-ping cream), thus allowing them to be secretly added without anything other than "milk" written on the ingredients label.[534,535]

MONOSODIUM GLUTAMATE (MSG)

Monosodium glutamate has become one of the most pervasive and potentially harmful food additives in the modern diet. This cheap and readily available compound gives big flavor to and provides instant satisfaction from snack foods such as chips, soups, sauces, salad dressings, fast food, takeout, frozen foods, TV dinners, marinated meats, and even baby foods and formulas. It is widely used in canned foods and in frozen and pre-prepared foods where the natural flavor is often lost; this is equally true with low-fat and fat-free foods, where avoiding or removing fatty oils leaves a void that MSG's ample flavor is apt to fill.[536]

In short, the chemical makes otherwise bland and inexpensive food vastly more palatable and often irresistible. It has been widely used from the early to mid-twentieth century onward.

The naturally occurring, seaweed-derived version of monosodium glutamate has been a favored part of the Asian diet for thousands of years, but its modern discovery and industrial patent in 1909 is credited to Japanese scientist Kikunae Ikeda, who readily helped its commercialization as a food additive through the Ajinomoto company. Ajinomoto's original MSG was extracted from kombu seaweed; however, since the late 1950s, most MSG is primarily produced in mass by fermenting starches, sugar beets, sugar cane, and molasses.[537,538]

Once marketed to housewives and quietly added to restaurant kitchens in Japan during the interwar period, MSG became a dominating and ubiquitous taste, also spreading into Chinese and other Asian cuisines.

It was touted as a quintessential flavor enhancer, the cornerstone of an additional taste sensation to the normal spectrum of sweet, sour, salty, and bitter, dubbed by its discoverer as "umami," or "deliciousness." MSG is a clever and useful additive, as its flavor intensity increases the consumer's perception of sweet and salty ingredients, as well as other prized tastes.[539,540] Because monosodium glutamate combines salt and glutamate, it allows foods like soups to be made with lower levels of sodium than typical recipes might ordinarily call for, yet still taste rich in flavor.[541]

Use of MSG eventually took hold in the United States and the Western world. The U.S. Armed Forces experienced MSG as a food additive in Japanese food rations, and after World War II, the Quartermaster Food and Container Institute for the Armed Forces officially enlisted it as a morale-boosting ingredient to add cheap and encouraging flavor to the rations of American servicemen.[542] Post-WWII domestic industries followed the Army's lead, making MSG a standard additive to frozen and ready-made dinners and canned foods. It was also "endorsed and encouraged" by the National Restaurant Association during the mid-1950s and became frequently used in diners, Chinese food restaurants, and the emerging fast food culture.[543]

Glutamates, including MSG, are salts and esters composed of the non-essential amino acid glutamic acid, and make up the most abundant form of amino acids found in the diet. Glutamic acid is one of the major components of most proteins and frequently shows up bound to other food compounds.

Glutamates are naturally found in many foods, although they are usually found together with natural fibers, oils, and other synergistic nutrients that vastly lessen the potency. Many foods rich in protein, including cheese, milk, eggs, tomatoes, mushrooms, and more contain high levels of natural glutamates. However, most glutamate contained in these foods is in bound form (and of course, as a nonessential amino acid, the human body produces its own glutamic acid and does not need food supplementation).

Unbound glutamate ("free" glutamate) is a different beast altogether. It only occurs in foods at a very low level; industrially produced MSG, however, adds high levels of unbound glutamate to the diet. (A few foods, including aged cheeses like Roquefort and Parmesan and soy sauce naturally have very high levels of free glutamate, with no additives, but by far, the food issue lies primarily with the highly processed, industrially produced MSG and glutamic acid derivatives used to add cheap and savory flavor.) In this unbound "free" form, studies show that glutamate is the "principal excitatory neurotransmitter in the brain."[544] The mechanism of free glutamate action as a toxin in the brain is not fully understood, but it has a significant interaction with specific neural receptors. Its chemical composition readily carries it from the bloodstream across the blood–brain barrier, where it meets glutamate transporters in both the neurons and the glial (important supportive structures to neurons). There, the free glutamate acts as an excitotoxin, reaching these specific glutamate brain receptors, hyper-stimulating brain activity, and eventually burning out the neurons through overactivity, causing damage to cells or even cell death.[545,546] Mice given MSG while in the womb experienced adverse effects on brain circuitry.[547] Free glutamate exhibits significant neuronal toxicity.[548]

Toxicity from MSG can induce not only brain damage and neurodegenerative disorders, but also endocrine disruption, irritable bowel syndrome, weight gain, reproductive issues, behavior disorders, and cancer. It also produces mild adverse reaction symptoms, most notably headaches, but also drowsiness; nausea; palpitations; chest pains; burning sensations in the neck, chest, or forearms; numbness; and general weakness.[549]

Its symptoms were first formally investigated in 1968 by Robert Ho Man Kwok, who conducted his studies after suffering from an illness derived from eating Chinese food, later dubbed "Chinese restaurant syndrome."[550]

Further studies were conducted by John Olney starting in 1969. Olney tested high-level doses of MSG on rhesus monkeys and found that it induced cell death in neurons.[551] In other tests, Olney injected newborn mice with high levels of monosodium glutamate, finding again the death of neurons along with impaired brain development—especially in the hypothalamus, where many important metabolic functions take place and the nervous system and endocrine system are connected. The study then found that these animals suffered from obesity, fertility issues, and damage to reproductive organs and skeletal structure as adults.[552]

John Olney's shocking findings stirred controversy, alerting the public to new dangers in the foods they had learned to take for granted, and putting the food industry on the defensive to dismiss the studies and maintain the safety of one of its most important commercial additives. Studies backed by big industry players were rolled out in an attempt to discredit the findings and give the impression that the toxic results could not be reproduced.[553]

Still, Olney's research continued to connect the accumulation of MSG with the destruction of neurons in the hypothalamus and other regions of the brain in his rat studies, which used large doses but demonstrated the potential for toxicity from ingesting the free form of this amino acid.[554]

In 1972, Olney presented his evidence on MSG toxicity in testimony to the Senate Select Committee on Nutrition and Human Needs, prompting discussion of an official FDA ban on MSG in baby foods. However, only a voluntary ban took place within the food industry, with additives companies opting to merely swap the monosodium glutamate in infant foods with other processed proteins such as hydrolyzed and autolyzed yeast, both of which contain high levels of free glutamate but look more "innocent" on food labels.[555]

Subsequent studies have found that free glutamic acid's role as an excitotoxin may cause additional or even wholesale damage to neurons in those already suffering from neurodegenerative diseases—including Alzheimer's, Parkinson's, Huntington's, ALS, seizures, stroke, and more.[556,557] Olney and his colleagues also believe that glutamate-reducing drug treatments could act as effective therapies for strokes and other degenerative disorders.[558]

In addition, monosodium glutamate has been thoroughly connected to neuronal damage in the retina, contributing to sharp decline in eye function, significant thinning of the retinal layers, and susceptibility to degenerative diseases. John Olney found severe retinal damage in studies on infant mice given high doses of MSG back in 1970, evidence of an "acute and irreversible form of neuronal pathology."[559]

Subsequent studies conducted by Harvard experimental eye researcher Liane Reif-Lehrer and her colleagues in 1975 and 1981 found morphological damage to the retinas of chick embryos given MSG, with damage becoming more severe over time.[560,561] A 1985 study by the National Eye Institute confirmed the retinal damage seen in newborn and embryonic rodents with adult rats, finding a progressive degeneration from MSG that began with swelling inside nerve cells and led to cell death and thinning retinal layers.[562]

Though it has previously been argued in academia and industry that dietary intake of MSG, unlike experimental lab doses given to animals, was not linked to actual damage, a study of retinal cell destruction in rats published in 2002 from Hirosaki University in Japan concluded that high dietary intake of monosodium glutamate could account for the dosage necessary to induce blindness or other degenerative eye diseases via MSG-induced cell death.[563,564]

A 2006 study conducted by the Lab for Development-Aging, Neurodegenerative Diseases in Guadalajara, Mexico, found that lab rats suffered liver and kidney damage after free glutamate in the brain severely impacted the neuroendocrine system, and subsequently organ function.[565]

Studies backed by the food industry, including the International Glutamate Technical Committee, have concluded, on the other hand, that MSG is completely safe, and that ingested glutamate does not cross the blood–brain barrier to create excitatory toxicity.[566] However, neither the hypothalamus nor the circumventricular organs are guarded by the blood–brain barrier, and the effects of MSG on glutamate receptors there have been found to disrupt the neuroendocrine system, which in turn has many important impacts on appetite and metabolism.[567]

Moreover, conditions such as hypoglycemia and a stressed immune system can compromise the blood–brain barrier and allow free glutamates to bypass and act as excitotoxins.[568,569,570] During pregnancy, MSG from the mother can cross the placenta barrier[571] and affect the development of the fetus, contributing to obesity and reduced energy levels during the vulnerable early years of life.[572]

Studies have shown that MSG can damage the hypothalamic regulation of appetite and contribute to obesity. Rats fed higher levels of MSG than found in the typical human diet expressed morbid obesity, with a propensity toward obesity found in rats fed levels similar to those found in average human consumption patterns.[573] Researchers believe that human consumption of MSG-laden foods at an early age is likely a significant—but largely invisible—contributor to the global obesity epidemic.[574,575]

Additionally, monosodium glutamate has repeatedly been shown in studies to produce insulin resistance, likely contributing to obesity, diabetes, and other detrimental health effects.[576,577]

Despite evidence of risk and adverse reactions in many in the population,[578] the manufacturers of MSG have been well protected by the supposed watchdogs of government. The FDA approved monosodium glutamate, giving it generally recognized as safe status, and helped push for the removal of labeling requirements to identify foods as "containing glutamate."[579,580] The Joint Expert Committee on Food Additives of the UN Food and Agricultural Organization and World Health Organization classified MSG as among the safest food additives—though it must be labeled with an E-number, a label that identifies all food additives in Europe.[581]

Negative press, ongoing since it began in 1968, has caused the food industry to bury MSG ingredients behind new and deceptive names because they are made of food products that contain MSG or other free glutamic acids. The FDA requires "added MSG" to be listed on the ingredients as monosodium glutamate; however, it does not require ingredients that contain MSG to state that they contain MSG as a molecular component.[582]

Thus, many confusing, cryptic, and coded ingredients are frequently found in processed foods of all kinds, contributing to high levels of MSG consumption by unwitting food consumers. Among these are "yeast extract," "hydrolyzed vegetable protein" (HVP), "textured protein," "torula yeast," "autolyzed yeast," "natural meat tenderizer," "soy protein isolate," "gelatin," "textured protein," "natural flavor," "amino acids," "proteins," and others.

Many of these, including hydrolyzed and autolyzed items, are proteins that are broken down by an enzyme to extract and isolate the MSG or other free glutamate used for flavor enhancing.

Some of these are considered different enough from monosodium glutamate to exempt them from E-number food additive–labeling laws in Europe and the United Kingdom that require MSG notifications. This technicality allows the addition of hidden sources of MSG-like flavor additives without clear disclosure.[583,584]

Additionally, related ingredients such as wheat and dairy hydrolysates, aspartame, and L-cysteine behave similarly to MSG in an excitatory neurotoxic manner.

Monosodium glutamate and other trade names for MSG/free glutamates are frequently used in common vaccines as an additive to preserve and stabilize the formulas.[585,586] By directly entering the bloodstream, these glutamic acid compounds may express even greater excitotoxicity.

Since the 1990s, MSG and other free glutamates have also been added to fertilizers, pesticides, and "growth-enhancement" products as agricultural inputs to increase yield, promote longer shelf life, and delay decomposition in produce.[587,588] One growth-enhancement product sold by Emerald BioAgriculture, AuxiGro, which contains a very high percentage of free glutamates, was approved for use on most vegetables by the EPA[589] and has been widely used across the United States since the late 1990s, though its application is thought to have been largely discontinued.[590,591]

In his paper, "A Short History of MSG: Good Science, Bad Science, and Taste Cultures," Jordan Sand, Georgetown University professor in Japanese history and culture, discusses how Ajinomoto and other glutamate producers were once proud to label their MSG products a "chemical seasoning" in both trade promotions as well as legal documents up until the point that the industry fell under great scrutiny over the safety of the food additive. Afterwards, it embraced "umami," the Japanese term for "deliciousness" or "savoriness," and frequently tied the product's image to its status as a "natural" flavor enhancer, shielding consumers as much as possible from its calculated mass production in giant industrial food-processing plants.[592]

People with MSG sensitivity who experience negative symptoms should thoroughly study the many trade names that can blind or disguise MSG or other excitotoxic ingredients, and make every effort to avoid them, as many people who react to MSG also react to aspartame and other similar ingredients.[593] As with avoiding many other potentially harmful ingredients, avoiding processed and preserved foods is a good start, while choosing foods with a minimal amount of recognizable ingredients is wise. Fresh cooked meals made at home with simple and known ingredients are a must as many cooking inputs also contain hidden MSG.[594]

While it is true that many people appear to demonstrate no acute sensitivity to MSG or free glutamates, my working theory is that certain individuals have diminished biochemistry potential to "clear" glutamates from their blood. It is these individuals, I believe, who experience the face flushes and intense, searing headaches that many people experience for up to twelve hours after consuming MSG. While I do not have scientific evidence to support this notion yet, I believe that nearly all people of Asian descent appear to be able to quickly and efficiently eliminate MSG through fortunate biochemical genetics. In my observations, Caucasians and even more so those of American Indian descent show unusually high susceptibility to MSG poisoning. This warrants further research, but there may be a genetic predisposition that either protects a person from MSG or creates a biochemical vulnerability.

Several vitamins and minerals appear to play a role in minimizing the effects of MSG, including vitamins C and E, as well as beta carotene and vitamins A, D, and K.[595] I've personally found that high-grade resveratrol appears to greatly diminish the duration and intensity of MSG headaches, especially when combined with L-Taurine (a common amino acid).

Magnesium plays a particularly important role in modulating MSG's toxic effects, as it is known in studies to block the neurotoxicity of glutamate and other excitatory amino acids, and it acts as a neuroprotectant.[596] Specifically, magnesium ($Mg2+$) maintains a voltage-dependent block on the

N-methyl-D-aspartate (NMDA) type of glutamate receptor; when the magnesium block is dropped, glutamate is able to "persistently" excite the NMDA receptor and damage neurons.[597] Thus, nutritional intake or supplementation of magnesium may be a viable safeguard against some effects of MSG.

ARTIFICIAL COLORS

Artificial colors have become a universal additive, typically found toward the end of the ingredients list for many packaged and processed foods. Food manufacturers add to snacks, meals, and beverages saturated colors that consumers have been trained to find appetizing and appealing, despite the fact that these additives also pose significant health dangers.

Dyes and lakes

There are two main types of artificial colors in use for food and cosmetics. *Dyes* dissolve in water and are used in most food-coloring applications, including beverages, dairy products, and even pet food.[598] *Lakes* are water insoluble and used in fats and oils and other foods lacking enough moisture to dissolve dyes, such as cake mixes, hard candies, gum, and coated tablets. Lakes are produced by mixing a color dye with aluminum hydroxide.

All FDA-certified dyes have traditionally been referred to as "coal tar dyes" that were originally produced from by-products of the coal-processing industry. However, according to Red40.com, a site created to raise public awareness about Red Dye No. 40, today's artificial food colorings are more likely petrochemical based. The full name for Red Dye No. 40, by the way, is 6-hydroxy-5-[(2-methoxy-5-methyl-4-sulfophenyl)azo]-2-naphthalenesul-fonic acid.[599]

The historical place of artificial dyes in modern foods

Controversy has surrounded artificial food dyes for nearly a century due to their known adverse health effects, which was instrumental in pushing for the legislation that founded the United States FDA.

Despite their associated health hazards, eighty color dyes were approved for use in foods and beverages by 1906 following the institution of the Wiley Act, otherwise known as the Pure Food and Drug Act, and the government hired Dr. Bernard Hesse to investigate which of these were truly still safe. By 1938, only fifteen of those colors remained. Over the years, as more information has emerged about these synthetic dyes and their effects on human health, more and more colors have been rejected from the list. For example, the FDA proposed a ban on Orange B in 1978, but the ban was never finalized. Orange B is still technically allowed for use in sausage casings up to 150 ppm, although batches of it have not been certified for over a decade now.[600]

Currently, only seven colors are still certified for use in foods in the United States, affirming the fact that the vast majority of artificial food dyes have been too unsafe to use in food for human consumption. Though the number of artificial dyes approved for use has declined, the ones that remain on the market today are still widely produced as more than 15 million pounds of dyes were certified by the FDA in 2009 alone. The top three dyes used by far are Red No. 40 (more than 6 million pounds) and Yellows No. 5 and No. 6 (nearly 4 million pounds each).[601]

Synthetic Color Dyes Currently Certified for Use in Foods

Name	Color Name	Shade	EU Code
FD&C Blue No. 1	Brilliant Blue FCF	Blue	E133
FD&C Blue No. 2	Indigotine	Indigo	E132
FD&C Green No. 3	Fast Green FCF	Turquoise	E143
FD&C Red No. 40	Allura Red AC	Red	E129
FD&C Red No. 3	Erythrosine	Pink	E127
FD&C Yellow No. 5	Tartrazine	Yellow	E102
FD&C Yellow No. 6	Sunset Yellow FCF	Orange	E110

Note: "FCF" in the name stands for "for coloring food."

According to an article posted on the FDA official website regarding its food-coloring regulatory process, "Color additives are important components of many products, making them attractive, appealing, appetizing, and informative. Added color serves as a kind of code that allows us to identify products on sight, like candy flavors, medicine dosages, and left or right contact lenses."[602]

Studies showing risk from artificial colors

Benjamin Feingold, a pediatrician, allergist, and clinical researcher with practices dating back to the 1920s, advanced a theory of dietary causes in the 1970s to explain the increasing effects of hyperactivity in children. As chief allergist for the Departments of Allergy he founded at the Kaiser Foundation Hospital and Permanente Medical Group in Northern California, Feingold identified artificial flavors and colors, as well as salicylates (an active ingredient in aspirin), as the culprits for a series of related developmental behavioral disorders, which are today known as attention-deficit/hyperactivity disorder (ADHD) and attention-deficit disorder (ADD).[603,604]

Seeing a rise in the prevalence of disease and behavioral problems in children coinciding with the advent of numerous artificial additives first available on the consumer market in the 1960s, Feingold focused his research on the growing intake of processed food additives and their increasingly adverse effects on mental and physical health.[605]

A series of studies by Feingold and his colleagues published between 1975 and 1982, when he died, found that these synthetic colors and flavors were connected to issues with nearly every system of the body, inducing causal effects on the respiratory system (asthma and cough), skin, gastrointestinal tract, and skeletal systems, as well as inducing allergies, headaches, and behavioral issues.[606,607,608,609,610]

Feingold's research also pointed to a connection between food dyes and asthma. A 1967 study at the Rhode Island Hospital Allergy Center documented a severe case of asthma caused by Yellow No. 5, also known as tartrazine, and other food dyes approved by the Federal Food, Drug, and Cosmetic Act[611] for use in food and other products.[612]

Artificial food colors and hyperactivity in children

The predominant issue from Feingold's research, however, is with developing children and hyperactivity, also termed hyperkinetic syndrome, where a state of overactive restlessness undermines attention, focus, learning, and behavior—ultimately a nervous system interaction.

As a result of research, the Kaiser-Permanente Medical Center recommended a detailed artificial additive elimination diet for treatment of these issues, which worked for other allergens, too.[613] Feingold successfully treated some six hundred children with this method, and found even greater effectiveness after also eliminating the synthetic antioxidant additives butylated hydroxytoluene (BHT) and butylated hydroxyanisole (BHA), which have been linked to possible cancer risk and genotoxicity.[614,615,616,617]

In 1978, Toronto researchers tested Dr. Feingold's theories on diet with twenty-six hyperactive children, validating much of his treatment.[618] They concluded that instituting a diet free of additives worked for three to eight of the twenty-six children, though the researchers favored pharmaceutical drugs as a more effective treatment to modify behavior (a conclusion the Big Pharma industry has been happy to run with in its ill-conceived bid to handle ADD/ADHD children).

A double-blind study published in *Science* in 1980 compared the behavior of twenty-two young children, half of whom were fed seven artificial colors along with a controlled diet. Parental observation confirms that toddlers who consumed artificial colors reacted dramatically to a challenge set by the study, as compared with a mild and temperate response by the children who did not consume these additives.[619]

Research in this area continued for decades, but a meta-analysis of the study data conducted by psychiatrist Dr. David Schab of Columbia University Medical Center in 2004 seemed to conclude once and for all that these artificial colors are indeed contributing to behavioral disorders, including ADHD.[620]

Schab told the Center for Science in the Public Interest that "The science shows that kids' behavior improves when these artificial colorings are removed from their diets and worsens when they're added to their diets. While not all children seem to be sensitive to these chemicals, it's hard to justify their continued use in foods—especially those foods heavily marketed to young children."[621]

A British study in 2004 put fresh focus on the problems of food additives, drawing a sample size of 1,873 three-year-old children from the general population, screening for hyperactivity as well as atopy, or the tendency to be hyperallergic. It concluded that artificial colors and the additive sodium benzoate did indeed agitate hyper behavior, and correlated in frequency with hyperallergic tendencies in the general population. Hyperactivity was significantly lower during the dietary withdrawal phase of these ingredients.[622]

Later studies have found a statistically significant positive association between atopic dermatitis and, separately, asthma, both hyperallergic conditions, and the prevalence of attention-deficit/dyperactivity disorder, increasingly connected with triggers from food allergies after eating common artificial food additives.[623,624]

In 2007, The Lancet published a study from the University of Southampton that led to bans of artificial colors in Europe and changed the landscape of the debate about color additives. The double-blind study exhaustively looked at two key developmental ages, using 153 three-year-old children and 144 eight- and nine-year-old children. Both age groups had adverse behavioral reactions to foods with artificial colors and sodium benzoate against placebo, giving great credibility to the earlier focus on the removal of food additives as a means of eliminating the cause of hyperactivity or other symptoms.[625]

This compelling study prompted hearings, removals, and warning labels for the use of artificial dyes in the United Kingdom and Europe, but when the U.S. FDA also held a public hearing and consulted its Food Advisory Committee about the study, it ultimately dismissed the information. Instead, the FDA determined that the risk of hyperactivity warranted no further action on its part—not a public information campaign warning of the dangers found in studies, nor a withdrawal of approval for the potentially harmful artificial colors being used in increasing quantities in thousands of food products consumed by millions.[626]

Artificial colors and cancer

Even more damning than the effects artificially created food colorings had on the behavior and neurodevelopment of children sensitive to the food additives, a number of dyes have been found in animal studies to be carcinogenic and genotoxic.

A seminal 2010 report by the Center for Science in the Public Interest titled, "Food Dyes: A Rainbow of Risks" made public the cumulative findings of studies on the remaining seven artificial color dyes still approved by the FDA for use in food and cosmetics, as well as Citrus Red No. 2, which is only used on orange peels, and Orange B, which was discontinued but never officially banned.[627]

The report revealed that the three most widely used color dyes—present almost ubiquitously in processed food products across the spectrum—contained known carcinogens and troubling results in animal studies, despite gaps in sufficient research.[628]

In fact, according to the U.S. Code of Federal Regulations regarding food coloring additives, a certain number of impurities are allowed in final batches of dye—impurities that include chlorides, sulfates, and carcinogens like azobenzene in addition to toxic heavy metals such as lead, arsenic, and mercury. For example, Red No. 40 is technically allowed to contain "not more than" 14 percent volatile matter (at 135 degrees Celsius) and chlorides and sulfates, 10 parts per million lead, 3 parts per million arsenic, and seven other substances; the total (actual) color in a batch may not be less than 85 percent.[629] That means that, according to regulations, up to 15 percent of each batch of Red No. 40 that is certified by the FDA can be made up of potentially dangerous impurities. Now consider for a moment that many foods contain multiple food-coloring additives that all have similar rules for the allowance of impurities. While 10 parts per million lead may not sound like a lot in one batch of Red No. 40, when added with other colors that contain their own small amounts of lead, it starts to add up fast.

Allura Red AC, better known as FD&C Red No. 40, is created with the use of the dye-processing intermediate p-Cresidine, which has caused urinary bladder cancer, nasal cancer, and liver cancer in mice and rats during feeding studies, and also produces allergenic effects. The U.S. National Toxicology Program classified p-Cresidine as "reasonably anticipated to be a human carcinogen" back in 1981.[630] Studies, though dismissed as flawed, found that Red No. 40 accelerated the growth of tumors.[631] Red No. 40 food coloring is used widely in the food industry, in such foods as gelatin, dairy products, artificially colored beverages, condiments such as ketchup, and baked goods.

Tartrazine, better known as FD&C Yellow No. 5, and Sunset Yellow (for coloring food), better known as FD&C Yellow No. 6, have both been found to contain benzidine and 4-amino-biphenyl, both known human carcinogens used in the production of azo dyes.[632,633]

The FAO/WHO Expert Committee on Food Additives set an acceptable daily intake level of 7.5 mg/kg for Yellow No. 5, the food additive thought to produce the most allergic reactions, inducing hives, eczema, and asthma, as it is a potentially carcinogenic nitrous derivative in the azobenzene class.[634]

Both benzidine and 4-amino-biphenyl, used as intermediates in Yellows No. 5 & No. 6 manufacture, have caused cancer through occupational exposure in workers engaged in dye manufacturing. Benzidine has caused at least 23 confirmed cases of bladder cancer among 198 workers exposed between 1935 and 1950. It can also induce nausea, vomiting, and liver and kidney damage.[635] Similarly, 4-amino-biphenyl caused bladder cancer in 19 dye workers at a plant of 171 employees between 1935 and 1955.[636] It also triggered blood-vessel cancer and liver tumors in mice. A 1977 study on rats fed high levels of Yellows No. 5 and No. 6 found toxic results, with slowed growth, bad fur, and the death of half of the rats over the course of a two-week period.[637]

Also, 4-amino-biphenyl was used as an intermediate to produce color dyes D&C Yellow No. 1—before D&C Yellow No. 1 was withdrawn—and D&C Red No. 33, which is still allowed. Its deliberate use in dye production has been discontinued, though it shows up as a contaminant in Yellows No. 5 and No. 6, as well as Red No. 33, at levels regulated by the FDA.[638]

Azo dyes and their derivative mixtures have long been flagged for allergenic potential and genotoxicity, due to known issues with several related formulas.[639] Many azo dyes have caused pulmonary and contact hypersensitivity in workers who manufactured them.[640] Coal tar hydrocarbons used as industrial solvents for dyes have been linked to a larger pattern of industrial dermatitis, in which workers exposed to occupational chemicals develop allergic responses.[641]

Other approved food dyes are not as widespread as Red No. 40 and Yellows No. 5 and No. 6, but further research is clearly needed to determine if they are truly safe. According to the Center for Science in the Public Interest, Blue No. 1 caused chromosomal aberrations in two studies, and another study suggested the dye had neurotoxic potential when it was found to act synergistically with L-glutamic acid.[642] Studies also show that up to 5 percent of Blue No. 1 is absorbed via the gastrointestinal tract, meaning it enters the blood stream and therefore has the potential to affect the body's neurological funtion, cellular function, and DNA.

No metabolism studies of Blue No. 2 on humans have ever been completed, but in rat studies, the dye breaks down in the gastrointestinal tract to 5-sulfoanthranilic acid, which gets absorbed and then excreted by the kidneys. Rat studies on this dye have also shown a statistically significant occurrence of brain gliomas and other tumors.[643]

Significant increases of certain types of tumors were found in male rats in high-dose Green No. 3 studies, and Red No. 3 is somehow still permitted for use in foods and drugs even though a 1990 FDA finding acknowledged that it is a known animal carcinogen.[644]

Removal of artificial dyes in the EU

Starting in 2010, the European Union began requiring a warning be placed on products containing Yellow No. 5, Red No. 40, and other dyes that reads, "May Have an Adverse Effect on Activity and Attention in Children."[645]

This requirement followed a 2008 European Food Safety Authority opinion based on a 2007 Southampton study on children published in *The Lancet*. It concluded that the safety of dyes was in question while they carried no nutritional value and were unnecessary. Thus, it recommended limiting the future use of the preservative sodium benzoate and artificial colors, both of which were connected with hyperactivity in children.[646]

This led to voluntary action by industry to begin removing these color additives by 2009 in the United Kingdom and the European Union. Following the Southampton study and subsequent EU requirements for warning labels on foods containing food dyes, the FDA performed another review of the safety of the remaining certified food colorings in the Southampton study in 2011. Ultimately, the review panel concluded that children with behavioral disorders who ingest artificial colors might find their conditions are "exacerbated by exposure to a number of substances in food, including, but not limited to, synthetic color additives," but typical children will be unaffected.[647] Considering the explosion of hyperactivity disorder diagnoses over the past several decades, it's hard to even determine what a "typical child" is anymore.

Notably, these artificial food colors still remain in use in the United States today.

Natural food colors as replacement additives

There are numerous natural food derivatives and extracts used for food coloring that could be substituted for artificial lakes and dyes, though a lack of water-soluble options may influence the ready use of artificial dyes in water-soluble food products.[648] Though these natural options are not completely free from adverse reactions, reports are very rare, and they clearly pose sharply less risk than do artificial color additives—enough to necessitate their adoption as a public health benefit in manufactured foods that use coloring processes.[649] One drawback is that natural colors are typically extracted using hexane, which has its own dangers; see the hexane section on page 107 for further information.

Currently it is primarily betacyanins—reddish to violet pigments—and anthocyanins—red, purple, and bluish pigments—that are widely used as natural food substitutes for artificial food dyes.[650] Betanin pigments, often derived from beet root, are perhaps the most widely used betacyanin, while colors from cactus flowers[651] and some flowers such as *Amaranthus caudatus*, which are all Caryophyllales, are often used as well. Anthocyanin pigments include a wide range of antioxidant-rich bioflavinoids mostly from fruits and berries, but commercially, the most used food color agents are grapes, elderberry, red cabbage, and roselle.[652]

Natural food colors that are water insoluble and show usefulness in fats, oils, cakes, and so on, include beta-carotene, chlorophyll, lycopene, and bixin, which are derived from beets, carrots, the *Bixa orellana* shrub, tomatoes, spinach, and cherries.[653]

Various spices are excellent alternatives for colors like yellow and orange. Turmeric, the iconic Eastern spice with legendary anticancer properties, is widely used as a food color.[654] Carmine, the bright-red aluminum salt harvested from cochineal scale, is very widely used as a natural alternative, as is paprika, derived naturally from red chili peppers and used to produce red, orange, or reddish brown food coloration.[655] Annatto, derived from achiote tree seeds, produces a natural yellow-to-orange coloration that is widely used in manufactured foods today, though there is some warning about its use. Annatto dye can cause an extremely rare but severe reaction in individuals with an uncommon level of hypersensitivity.[656]

Despite the availability of these options, artificial coloring has been a mainstay of processed foods for many decades, mainly due to its cheap cost, giving a mass marketing competitive edge, as well as its consistency and character in giving processed food an often alluring appearance.[657] Though slowly, the pervasive use of artificial colors in food products of almost every kind is finally beginning to change in response to scientific research, bans in other Western nations, and consumer demand for cleaner foods with fewer chemical additives.

Kraft Foods leads industry backpedal on artificial dyes

In the spring of 2015, Kraft Foods announced its decision to remove the artificial dye Yellow No. 5 from its flagship macaroni and cheese products, but only for certain markets; however, the move marked a potential turning of the tide. Consumer advocates vociferously critiqued the food mega-conglomerate for replacing the artificial dye only in its European recipe with the natural food color additives paprika and beta-carotene after regulations there required a warning label stating that "This product may have adverse effect on activity and attention in children"—an obvious buzzkill to its image as a staple part of many children's diets.

Eventually, Kraft caved and agreed to remove Yellow No. 5 from some of its American macaroni and cheese products, but only from three varieties marketed directly at children—one featuring the SpongeBob SquarePants character, a Halloween design, and a winter design—not from the main elbow-shaped classic, or most of its other processed food products.[658]

Kraft defended its use of the dye on grounds that other entities embedded in the modern profit-driven, mass-scale food industry understand: Consumers have been accustomed to the appearance, coloration, and texture provided by these artificial colors and other food additives. Without artificial colors, much of the appeal of processed foods in general would be lost. "All of the ingredients must work together to deliver the distinctive taste, appearance and texture consumers expect and love from Original KRAFT Mac & Cheese. Our fans have made it clear they won't settle for anything less," Kraft spokesperson Lynne Galia stated.[659]

For Kraft and most other Big Agra conglomerate players, food marketing is a tireless business based on providing a consistent food-flavored product that looks and tastes larger than life. Expectations for the image of the product and a race to the bottom for the lowest price of ingredients frequently supersede concern for nutritious or safe ingredients.

Michael Jacobson, executive director of the Center for Science in the Public Interest, who has been at the forefront of publicizing the risks of artificial dyes and lobbying for their elimination, pointed to larger implications of the significance of these widespread yet unnecessary food additives.

"The continued use of these unnecessary artificial dyes is the secret shame of the food industry and the regulators who watch over it," Jacobson said. "The purpose of these chemicals is often to mask the absence of real food, to increase the appeal of a low-nutrition product to children, or both. Who can tell the parents of kids with behavioral problems that this is truly worth the risk?"[660]

Artificial colors, then, not only pose significant and well-documented health risks, but they predispose the population to other health risks, including obesity and diabetes, drawing impressionable children and many child-like adults to bright and colorful foods that feed illusion rather than nutrition.

CHEMICAL PRESERVATIVES

Preservatives are used by food manufacturers to extend the shelf life of foods and to prevent food products from spoiling or going rancid, allowing for the most impersonal of mass-scale industrial production. In this form, foods become chemical widgets that must, for commercial purposes, appear exactly like the food is expected to appear and taste exactly how the food is expected to taste—every time.

Many preservatives help maintain the consistency and volume of chemical food blends as emulsifiers and thickening and bulking agents, in addition to enhancing their perceived qualities of fullness and flavor. Some food additives, like MSG and its derivatives, add flavor to a food product while simultaneously preserving its shelf life. In effect, this allows low-quality, bland, and even stale foods to be perceived by the consumer as fresh, tasty, and wholesome while hiding behind layers of cosmetic food treatment, no matter how nutritionally void a substance it might be.

At its extremes, a careful combination of preservatives, artificial flavors, colors, and other chemical additives can dress up a Frankenstein recipe to look like a debutante beauty queen. In return, consumers are subjected to little-known dangers from hard-to-pronounce and unfamiliar chemical ingredients, potential carcinogens, harmful toxins, and junk food likely to contribute to ill health and degenerative disease conditions such as obesity, diabetes, heart disease, cancer, reproductive issues, and more.

Avoidance of these chemical food additives—where nutritional labeling is transparent and accurate—is overall the best strategy for eliminating potential toxins, inflammatory ingredients, and harmful additive interactions.

Benzoic acid and its salts as food preservatives

Benzoic acid (labeled as E210 in the EU) and its salts and esters are commonly used in food production as preservatives and stabilizers, despite known risks to human health. Benzoic acid is used to prevent decay in common foods such as reduced-sugar products, certain meats, cereals, and beverages.[661]

Potential adverse effects for benzoic acid and its commonly used salts—sodium benzoate, potassium benzoate, and calcium benzoate—include temporary impairment of digestive enzymes and depleted glycine levels, as well as allergenic triggers for hay fever, hives, and asthma. Controlled studies on piglets found that benzoic acid increased feed intake and body weight gain.[662] Through its strong antimicrobial properties, benzoic acid reduced the number of bacteria in the gastrointestinal tract, including many beneficial strains that could potentially affect digestion and immunological factors.

Sodium Benzoate (E211)

Sodium benzoate, the sodium salt of benzoic acid, is one of the most pervasive food preservatives from this class, used to stave off microbial growth and frequently added to acidic foods and beverages, including carbonated sodas, fruit juices, margarine, vinegar-preserved foods, and jellies. Exposure through food has been linked in a number studies to attention-deficit/hyperactivity disorder symptoms as well as to worsening asthma and eczema, especially in children.[663]

The widely reported and scientifically confirmed allergic sensitivity to sodium benzoate (along with artificial food colorings) among children was most famously investigated in the 2007 Southampton, England, study that found consistent adverse behavior effects and hyperactivity among hundreds of randomly sampled children from the general population in two distinct developmental age groups—three-year-olds and eight- and nine-year-olds.[664]

This study, compounded with previous data, prompted a UK and European Union mandatory warning label that sodium benzoate (E211) "may have an adverse effect on activity and attention in children."[665] However, unlike the artificial colors subjected to a "voluntary ban," sodium benzoate was not put under further regulation because its use as a preservative was determined to distinguish it from non-functionary colorings.[666]

Since the end of the twentieth century, researchers have chronicled sodium benzoate's potential to damage DNA through mutagenesis and promote oxidative stress in the gastrointestinal tract.[667]

The FDA and the soda and beverage companies have known since the early 1990s that certain formulas with sodium benzoate or other forms of benzoic acids and ascorbic acid (vitamin C) as ingredients were converting into benzene, a known carcinogen and elementary petrochemical classified as a hydrocarbon—but the public was never told.[668] More than fifteen years later, the issue resurfaced as a hard-hitting contamination scandal on a global scale, with findings that benzene had formed in beverages during production, particularly those with orange flavoring, including citric acid, leading to subsequent recalls and reformulation in many markets.[669,670] Further, a Belgian study found that plastic soda containers were contributing to the acidic reaction that produced benzene in trace amounts, as demonstrated in approximately 47 percent of samples.[671]

In 2008, Coca-Cola Great Britain hailed plans to remove sodium benzoate from its UK Diet Coke formula,[672] while seeking a replacement preservative for sodas with fruit content such as Sprite and Fanta Orange.[673] The switch was also made in the United States and Europe, while soft drinks such as Pepsi Max and diet sodas produced by the name brands Sunkist Orange, Mountain Dew, and Nestea continues its use today.

Potassium Benzoate (E212)

Potassium benzoate is also a salt of benzoic acid, frequently used in acidic food and beverages as a preservative to protect artificial flavor enhancers. It is often used as an alternate to sodium benzoate. It is an ingredient in popular low-calorie soft drinks, including Diet Coke, Diet Pepsi, and many of their variants such as Coke Zero and Diet Pepsi Wild Cherry,[674] as well as Lipton Diet Iced Tea, all of which reportedly transitioned to potassium benzoate in reaction to the controversy over sodium benzoate.[675]

Like other benzoic acids, potassium benzoate has been linked to triggering or worsening allergic reactions and contributing to ADHD and hyperactivity, and it also poses a risk of producing benzene when formulated with ascorbic acid.

Calcium Benzoate (E213)

Calcium benzoate is yet another benzoic acid salt used as a beverage and food preservative—appearing in low-sugar products, cereals, and meats—that is connected with allergic reactions and hyperactivity. It has been listed as one of the top ten E numbers to avoid.[676]

Parabens (E214, E215, E218, E219)

Parabens are chemicals most commonly known for their use as antimicrobial preservatives in cosmetics and skin-care products. Their variants include methylparaben (E218—methyl p-hydroxybenzoate), sodium methyl p-hydroxybenzoate (E219), propylparaben, isopropylparaben, ethylparaben (E214—ethyl p-hydroxybenzoate), sodium ethyl p-hydroxybenzoate (E215), butylparaben, isobutylparaben, and benzylparaben. Methylparaben and propylparaben are the only two parabens classified as generally recognized as safe by the U.S. FDA.[677,678] (Paraben food additives approved for use in the EU include the E numbers in the parentheses following the variants above.[679])

Parabens are found in tens of thousands of personal care products on the market today, even though the mechanism by which parabens are antimicrobial "is not fully understood."[680]

A storm of controversy began brewing in the 1990s surrounding parabens when researchers discovered they acted as xenoestrogens, or chemicals that mimic female hormones, and as endocrine disruptors.[681] Fast forward to 2004, when a study published in the *Journal of Applied Toxicology* found five types of parabens in eighteen of twenty breast tumor tissue samples tested.[682] Methylparaben was discovered at the highest levels, comprising 62 percent of total paraben discovered. Of the six parabens analyzed (isopropylparaben was not included in the study), benzylparaben was the only paraben not found in any of the tissue. The research team concluded that some of the paraben absorbed through skin-care products or food is able to be retained, although they could not identify the specific route—oral or topical—in which the parabens entered the body. Nor could the study provide conclusive proof that the parabens in the breast tumor tissue actually caused the tumors in the first place. Of the study, *Discovery Fit and Health* noted, "Paraben may very well be found in all tissue, due to widespread use."[683] Still, the study was cause for alarm and further research, given that parabens are ubiquitous in cosmetic, skin-care, pharmaceutical, and even food products.

Despite their established action as xenoestrogens and endocrine disruptors, the FDA has classified both methylparaben and propylparaben as GRAS for use in food. Parabens can be found in processed foods, cakes, pie crusts, pastries, icings, dried meat products, coated nuts, liquid dietary food supplements, and more.[684] Regarding parabens as food additives, the FDA says, "There is no evidence that consumption of the parabens as food ingredients has had an adverse effect on man in the 40 years they have been so used in the United States."[685] While public concern mounted following the 2004 study previously discussed, causing a flurry of companies to remove parabens from their cosmetic and skin-care products altogether and openly noting "paraben-free" on the packaging as a selling point, many health-conscious U.S. consumers may likely remain unaware that paraben is used as a food ingredient and that it has been for over four decades.

In a follow-up to the 2004 study, a study in 2012 analyzed 160 breast tissue samples from forty women with breast cancer for five different parabens. This time, parabens were detected in a whopping 99 percent of samples. Propylparaben and methylparaben were found in the highest concentrations, respectively, but over 60 percent of samples analyzed contained all five parabens considered. While many underarm deodorants contain parabens that have been postulated as a potential breast cancer agent due to the close proximity of the underarms to the breasts and the typical daily usage of deodorants, the researchers noted that parabens were even present in the breast tissue of women who do not use deodorant.[686]

Even though only methylparaben and propylparaben are listed by the FDA as GRAS, two 2013 studies discovered those weren't the only types found in food samples tested. The first study involved 267 food samples, including meat, grains, fruits, vegetables, fish, fats/oils, dairy products, and beverages taken from Albany, New York.[687] Over 90 percent of the food samples tested positive for parabens, and all five types the researchers were testing for were present: methyl, ethyl, propyl, butyl, and benzyl. The highest concentrations were methyl, ethyl, and propyl. The abstract noted that, to the researchers' knowledge, it was the first study of its kind on paraben levels in foods.

In a follow-up study, the same researchers looked at food samples from China and determined that, out of six parabens, 99 percent of 282 food samples from thirteen categories collected from nine cities in China contained the chemical preservative. According to the study abstract, "Methyl paraben (MeP), ethyl paraben (EtP), and propyl paraben (PrP) were the major paraben analogs found in foodstuffs, and these compounds accounted for 59 percent, 24 percent, and 10 percent, respectively, of paraben concentrations."[688] The researchers also determined that estimated daily intake levels for the foodstuffs from China for parabens were approximately three to, in some cases, ten times as high as was found in the U.S. study.

The public's avoidance of parabens in foods is difficult because awareness of the issue is so low; many do not realize they are in foods and not just limited to cosmetics and skin-care products. Worse, when parabens are in foods, they are deceptively labeled as *methyl p-hydroxybenzoate* or *propyl p-hydroxybenzoate* instead of methylparaben and propylparaben.[689] Knowledge is half the battle and, unfortunately, the public is simply not well informed.

Propyl gallate (E310)

Propyl gallate is the ester of gallic acid and propynol used as a synthetic antioxidant food preservative to keep oxygen from turning the oils in some food rancid. It is commonly found in microwave popcorn products, mayonnaise, chewing gum, soup mixes, frozen TV dinners, and other foods containing oils and fats. It's also used in personal care products, cosmetics, adhesives, and lubricants. Propyl gallate is commonly used in conjunction with the preservatives BHA and BHT (see pages 149 and 163).

Propyl gallate has been shown in studies to be both genotoxic and cytotoxic, to inhibit and kill human endothelial cells, as well as to cause everything from allergic reactions such as seborrhoeic dermatitis and depigmentation of skin to liver damage. It was also recently identified as a xenoestrogen.[690,691,692,693,694,695] Propyl gallate can cause stomach irritation and asthma attacks, and it can negatively affect aspirin-sensitive people; in addition, some countries such as South Africa ban it from use in foods for babies and young children.[696] The FDA still considers it generally recognized as safe and has determined there are no safety hazards when used at appropriate levels.[697]

As it is not used as frequently as some preservatives, it is easiest to avoid by simply reading food labels.

TBHQ (E319), BHA (E320), and BHT (E321)

TBHQ, or tertiary butylhydroquinone, is a phenolic antioxidant-based preservative created with coal tar and the petroleum derivative butane. It is added to bread, pasta, margarine, potato chips, condiments, and other processed foods including fast food to prevent oils and fats from turning rancid. It's also used in a wide array of manufacturing capacities, including varnish and lacquer production, as well as for the stabilization of explosives. A five-gram dose of TBHQ is known to be fatal.

TBHQ is often used in combination with other preservatives, specifically BHA (butylated hydroxyanisole), and BHT (butylated hydroxytoluene). These three preservatives are commonly used together. In dozens of studies that have been done on all three since the mid-1970s, a wide range of adverse health effects—including reproductive, gastrointestinal, cardiovascular, liver, lung, and skin problems—have been demonstrated, along with severe allergic reactions, nausea, and delirium.[698]

These preservatives have also been linked in studies to behavioral problems in children, including ADHD. In the 1970s, before TBHQ existed, Dr. Ben Feingold was able to reduce behavioral issues in six hundred children just by removing BHA and BHT (in combination with the removal of artificial colors and flavors) from their diets.[699] While behavior improved in 30 to 50 percent of children after the colors and flavorings were taken out, the removal of BHA and BHT improved behavior in 60 to 70 percent of them. That same decade, in 1974, a study found that chronic ingestion of BHA and BHT by pregnant mice resulted in adverse behavior patterns, including insomnia, cognitive deficits, decreased self-grooming, and increased aggression.[700] According to the New England Health Advisory, not only has TBHQ been linked to ADHD, but studies have shown it affects estrogen levels in women as well.[701] According to the International Programme on Chemical Safety, TBHQ has damaged DNA in vitro and produced stomach tumor precursors in lab animals.

Both the FDA and the European Food Safety Authority have determined that BHA, BHT, and TBHQ are safe at the permitted acceptable daily intake levels. Despite evidence to the contrary, the EFSA ruled in 2004 that TBHQ is not a carcinogen and no further genotoxicity studies would be necessary.[702] Consumers eat an estimated 20 milligrams of BHA and BHT daily.[703]

TBHQ, BHA, and BHT are all required to be listed on food packaging, but unless someone specifically requests the ingredients list at a restaurant, they aren't going to know whether or not these preservatives are present in the food. In addition, as with many additives, federal laws do not require food manufacturers to disclose if ingredients were already preserved with BHA or BHT prior to being made into a final product. Vitamin A palmitate, used to fortify foods such as dairy products, may contain small amounts of undisclosed BHA and BHT, for example.[704]

Sulfites (E223)

Sulfites are common preservatives and antimicrobial agents added to foods, medicines, and especially wines to stop the fermentation process. Sulfites also prevent spoilage and can stop the browning process in some fruits and vegetables. They can be found in alcoholic beverages, condiments, modified dairy products, fish, gelatins, puddings, jams and jellies, shredded coconut, processed vegetables, dried fruits, and some snack foods and soup mixes.[705] According to natural wine promoter More than Organic, sulfites are present at concentrations of up to 10 mg/L even in unsulphured wine, but conventional wines on the market today contain an average of ten to twenty times that much. In addition, conventional winemakers typically add sulfites to red wine even though its antioxidant properties are such that it is an unnecessary step.[706]

While sulfur is an essential element found in all animal and plant cells—some foods, such as eggs, onion, garlic, and cabbage, naturally contain high amounts of sulfur—the inogranic chemical compound sulfite created from sulfur can cause adverse reactions in sensitive people, including autistic children who have issues ridding themselves of excess sulfur.[707] Some studies show that the sulfites regularly added to wine can actually trigger wine-induced asthma.[708] Research has also linked sulfite exposure to an increased risk of liver disease due to the oxidative damage it can cause.[709]

The FDA requires food manufacturers to list sulfites if 10 ppm or more are present in the finished food or beverage product.[710] Still, a rash of twenty-seven deaths between 1985 and 1990 were blamed on sulfite-induced anaphylactic reactions; at least six of those occurred in restaurants, where ingredients lists are either not readily available or not double-checked.[711,712]

Consensus was the deaths were due to sulfites on potatoes, which prompted the FDA to stop allowing sulfites on fresh fruits and vegetables in addition to establishing the 10 ppm-labeling limit. Still, sulfites are one of the few approved food preservatives that even the government has acknowledged has killed people—and yet, they are still allowed to be added to food with only limited regulation.

A meal consisting of a regular green salad, three ounces of dried apricots, and a four-ounce glass of wine would contain approximately 375 milligrams of sulfite, an amount far in excess of the World Health Organization daily limit of 42 milligrams for a 132-pound adult.[713] It's very likely the average diet contains far more sulfites than recommended, but because they are ubiquitous, it may be unavoidable.

EMULSIFIERS AND THICKENING AGENTS

A number of food additives are used to structure or blend otherwise incompatible mixtures of oil and water, or water-soluble ingredients. Thus, these chemicals and naturally occurring ingredients help hold many processed food concoctions together and maintain appearance, texture, and freshness, acting secondarily as preservatives. Ingredients such as cellulose and various gums—including gum arabic, furcelleran, guar, locust bean, and xanthan—frequently serve functions in processed foods as thickening agents, emulsifiers, and/or preservatives with little or no known risks and, in some instances, certain benefits. Guar gum, for one, has numerous positive interactions and possible health benefits.[714,715,716,717,718] However, certain other emulsifiers and thickeners may pose significant health risks.

Carrageenan (E407)

Carrageenan, a red seaweed extract, is one of the most commonly added emulsifiers and thickening agent preservatives in numerous cheese and dairy products, alternative nondairy and low-fat products, desserts, cereals, drinks, baby formulas, gums, and other snacks. In many cases, carrageenan works as a fat substitute to bind ingredients together and establish texture.

Carrageenan is even a favorite additive in many USDA-certified "organic" and "natural" foods, despite its unhealthy connection to gastrointestinal inflammation and experimental cancer in lab animals.

166

Its health risks center on the fact that degraded forms of carrageenan, which have lower molecular weights, have been found to cause inflammation of the gastrointestinal system and colon. The native carrageenan used in foods begins not degraded, but a certain percentage becomes inadvertently degraded through the alkaline-based extraction process. This process contributes a potentially dangerous and inflammatory form of carrageenan seeping into foods.[719]

In numerous scientific studies, researchers have administered degraded carrageenan to rats as a way to induce adverse health effects for tests, including pain, chronic prostatitis, arthritis, synovitis, pleurisy, insulin resistance, Achilles tendinitis, and edema, just to name a few.[720,721,722,723,724,725,726,727,728,729,730]

The Cornucopia Institute details how a working group formed in 2005 by the carrageenan industry trade group Marinalg tested samples of food-grade carrageenan produced by its industry members, finding degraded carrageenan in every single sample. Two-thirds of these samples contained levels of this dangerous derivative above 5 percent, the amount considered by the industry as a working limit. However, by 2012, Marinalg was reportedly unable to establish a reliable testing procedure that would allow limits to be set or met, meaning that there is no guarantee of consumer safety of this food additive in spite of its widespread use.[731]

The U.N. WHO's International Agency for Research on Cancer identified degraded carrageenan as a Group 2B "possible human carcinogen" back in the early 1980s.[732] The U.S. FDA considered restricting degraded carrageenan, as defined by molecular weight under 100,000, back in 1972, but no action was ultimately taken.[733]

The FDA has approved carrageenan as safe in its not-degraded food grade form, along with several of its salts and also formulas combined with Polysorbate 80 as stabilizers.[734]

In a controversial move, the USDA's National Organic Standards Board first approved carrageenan for use in organic foods in the mid-1990s. The Cornucopia Institute reported that when the food additive came up for periodic review in the spring of 2012,[735] one of the NOSB board members overemphasized the claims of the carrageenan lobbying group Marinalg (whose member companies include Cargill Texturizing Solutions and DuPont Nutrition Biosciences).[736]

That board member—supposed to be representing public interests—reportedly spent floor time reading direct passages from Marinalg-sponsored studies that asserted carrageenan's unequivocal safety,[737,738] but passed the claims off as if they were authored by United Nations' Joint FAO/WHO Expert Committee on Food Additives. Despite strong opposition from every public interest group in attendance, carrageenan was reapproved for use in organic foods for another five years by a slim one-vote margin.[739]

Dr. Joanne Tobacman has published more than twenty peer-reviewed studies on the health effects of carrageenan, and she has not only studied carrageenan as an associate professor at the University of Illinois College of Medicine,[740] but has also used her acumen to be a public advocate for the removal of carrageenan, initiating a petition to the FDA back in 2008, addressing the USDA National Organic Standards Board in 2012, and informing the public through various media outlets.[741,742]

According to Dr. Tobacman, the food additive is capable of causing inflammation in any of its forms,[743] making it not just another inert ingredient but also cause for significant alarm. Chronic inflammation can trigger a perpetual inflammation cycle that invites everything from Parkinson's to coronary artery disease to rheumatoid arthritis to cancer.

Significantly, Dr. Tobacman found that chronic, low-dose exposure to carrageenan contributed to glucose intolerance, insulin resistance, and impaired signaling, all precursors to diabetes and obesity.[744]

Another study by Tobacman and her team described how colon cells interact with carrageenan promoters to prolong inflammation caused by the food additive.[745] Most recently, Tobacman and her colleagues published a study in January 2014 demonstrating how carrageenan contributes to colon cancer.[746]

Despite all this, the public is largely unaware of carrageenan's risks, unlike the attention paid to high-profile ingredients such as aspartame, MSG, high-fructose corn syrup, and other controversial additives.

Ultimately, carrageenan has no nutritional purpose and is nonessential as an additive because it could easily be replaced by alternatives such as locust bean gum or guar gum. Moreover, many foods do not require an emulsifier in the first place but could instead simply prompt consumers to "shake" before eating or drinking.

The Cornucopia Institute has published a shopping guide to aid buyers in avoiding carrageenan.[747]

Soy lecithin (E322)

With soybeans as one of the most heavily subsidized, widely used, and cheapest sources of raw food material, soy lecithin is one of the most common components of modern mass-produced processed food products. It is relatively non-toxic, inexpensive (due to government soy subsidies), and reduces viscosity while preventing separation and keeping ingredients—such as oils and chocolate—evenly mixed inside product formulas. Made up of the phospholipids phosphatidylcholine (PC), phosphatidylethanolamine (PE), and phosphatidylinositol (PI), lecithin is separated from soybean oil through industrial production and can be found on a significantly large portion of product ingredient labels for foods of nearly every kind.[748]

In almost every case, that soy lecithin is also derived from genetically modified soy, unless the label specifically says it is made from non-GMO or organic soy. About 93 percent of soy grown in the United States and 81 percent of soy grown globally is genetically modified, although food producers prefer to keep that fact off of labels.[749]

The sordid details of soy lecithin's history as a food staple was taken on by author Kaayla Daniel in her 2005 book *The Whole Soy Story: The Dark Side of America's Favorite Health Food*. Though lecithins are naturally occurring in all organisms and can be extracted from many sources, soy lecithin came to dominate the market due to its cheap cost and surplus abundance as a foul-smelling industrial waste sludge that remains from the degumming processing of crude soy oil.[750,751]

Daniel cites historian William Shurtleff, who wrote an unpublished history on soy with coauthor Akiko Aoyagi, claiming that German soy oil refiners of the early twentieth century were seeking ways to dispose of this industrial sludge and turned to a vacuum drying method that led to the patenting and marketing of soybean lecithin as a major commodity.[752] Reportedly, the German industry hired scientists to develop hundreds of new commercially viable applications; several of the new applications it developed for the food industry now heavily affect the diet of the global consumer.

Shurtleff and Aoyagi further detail how Archer Daniels Midland (ADM), now a massive Big Agra conglomerate, became the first American manufacturer of soy lecithin in 1934, and by 1935, the company had patented a new process for oil extraction—using hexane.[753] This displaced the dominant ethanol-benzol extraction method and allowed for a more palatable and appealing soy lecithin product. ADM's aggressive marketing allowed soy-derived lecithin to overtake egg-derived lecithin and unleashed a whole new era of processed foods into the Western diet.

Using hexane as a solvent to extract soy lecithin underscores the pressing concern in weighing the potential health risks and contamination issues for this industry standard food emulsifier.[754] Hexane, a constituent of gasoline and jet fuel, poses significant chronic toxicological health hazards, including damage to the nervous and muscular systems and vision impairment. Hexane is also a known potential carcinogen.[755] The Cornucopia Institute found that hexane is persistent in soy lecithin production and thus poses a legitimate health concern.[756] This issue is more thoroughly covered in the hexane section of this book (see page 107).

Soy lecithin production supposedly eliminates soy proteins and, with it, the potential for allergic reaction. However, the expectation that mass production and mass consumption of soy lecithin does not carry with it the risks of soy-related allergies is not based on any long-term dietary studies, so it warrants further study. Nevertheless, aside from the hexane, soy lecithin likely carries a low allergenic risk.

Polysorbate 80 (E433)

Polysorbate 80, which is also known as polyoxyethylene (80) sorbitan monooleate, (x)-sorbitan mono-9-octadecenoate poly (oxy-1,2 ethanediyl), Tween 80, and POE (80) sorbitan monooleate, and its fellow polysorbates (including -20, -40, -60, and -65) are emulsifiers traded under brand names such as Tween, Alkest, and Canarcel. Polysorbates are made up of sorbitol, a sugar alcohol, esterified with fatty acids. Polysorbate 80 and polysorbate 60 are widely used in foods, while polysorbate 80 has become a common (and controversial) adjuvant and excipient in vaccines and pharmaceutical drugs, included increasingly in the nanoparticle delivery of medication.

Polysorbate 80 has GRAS status from the FDA and is accepted as safe in Europe as well; it is very frequently found in whipped dessert toppings, ice cream, shortening, desserts, and condiments.

However, few studies have been done on the actual safety of this processed food ingredient in the human diet. While no great potential for harm has yet been demonstrated, and no evidence exists in regards to carcinogenicity or neurotoxicity,[757] there is some emerging evidence to cast doubt on the overall safety of dietary polysorbate 80.

Gastroenterology research into the causes and rising prevalence of Crohn's disease and other gastrointestinal inflammatory diseases has raised significant dietary questions about the developed world's modern diet of highly processed foods. Does polysorbate 80 play a role?

In 2010, researchers probed the impact of foods on aiding or inhibiting invasive disease bacteria across the gastrointestinal barrier through transportation on M cells (microfold cells),[758] which play a role in immune response and in breaching this barrier during intestinal inflammation.[759]

The study found that high-fiber foods like broccoli and plantains inhibited the translocation of invasive disease-carrying bacteria, while emulsifiers such as polysorbate 80 from processed food diets facilitated the transport of pathogens, increasing the rate across M cells fivefold.

Researchers now believe that emulsifiers generally may play a significant role in increasing intestinal permeability in patients with Crohn's disease, particularly as emulsifiers are detergents—and amphiphilic (friendly with water and fats)—which are known to increase intestinal permeability. These researchers noted that in previous studies, "Polysorbate 80 has been shown to integrate within cell membranes, altering their microviscosity."[760,761]

This research would support evidence that polysorbate 80 could be affecting transport of disease-causing agents across the intestinal barrier. So far, there has been little investigation into the effect of emulsifiers like polysorbate 80 on gut permeability, but the implications of these initial findings for the emerging rainbow of gastrointestinal disorders is immense.

A 2003 study found that injected polysorbate 80, frequently used as a vaccine adjuvant, was found to increase digestive efficiency, but at the same time, it also caused a toxic irritating effect on the gastrointestinal system at a high dosage.[762]

In commercial food production, polysorbate 80 has also been combined with carrageenan into a single food additive, which has been approved by the FDA for use in foods.[763,764] Now, it need only be labeled "carrageenan" even when it contains up to 5 percent by weight of polysorbate 80. The FDA currently limits the concentration of polysorbate 80 in the final food product to 500 ppm.[765] Given the results of the Crohn's study with polysorbate 80 and the significant and toxic effects of carrageenan with gastrointestinal inflammation (see the earlier section on carrageenan on page 166), there is reason to be concerned that low concentrations of polysorbate 80 may now be hidden in foods.

A study published by *Nature* in 2015 highlighted the negative impact of dietary emulsifiers on digestive disorders and even metabolic syndrome. Entitled "Dietary emulsifiers impact the mouse gut microbiota promoting colitis and metabolic syndrome," this study reported:

> [A]gents that disrupt mucus–bacterial interactions might have the potential to promote diseases associated with gut inflammation. Consequently, it has been hypothesized that emulsifiers, detergent-like molecules that are a ubiquitous component of processed foods and that can increase bacterial translocation across epithelia in vitro2, might be promoting the increase in inflammatory bowel disease observed since the mid-twentieth century.[766]

Avoiding polysorbate 80, along with other synthetic preservatives and emulsifiers, may be prudent under the prevailing wisdom of the precautionary principle and a little common sense, as polysorbate 80 was not an ingredient in anyone's diet a century ago. It may well be confounding or aggravating to our digestive systems, regardless of how readily it is sold to us in cleverly marketed food products with trendy and inviting packages.

A feeding study from 1956 testing for fertility effects on mice from partial ester emulsifiers found no effect from eating a 5 percent diet of polysorbate 80, but it did find a slight reduction in fertility at the extremely high dietary intake level of 20 percent.[767] Though this level of consumption is unrealistic in terms of typical human diets, it may warrant further investigation, as polysorbate 80 has been connected with lowered fertility and birth defects when used in vaccines in animal studies[768] and accompanied a spike in fetal loss reports across three consecutive flu seasons while it was in the flu vaccine.[769] It is in current versions of vaccines for influenza, HPV, Pneumococcal, Rotavirus, Tdap, and DTap.[770]

Medical administration of polysorbate 80 in vaccinations caused anaphylactic shock in at least one man, according to a 2005 paper.[771] Rats given injections of polysorbate 80 (Tween 80) in saline experienced convulsions and death within minutes.[772] It should be noted, when administered capsaicin, the compound that makes hot peppers hot, prior to the Tween 80 shot, the rats' lives were saved from the Tween 80.

Carbon monoxide

Carbon monoxide (CO), the notorious odorless killer gas, is used as a color preservative in meat and seafood products to maintain a reddish color that gives the appearance of fresh meat and lasts for up to three weeks.[773] Retail cuts are packaged in gas mixtures containing less than 0.5 percent CO. It is approved for use and generally recognized as safe by the FDA.[774]

Though studies have claimed that the low level of gas used is safe and a highly improbable toxic threat,[775] critics have called attention to its use on the basis of potential consumer fraud, by potentially making old foods seem fresh.[776,777] If meats or fish appear fresh even after they are past their prime, shoppers could be duped into purchasing spoiled meats, filled with dangerous microbes, which could be hazardous if consumed.[778]

In addition, low-level chronic exposure to breathing carbon monoxide can cause amnesia, headaches, memory loss, behavioral issues, loss of muscle and bladder control, and vision impairment, although no studies have considered the effects of long-term CO ingestion.[779]

All in all, the use of such a well-known toxin in food preservation holds a creepy overtone—one more cosmetic agent of food mummification.

Potassium bromate (E924)

Potassium bromate is a preservative and bleaching agent that strengthens glutens and was widely used across the globe in nearly every type of enriched bread for many decades, until it was confirmed to be a carcinogen targeting the kidneys[780] and thyroid with oxidative damage.[781,782]

The food additive has since been banned in numerous countries, starting in Europe and the United Kingdom in 1990, in Canada by 1994, in Sri Lanka and parts of Latin America by 2001, and even in China by 2005,[783] while the state of California requires a warning label listing it as a carcinogen under Prop 65.[784]

Nevertheless, it remains approved for use in the United States by the FDA as an optional ingredient in standardized foods at levels less than 75 ppm in whole wheat flour and 50 ppm in white flour, though use has reportedly declined.[785] It remains legal because it was approved by the FDA back in 1958 before the Delaney Clause took hold.[786] The EPA classified it as a

Group B2 carcinogen in 1993 and established a final rule by 1998, stating that "there is sufficient laboratory animal data to conclude that bromate is a probable (likely under the 1996 proposed cancer guidelines) human carcinogen."[787]

In addition to potassium bromate's direct effects, bromism, or bromide dominance, can develop in the human body, in which long-term chronic exposure to bromide can inhibit iodine absorption, leading to a deficiency that can trigger cancers of the thyroid, prostrate, and ovaries after significant accumulation.[788,789]

Potassium bromate, where still in use, remains a largely hidden ingredient, typically only listed on the label as "enriched flour," but occasionally appearing as "bromated flour." Several fast food chains continue to use it in buns and breads, despite the clear risks.

Used in the United States since the early 1900s, it is added to the brew and dough recipes for enriched flours, particularly after new requirements called for nutritional enhancements and constant refinement to maintain texture, volume, and a palatable taste in industrial scale breads produced for the commercial market. Potassium bromate was considered essential at trace levels to solidify the addition of soy or wheat-gluten proteins. *Cereal Chemistry* journal articles detail the sometimes disastrous recipe revisions[790] and workarounds in 1970s-era cereal-enrichment formulations based on low-quality ingredient mixtures with potassium bromate as a stabilizing agent.[791]

Documented industrial recipes describe its continued use in 2002 with a raw wheat germ and vital wheat gluten formula.[792] Organic flours and baked goods typically avoid the use of this chemical, and are the best bet to avoid intake.

Studies have shown that glutathione, cysteine, and vitamin C protect against the cytotoxic carcinogenic effects and oxidative DNA damage of potassium bromate by blocking its ability to induce oxidative stress.[793,794]

Brominated vegetable oil (BVO) (E443)

A related bromide preservative is controversially used in the soft drink and beverage industry in sodas and sports drinks with citrus flavors. Brominated vegetable oils (BVO), composed of bromine and corn or soy oils, emulsify these citrus flavor agents and allow them to remain suspended in a cloudy mixture in drinks, including Mountain Dew, Gatorade, Powerade, Amp, Squirt, and Fanta Orange.

BVO was originally approved for use as a flame retardant. According to the Center for Science in the Public Interest, safety concerns over brominated vegetable oils led the FDA to remove the additive from the generally recognized as safe list back in 1970. However, with behind-the-scenes pressure from the beverage industry, the toxic ingredient continued to be allowed as part of the FDA's Interim List pending further safety studies.

Decades later, the FDA has indicated it believes the ingredient to be "safe," while Europe and other countries have banned its use and embraced safer alternatives.[795]

Long-term exposure to BVOs can cause inhibited growth, adverse behavioral and reproductive effects, heart lesions, and liver damage, according to rat studies, and several isolated cases of human toxicity after extreme overconsumption of sodas, triggering a severe case of bromism.[796,797,798,799]

Sodium nitrite (E250)

Sodium nitrite, used in everything from pesticides to dyes to pharmaceuticals, is an inorganic compound perhaps best known for its role as an additive in processed meats. The FDA has approved sodium nitrite for use in foods to prevent the growth of botulism spores and as a color fixative.[800] Sodium nitrite is added to give meat that seemingly "fresh," vibrant red or pink color that will make it more appealing to consumers for its potentially lengthy shelf life.

While it may be visually appealing—causing cured deli meats, pepperoni, salami, jerkies, bacon, hot dogs, and sausage to look the way people expect them to—sodium nitrite in processed meats doesn't look quite so pretty otherwise.

A multitude of studies have associated processed meats with a bevy of cancers and health issues due to the nitrites used to cure them. In just the last decade, researchers have linked sodium nitrite in processed meat to a 74 percent increase in leukemia;[801] a significant increase in the risk of esophageal carcinoma according to a thirty-year cohort study;[802] reproductive toxicity and interference with normal embryo development;[803] the parallel rise of Parkinson's disease, Alzheimer's disease, and diabetes;[804] an increased risk of gastric cancer;[805] a 31 percent increase in ovarian cancer risk with high intake of dietary nitrite;[806] obstruction of lung function and increase in risk of chronic obstructive pulmonary disease (COPD);[807] formation of a hepatocarcinogen;[808] a 67 percent increase in pancreatic cancer risk;[809,810] a positive association between red meat intake and bladder cancer;[811] nephrotoxicity and oxidative damage in the kidneys of rats;[812] a twofold higher risk of thyroid cancer in women with the most dietary intake of nitrite, particularly from processed meats;[813] and the list goes on and on.

When sodium nitrite hits the human digestive system, all hell breaks loose. At high temperatures, nitrites in processed meats combine with the proteins in meat called amines, forming toxic, carcinogenic nitrosamines in the stomach that can enter the blood stream and wreak havoc on the body. Nitrosamines were first outed as cancer-causing agents in 1956 when two scientists discovered dimethylnitrosamine gave rats liver tumors, so the dangers have been known for some time.[814]

In her book *Eating May Be Hazardous to Your Health,* former FDA aspartame panel member–turned–whistleblower Jacqueline Verrett talks about how Dr. William Lijinsky, a scientist at Oak Ridge National Laboratory, reported that 100 percent of his lab rats fed combinations of nitrite and amine (found in meat, wine, fish, and many prescription drugs, as well as other products) developed malignant tumors in nearly every organ system within six months.[815]

Although some may try to argue that nitrites in processed meats are safe because some vegetables naturally contain nitrites, those vegetables do not contain the amines that meat does; neither are vegetables heated to the same range of temperatures as meats, so the likelihood of vegetables creating nitrosamines is much lower.

The bacteria found in meat reduces nitrates into nitrite, which in turn becomes the nitric oxide that actually cures the meat. Environmentally relevant concentrations of nitric oxide have been found to induce everything from reproductive and developmental toxicity to colon cancer.[816,817]

Is sodium nitrite toxic? Without question, it is.

In 2008, a Missouri woman who worked at a meat processing plant filled a capsule with sodium nitrite and gave it to another woman under false pretenses, allegedly in order to hospitalize her so she could have a chance to get close to the woman's husband. The victim of this poisoning collapsed twenty minutes after taking the pill and was rushed to the hospital. She survived, but only because the quantity of sodium nitrite was deliberately chosen to be nonfatal.[818] Sodium nitrite is currently being developed as the main ingredient in a feral hog toxicant for population control purposes.[819]

As stated on the USDA website in a document extolling the virtues of a sodium nitrite–based feral hog killer:

> The toxin, sodium nitrite, a common meat preservative that prevents botulism, had previously been shown to be a quick-acting and low-residue toxicant for feral pigs in Australia and has since been patented. Pigs are particularly sensitive to nitrite-induced methemoglobinemia because they have low levels of methemoglobin reductase, the enzyme required to reverse the effects of nitrite toxicosis.

It raises the question: If sodium nitrite is toxic to feral hogs because they have "low levels of methemoglobin reductase," isn't it also possible that some humans may also share that enzyme deficiency due to natural genetic variation?

There's no question that sodium nitrite is a toxin in both humans and feral hogs. The solution to this chemical contaminant is to stop consuming it. The best way to avoid sodium nitrite is to stop buying processed meat products containing the ingredient all together, and if meat is on the dinner menu, look for fresh meat and meats that explicitly state "no nitrites" on the packaging.

Vitamin C has been shown in studies to protect people from the damaging effects of nitrites.[820] In addition, consuming large amounts of vitamin C, as well as E, will reportedly protect one from the cancer-causing nitrosamine conversion process if taken before processed meats are to be consumed.[821] Cod liver oil was also recently proven to protect the liver against sodium nitrite, significantly reducing nitrite-based liver inflammation in rat studies.[822]

MOLECULAR ALTERATION OF FOOD

Food manufacturers routinely alter foods at the molecular level, processing them in ways that have enormous consequences on human health. Typically, these alternations are carried out in order to boost product sales through increased shelf life or improved cosmetic appearance of the finished product.

Homogenized milk fat

Whole, raw milk is not molecularly homogenous. When milk comes from a cow, it contains cream that typically separates and rises to the top. This cream is made of intact, whole fat molecules that are perfectly formed for the nutritional needs of a baby cow. Before drinking or using the milk, people typically shook the milk bottles or jugs to reintegrate the cream and fat into the more predominant liquid.

Most commercially available milk on the market today is homogenized, sometimes labeled "homo" for short. This means the fat in the milk has been subjected to a mechanical process that uses heat, then high pressure (estimated at 4,000 pounds per square inch) to push the milk through tiny tubes that break the fat molecules into far smaller pieces, from up to 15 micrometers down to less than 2 micrometers.[823] When complete, this process keeps the milk fat evenly distributed throughout the finished product, so it does not rise to the top and the milk doesn't have to be shaken up.

While that sounds convenient, some research has pointed to the theory that homogenization is dangerous to health. Why? Because it ends up producing fat globules so tiny that the particles of proteins that would normally be digested instead pass through intestinal walls and enter the bloodstream, undigested. There, they may interfere with healthy cardiovascular function and arterial lining. They also release the enzyme xanthine oxidase (XO), potentially causing damage to arteries and inducing arterial plaque formation, ultimately leading to heart and circulatory disease.

Like many other processed foods, then, milk begins as a wholesome, nutritionally intact product. But through homogenization and pasteurization, it is artificially modified into a beverage that serves the interests of the industry producing it while simultaneously exposing consumers to health risks that aren't present in the pre-processed form of the product.

The theory of the XO enzyme posing significant risk to human health was first published by Dr. Kurt Oster, a cardiologist, and his coauthor, Dr. Donald Ross, in 1973,[824] and it was immediately attacked. Andrew Clifford and Charles Ho published their oppositional study on bovine milk xanthine oxidase in 1977, claiming that large intravenous doses of XO administered to rabbits did not lead to arterial plaque formation, nor did it deplete plasmalogens.[825] The study, however, was funded by the National Dairy Council.[826] Oster and his colleagues continued undeterred, in 1981 publishing further research that evaluated the blood of 300 heart attack victims over a five-year period and discovered significantly elevated XO levels in every single one.[827]

Researchers have also debated whether the cause of the damage in question came from the XO naturally occurring in the human liver or in the cow's milk, but XO in cow's milk is fifteen times more prevalent.[828] Cow's milk is the largest source of dietary XO, and while pasteurization destroys about half of it, the other half is still being ingested by homogenized milk drinkers.[829]

The debate over XO's role in artery damage and heart disease rages on today, but further evidence has emerged that Oster and Ross were right. A study in 1997 reported that XO was at least partly responsible for impairing heart function, and inhibiting XO in patients with high cholesterol reversed these effects, though not entirely.[830] Two years later, another study concluded that "circulating XO can bind to vascular cells, impairing cell function via oxidative mechanisms."[831] A 2002 study concluded that the presence of increased XO was closely associated with increased vascular oxidative stress in people suffering from chronic heart failure.[832] Overall, XO has been linked in research to more than fifty inflammatory and autoimmune diseases.[833]

Researchers have also noted a positive correlation between milk consumption and coronary heart disease death rates.[834]

Steve Bemis, lawyer and board member of the Farm-to-Consumer Legal Defense Fund, notes that in the years prior to World War II, milk competition was based entirely on the "cream line" in the milk—the more cream, the better. Bemis argues that milk cream was considered the premium portion of milk, which could be used in other products such as cheese; so having to leave more cream in milk undermined potential profits in what would become an exploding grocery industry. The homogenization process helps solve this problem because it removes the cream line, leaving producers free to use much of the cream in other dairy products.[835]

Homogenization was important to the burgeoning cheese industry because it improved texture, flavor, and softness in some cheeses.[836] Just as we've seen in other food additives and processed food commodities, the production of cheese was altered to (excuse the pun) milk it for all it was worth; improving appearance, taste, and shelf life were valued far above nutrition or tradition. Additionally, standardizing the production and redirecting valuable cream from milk to other dairy products was a boon for the dairy industry because it resulted in a new ingredient supply for cheeses, dessert products, and more, substantially increasing revenues.[837]

Additionally, homogenization led to the standardization of both milk and cheeses, which further lowered production costs and increased profits. Once milk is homogenized, it must be pasteurized, or treated with high heat; otherwise, it will spoil within hours. (The reverse, however, is not true—pasteurized milk does not require homogenization.) Another issue with homogenization is that it's paired with pasteurization, and more broadly, the large-scale commercialization of the dairy industry that began in the post–World War II period. Prior to this, milk was locally pasteurized in urban areas, while many rural areas offered fresh, raw milk. This allowed consumers in cities who still wanted fresh, whole milk to acquire it from rural areas if they chose.

Dairy processing for mass commercialization became mutually beneficial for milk, cheese, and other dairy products, making homogenization and pasteurization a necessary, standard combination process for producing finished products. This is despite the fact that these heat treatments destroy valuable nutrition in the milk, including an array of essential amino acids and import-ant enzymes. One reason so many people are allergic to lactose in milk, for example, is because pasteurization destroys the lactase enzyme that helps the human body break down the proteins. Undigested proteins lead to allergic reactions.

Homogenized milk can be avoided by looking for nonhomogenized products typically carried in natural, organic food stores and buying local, raw milk and milk products. While some countries such as France sell raw milk in vending machines, strict raw milk purchasing laws vary by state in the United States, with some completely banning sales and others allowing retail sales, while others will only let people buy it directly from the farm.[838]

Hydrogenated fatty acids

Hydrogenated fats are commonly found in fast foods, from French fries to chicken fingers; in dairy products such as margarine; in breads, cakes, and biscuits; and in TV dinners and sweets. Because it isn't a saturated fat, con-sumers who do not expressly know what it is may mistakenly think it's some-how healthier for them. Unfortunately, they would be wrong.

Hydrogenated fat is liquid vegetable oil that has been treated with hydro-gen. These are normally healthy oils that undergo manufacturing processes that ultimately turn them into poisons. First, the oil is heated from 500 to 1,000 degrees under high pressure; then it is exposed to hydrogen gas. Finally, a catalyst, typically a metal such as nickel or aluminum, is injected into the oil for several hours to change the molecular structure, increasing the oil's density. The finished product is either semi-solid, known as partially hydro-genated oil, or solid, known as hydrogenated oil.

The resultant synthetic fat is heavier. It actually thickens the blood after it's been eaten, forcing the body's circulatory system to work harder to push blood around inside it. Hydrogenated oil sticks to and readily clogs arteries, leading to an increased risk of high blood pressure, blood clots, and heart attacks (and that's just for starters).

The thicker blood also has a hard time circulating through the brain, leaving a person open to everything from muddled thoughts to Alzheimer's disease.[839] The use of the metal aluminum as a hydrogenation catalyst can't be helpful for brain-related disorders either.

Hydrogenated and partially hydrogenated oils are the main sources of trans fats, and an abundance of scientific studies litter medical journals declaring one after another just how bad trans fats are for people. One 1997 Harvard University study involving more than 80,000 women found that simply decreasing one's dietary trans fat intake by a mere 2 percent reduced the risk of coronary heart disease by a whopping 53 percent.[840] Researchers in another study discovered that eating just 5 grams of hydrogenated fat a day—particularly partially hydrogenated fat—increased coronary heart disease risk by 29 percent.[841] Yet another study showed that eating foods considered major trans fat sources, such as biscuits, cookies, cakes, and margarine, was significantly tied to higher coronary heart disease risks.[842] The FDA admitted in 2013 that ridding the U.S. food supply of industrial trans fats could prevent approximately 20,000 heart attacks and 7,000 heart-related deaths a year.[843]

"Trans is a secret killer. Labels tell you how much saturated fat you're eating. With trans, it's anybody's guess," Dr. Walter Willett, chairman of Harvard School of Public Health's nutrition department, told the Center for Science in the Public Interest in 1996.[844] Dr. Willett is a pioneer in establishing a link between heart disease and trans fats.

Trans fats in the United States were not required by law to be listed on nutrition information labels until January 2006, but even after that point, the product had to contain 0.5 grams or more.[845] This standard means that up to half a gram of trans fats could be in a product and the manufacturer could still claim it had zero trans fats. Yes, to the FDA, 0.5 grams rounds down to zero. Maggie Stanfield, author of *Trans Fat: The Time Bomb in Your Food*, explained to the British *Independent* that, due to their synthetic nature, hydrogenated vegetable oils confuse the body's cells. "They identify the fat as unsaturated—it comes from vegetable oil, after all—but because of the industrial process involved, they can't handle the fat as they would a truly unsaturated one."[846] Industrially produced hydrogenated oils are not natural and the body cannot process them as such.

Hydrogenated oil shares a molecular resemblance to plastic. It's nonessential, serves no nutritional purpose, and has no known human health benefit whatsoever.[847] So why would food manufacturers knowingly put it in people's food? Simple. It's a cheaper, taste-free butter alternative that extends a product's shelf life. How long? One American television program featured a cake that appeared to have just been made but was actually baked over two decades ago.[848] Consider what happens when this substance enters the body. Unless governments force the issue, it's a safe bet companies will continue to show more concern for their financial bottom lines than the health risks their trans fat–laced products pose to the public at large.

Some countries have forced the issue. Denmark became the first country to initiate strict regulation on the sale of trans fat foods in 2003, effectively banning partially hydrogenated oils with its limit of 2 percent oil and fat ingredients destined for human consumption.[849] Iceland followed suit, as did Switzerland.[850,851] Canada passed a Commons motion similar to Denmark's ban, although it is not binding (as no motions that pass through Commons are legally binding on the Canadian government). The European Food Safety Authority released an official scientific opinion on the effects of trans fatty acid consumption in 2004, admitting the link between eating hydrogenated fats and an increased risk of cardiovascular disease.[852] In 2013, the EU margarine and vegetable fat trade association tightened the Code of Conduct for the third time since 1995 in a bid to reduce retail food's trans fat levels.[853]

Due to the overwhelming preponderance of the evidence that trans fats are detrimental to health that has continued to pour out, the FDA finally took action, announcing in November 2013 that it was going to revoke the generally recognized as safe status for partially hydrogenated oils[854] in a move FDA Deputy Commissioner for Foods and Veterinary Medicine Michael Taylor called "the next step" in removing these oils from the nation's list of approved food additives.[855]

Because more and more evidence is stacking up against hydrogenated vegetable oil all the time, some food manufacturers are turning to palm oil instead. Unfortunately, a 2006 *American Journal of Clinical Nutrition* study and a 2009 U.S. Agricultural Research Service–supported study both found that palm oil impacted the lipoprotein profile even more negatively, resulting in higher levels of bad cholesterol.[856,857]

Many people are still laboring—and chewing—under the false assumption (strongly pushed by the media and major health outlets) that *all* saturated fats are bad and that butter is much worse for their health, when in reality butter from grass-fed cows is actually a much more nutritious option than trans fats, because butter contains vitamin K2 and omega-3s and has been shown to raise good cholesterol levels while lowering bad cholesterol. Although it's been the health mantra for years, scientific studies have found no reliable association between saturated fats and coronary heart disease.[858,859,860]

Until more governments get on board with banning trans fats altogether, the best strategy for avoidance is to steer entirely clear of fried foods, avoid processed foods as much as possible, and double-check ingredients lists thoroughly when unsure. Partially hydrogenated oil or hydrogenated oil will sometimes show up on ingredients lists as "shortening." Always look for the word "hydrogenated"—even if a product says zero grams of trans fat, it may still contain up to 0.5 grams. Remember, it doesn't take much hydrogenated oil to have a disastrous effect on health.

ANIMAL FEED CONTAMINANTS

"You are what you eat."

One of life's oldest adages holds remarkably true in this modern age of industrialized, centralized, globalized, and profit-driven food production.

The principles of bioaccumulation in the food chain expose us to environmental toxins consumed by fish and wildlife that end up on our plates. They simultaneously expose us to the toxins, hormones, drugs, and pathogens absorbed by the livestock raised to feed our growing population, who have less transparency, control, or input into the food we eat than ever before.

Even when consumers are meticulous about the choices they're making when purchasing fresh produce, grains, and food staples at the grocery store, they often have little or no information about what ingredients and feed practices have gone into the meat they're buying. Thanks to industry collusion with government regulators (USDA and FDA), nearly all meat sold in America today can be described as "mystery meat" with unknown origins, undisclosed feed practices, and unwise treatment with aggressive antibiotics and hormones.

For consumers, buying meat is much like a game of "nutritional Roulette," where every meal comes with a regrettable list of unknowns that rightly make people concerned. And while producers of mass-marketed foods have taken shortcuts with the food we eat, adding cheap, artificial, and nutritionally void ingredients to our detriment, they have taken even more questionable liberties with the animal feed that is forced on the cattle, poultry, swine, and farmed fish that we in turn ingest.

Feed sources

The U.S. Department of Agriculture administers vast amounts of subsidy dollars through the U.S. Farm Bill, a massive piece of legislation that is renewed every five years with billions of dollars annually for the biggest producers of corn, soy, and other crops as well as beef, poultry, fish, and other livestock. More than $295 billion in subsidies were issued between 1995 and 2012, including $177 billion in direct commodity subsidies.[861]

According to the Environmental Working Group, the top ten most subsidized crops are corn, wheat, cotton, soybean, rice, sorghum, peanuts, barley, tobacco, and sunflower, with corn subsidies topping $84 billion between 1995 and 2012. Another $27 billion in subsidies went to soybean crops during the same period. Corn and soybeans are predominantly cultivated as genetically modified crops and frequently used in processed foods of all kinds, and are a major source of livestock feed. The economics of raising meat with cheap inputs has made heavy grain diets standard with most livestock, even when other diets are more natural choices. The drive to find the cheapest inputs for feed rations makes subsidized, genetically modified corn and soybean feed the most popular choice, with smaller amounts of hay, forage, or by-product added in. Typically drenched in pesticides and cheaply produced for fodder, ethanol, or junk food production, these crops often have contaminants that, when consumed by livestock, enter into our food supply.

The rules of agriculture and livestock production are such that smaller-scale, independently operated, local, and/or holistic food producers work at a strong economic disadvantage, which creates difficult market conditions for those raising livestock responsibly.

The vast majority of meat and animal protein raised and sold in the United States and much of the global market is dominated by remarkably few companies. This is not strictly a problem of too few producers, as there are many hundreds of thousands of farmers who raise poultry and swine or own cattle operations, many of whom are relatively small-scale. The problem is a monopolized market dominated by irresponsible and inhumane animal husbandry.

The U.S. Justice Department held a workshop in partnership with the U.S. Department of Agriculture to address antitrust issues with respect to agricultural competition and market concentration, frequent mergers, bid rigging, and market manipulation—including through captive ownership and supply—as well as the lack of transparency and issues of oligopoly and monopsony, a problem where many producers are impacted by one or few buyers or marketplaces, forcing low prices and often unfavorable conditions.[862]

Contamination of animal feed and bioaccumulation in livestock

One predominant strain of genetically modified crops—Bt corn and Bt soy, which are engineered to produce their own pesticides—have been found to contain higher lignin content than natural varieties.[863] This is a clue that genetically modified crops are not nutritionally equivalent to non-genetically modified foods, and it could have a significant impact on animal feed, as ruminated animals—such as cows, goats, and sheep—are less able to digest lignin than humans.[864] The widespread use of genetically modified crops could have far-reaching implications for livestock and the economy that surrounds them. For example, long-term feeding on such crops could create nutritional imbalances in livestock that ultimately require treatment with an additional burden of medications and antibiotics.

China is officially the world's largest agricultural product producer and consumer, while the nation is home to some of the most polluted cities on the face of the Earth.[865] Pollution from China's vast industrial complex rains down from the sky, where it is deposited into soils that are then used by the vegetation that is eaten by the livestock. The vicious cycle continues, as the manure from heavy metal–contaminated livestock contributes to toxic runoff and further concentration of toxins in already contaminated water and soils. As if that weren't enough, Chinese farmers also regularly add heavy metals, such as copper and arsenic, purposefully to farm animal feeds for their antimicrobial and growth-stimulating properties.[866,867]

A 2004 survey of heavy metal pollution from thirty-one farming plants in ten major Jiangsu cities found that the majority of feed samples surveyed contained high concentrations of toxic heavy metals; manure from the animals also contained alarming levels of cadmium, lead, zinc, copper, and chromium.[868] A similar survey of 104 livestock feeds and 118 manure samples from farms in Northeast China in 2012 also found high levels of arsenic, cadmium, and copper in poultry, pigs, and cattle.[869] In addition, pig and poultry feeds contained higher levels of heavy metals than cattle feed, a finding mirrored by an analysis of farm and animal feeds in England and Wales.[870]

These heavy metal constituents, which readily bioaccumulate, are often stored in animals and then passed along to humans who consume their meat.

Beef

Beef cattle represent about a third of global meat production, with the United States, Brazil, and China leading the way. The United States has the largest beef production industry in the world, with the vast majority raised on a grain feed diet, typically composed in large part of corn, soy, and alfalfa, most of which is produced from genetically modified crops.[871]

According to the USDA National Agriculture Statistics Service and the National Cattlemen's Beef Association, about 26 billion pounds of beef is harvested annually in the United States from about 33 million head of cattle, with about 2.5 billion pounds of beef sold in export markets. Additionally, about 34 million cattle are raised in calf operations.[872]

Though the industry is quick to point out the large number of independent farmers with fewer than one hundred head of cattle, it remains true that Big Agra players control the market by volume. Eighty-five percent of the fed cattle market is dominated by large-scale feeding operations, where operators like JBS and Cargill feed cattle owned by independent ranchers as well as bigger players. They also frequently coordinate or arrange for fed cattle to be sold to slaughterhouses and/or meat packers.[873]

Large-scale feeding operations are a function of powerful conglomerates that own and operate the feedlots, slaughterhouses, packing facilities, calving operations, and seed stock, where genetic varieties of cattle are licensed. Small-scale producers are often dependent on the feeding services and the market connections of Big Agra players—many who have major interests in each step of cattle production—who heavily influence price and conditions of sale, including the use of drug and feed inputs.[874] In many cases, these smaller ranches, such as those with fewer than one hundred head of cattle, contract with large operations for feeding, which might take place in a large lot over several months of the year. These large-scale feedlots typically provide feed, veterinary service, transportation, and market access to herd owners, who contract with them or other large feeding operations.

Top players in feedlot operations are JBS Five Rivers in Colorado with a one million head capacity on twelve yards in seven states including Kansas,

Texas, Colorado, and Oklahoma. The second largest is Cactus Feeders, with the capacity for over half a million head of cattle on nine yards in Texas and Kansas. Third, Cargill Cattle Feeders, LLC, in Kansas, has the capacity to feed more than 350,000 head of cattle at a time on five yards in Kansas, Texas, and Colorado. Friona Industries, LP; Cattle Empire, LLC; J.R. Simplot Co.; Irsik and Doll Feed Services, Inc.; Four States Feedyard, Inc.; Foote Cattle Co.; and Agri Beef Co. round out the top ten feeding operations in the United States. Many of these companies are transnational, with major beef and livestock operations in Brazil and Australia, among other locales.

J.R. Simplot, which is also the largest potato supplier to fast foods and the sole supplier of potatoes for French fries to mega fast food chain McDonald's, is emblematic of the vast control industry has over food production, from beginning to end. Based in Idaho, Simplot ranks as the largest Western-based cattle producer, and is among the largest feedlot operators as well as one of the top calf-producing firms. Simplot operates one of the largest feeding lots in the world, with the capacity for 150,000 head of cattle at one time, in Grand View, Idaho,[875] and maximizes synergy by supplying beef and dairy cattle—largely destined to produce fast food meals—with "custom" feed composed in part of by-products from its potato empire.[876,877] Simplot feedlots also ship in corn for grain-feed bulking on its specialized rail delivery system and can store some 1.5 million bushels of corn on site.[878] It further operates meat processing plants that supply fast food restaurants, including McDonald's, across the United States and in global markets.[879]

Tyson Foods is the giant of the slaughter and meat packing industry, doing even more business in beef production than it does in the chicken production with which its namesake is so closely associated. Tyson ships frozen carcass, boxed, case-ready, and value-added (processed) meats to nearly every major fast food, grocery store, restaurant, and retail chain, as well as to school cafeterias, hospitals, and prisons throughout the United States and elsewhere.[880,881,882] JBS and Cargill Meat Solutions are the next largest packers, working synergistically with other branches of their operations and significant inroads with other sizeable firms throughout the food industry that they supply.[883]

Cow diets: grains, candy, and hormones

Despite the fact that cows are biologically predisposed to eat grass and prefer to spend their time out in pastures eating it when given access to do so (except during freezing weather),[884,885] the cattle bred for the beef industry are primarily bulked on corn, soy, alfalfa, and other grain mixtures and vitamin concentrates, with hay as only one component of their diet. The primary motivation here is weight gain and the maximized use of limited land.

With grain composing the vast majority of their diets, livestock health is significantly affected.[886] High levels of corn and other grains lower the pH of the cattle rumen, the primary stomach where microbial fermentation of feed takes place, causing acidosis.[887,888] This condition then requires the use of antibiotics to manage cattle diseases that become common under feedlot conditions.[889,890]

However, cattle diets have leaned toward even more outrageous extremes in recent years when corn prices more than doubled.

CNN declared in 2012, "Cash-strapped farmers feed candy to cows," noting that the rising price of corn due primarily to ethanol demands had farmers literally turning to candy—gummy worms, marshmallows, chocolate bars, and ice cream sprinkles—for lower cost feed. "Cut-rate by-products of dubious value for human consumption seem to make fine fodder for cows," CNN's Aaron Smith reported.[891]

Worse, the crowded conditions in factory farms has led to serious veterinary intervention to administer an array of pharmaceuticals, antibiotics, and hormone treatments all designed to make every type of livestock gain weight, remain healthy enough to survive until slaughter, and prevent the spread of diseases that could compromise other commodity creatures in confinement among close quarters. Factory farm conditions involve frequent encounters with feces and microbes of all kinds, including new strains of antibacterial resistant superbugs.

Antibiotics such as macrolide are used to fight pathogenic bacteria and the spread of disease. The antibiotics ractopamine and Zilmax (beta-agonists) double as growth hormones, helping cattle and pigs metabolize their unnatural, grain-heavy diets while promoting the conversion to lean muscle weight gain. The intention is to add a marbling effect, where fat deposits are interspersed in lean muscle tissue, adding to desirable texture and flavor in marketed meat cuts.[892] However, their use has come into question in light of their studied effects on human health and behavior.[893]

Already, the drug ractopamine has been banned in the EU, Russia, China, and elsewhere over concerns about its adverse effects on humans, as studies have shown rapid heartbeat in animals and other human health issues, though it continues to remain legal in the United States. Even the tenderness of meat and its taste is perceptively affected by the weight gain drug.[894] Refusals and seizures by Russian and Chinese authorities have added tension to international trade and impacted export profits over issues concerning the regulation of animal growth hormones. A Consumer Reports study found traces of ractopamine in 20 percent of U.S. pork products.[895] The Animal Legal Defense Fund (ALDF) and Center for Food Safety (CFS) have sued the FDA over allegedly withholding records about the human health effects of ractopamine, which other studies signify may include information about harmful behavioral and neurological effects.[896]

Zilmax, a drug produced by pharmaceutical giant Merck & Co., can reputedly add up to 33 pounds of salable beef to each head of cattle, making for significant market profit while ensuring corn-fed life in a confinement operation is manageable until slaughter.[897] It has become controversial ever since reports that it may be contributing toward health decline in cattle, preventing harvest and requiring negatively affected animals to be euthanized. Merck halted sales in August 2013 after meat processors began refusing cattle treated with Zilmax due to concerns about debilitated cattle. Reports emerged of some fifteen cattle headed for a Tyson processing plant whose hooves were all but disintegrated, losing the ability to walk after treatment with the drug. Video recordings were reportedly circulated within the industry, raising concerns about impacts on market share if antibiotic-treated cows were thought to be unfit for consumption.[898] According to FDA data, at least 285 cattle have had to be euthanized after taking Zilmax since the drug was first marketed in 2007. China and other nations have since banned Zilmax imports. U.S. processors have further expressed fear that accidental exports in violation of this ban would hamper overall trade shares.

Clenbuterol is another drug that has been used to promote enhanced muscular meat, though it is not approved for use in the United States, EU, or China in food-producing animals due to its potential to impact human health if consumed from treated animals.[899] Explosive contamination scandals of unauthorized clenbuterol in pork sold in China have made major headlines, while residues reportedly made several hundred people sick in

Shanghai.[900] Past cases of food poisoning in humans linked to clenbuterol have also contributed to unease with the chronic effects of antibiotic use in livestock.[901]

The U.S. milk and dairy beef industry has been required to screen for the presence of lingering antibiotics in marketable products, finding violations in the form of trace levels of penicillin, gentamicin, sulfadimethoxine, and other antibiotic residues that could wind up in food destined for human consumption.[902]

Many other pharmaceutical drugs are given to livestock simply to help mitigate the hazardous conditions of living in what amount to little more than animal concentration camps. Such antibiotic treatments as efrotomycin, bambermycin, avilamycin, salinomycin, narasin, and lasalocid are widely used, despite evidence of a rise in drug-resistant strains of bacteria.[903]

One of the most widely administered antibiotics in the cattle industry is monensin, which is added to feed to control diseases that can contribute to further contamination and loss of herd potential.[904] It was synthesized in 1979 to deal with coccidiosis, a pathogenic disease derived from contact with infected feces.[905] The disease is most frequently a result of animals being tightly packed into contaminated pens or pastures in calving operations or crowded feedlot conditions.[906] Coccidiosis can infect humans, and drug resistance to monensin has been observed in both cows and humans since 1983.[907]

Appearing with symptoms of bloody diarrhea, coccidiosis is considered "one of the most economically important diseases of the cattle industry,"[908] as it can disqualify calves or fed cows from harvest potential. Cattle salmonellosis, as a result of contaminated feed and transmission of bacteria through feces and tainted feed in crowded confinement lots, can lead to salmonella in food-grade beef products winding up in grocery stores.

The occurrence of salmonella contamination is found in cattle lymph nodes in only about 1 percent of tested cull cattle, but averages an astonishing 11 percent among feedlot cattle,[909] and this number can range much higher, from 30 to 60 percent, depending on the season and climate. Not all salmonella is dangerous, but outbreaks tied to ground beef have made headlines after sickening dozens of people who have eaten undercooked meat.[910]

The beef industry credits the 1997 law prohibiting the addition of animal waste—including cow remains—to ruminant feed for curbing the incidence of salmonella contamination, as well bovine spongiform encephalopathy, better known as "mad cow disease," for which the law was passed.[911,912,913] However, laws still allow plate waste from human food, which includes animal protein, pork, poultry, horse meat, gelatin, and blood, to be added to livestock feed, which can also lead to salmonella and other cattle diseases.[914]

Despite rhetoric to the contrary, the actual welfare of the animals is secondary to the profit potential of the herd, which is first and foremost focused on marketable meat weight. Livestock pharmaceuticals play a huge role in lowering the cost of food production, pound for pound, and these drugs have played a particularly important role, along with genetics and feed, in raising super-sized cattle. The use of antibiotics in farming has become increasingly controversial among conscious consumers who seek out more expensive meats labeled "hormone-free" or "antibiotic-free."

Though meat consumption remains high in Western countries, it is overall on the decline. New markets may be developing globally—particularly in China where a large population of middle class workers are adopting Western consumption patterns—but the beef industry has experienced shrinking herds, higher costs, and drastic changes in business structure.[915] These conditions, along with the lure of biomedicine and technology, have driven a focus on increasing the weight of the cow rather than simply raising more cows.

The American beef industry produced some 26 billion pounds of beef in 2012, as compared to about 24 billion pounds in 1975. The difference is that there were more than 135 million head of cattle then; now herd totals are down to approximately 91 million head of cattle. Today, the industry harvests approximately 150 percent more marketable meat from each cow than it did per head in 1975—simply by artificially growing larger cattle. Current live weights at slaughter now routinely top 1,300 pounds, with the aid of growth hormone drugs and intensive feeding practices.

Drought conditions have played a significant role in herd decline, forcing many farmers to reduce herd size and keep fewer cows. Market consolidation has absorbed former competitors, while approximately 20 percent of the feedlot industry has gone out of business over the past decade or so due to soaring costs, including competition for feed crops such as corn and soy with producers of ethanol and other biofuels. Demand in Europe and other markets for beef raised without genetically modified corn and artificial growth hormones also has some measured impact on export markets and potential future consumer confidence.[916]

Poultry factory farms and environmental pollution

Foul odors. Flies. Rats. A landscape lined with chicken litter, and miles of polluted rivers.

That's how locals have described factory farm chicken operations, full of millions upon millions of birds often packed into tiny cages, and run by "growers" who execute independent contracts to raise broilers and eggs for some of the biggest global food companies.[917]

Tyson Foods is far and above the giant in this industry, slaughtering and packaging more than 40 million chickens per week, in addition to gigantic volumes of beef and pork. Tyson, like other Big Agra poultry players, has a long environmental record of alleged damage and plenty of controversy.

In 2003, Tyson pled guilty to twenty felony charges and forked over $7.5 million in fines, for violating the Clean Water Act by illegally dumping untreated wastewater from a processing plant outside of Sedalia, Missouri, on the level of thousands of gallons per day over the course of at least six years.[918,919]

A 2004 decision by a federal judge held Tyson responsible for the ammonia emissions and waste dumped by "growers" it contracted with and directed throughout the chicken production process, leading to the shutdown of a plant in Kentucky, after the activism of the Sierra Club and many concerned locals forced the issue into the spotlight.[920,921]

Tyson Foods was sued by the state of Oklahoma in 2009, along with Cobb-Vantress, Inc.; Cal-Maine Foods, Inc.; George's, Inc.; Peterson Farms; and Cargill, Inc., for contaminating public drinking water through the use of poultry litter, which included salmonella- and E. coli-tainted fecal waste, on some one million acres of land as a fertilizer.[922,923]

Along with the fecal waste and potential bacterial contaminants is a significant amount of the poisonous heavy metal arsenic, known for its deadly potential for and incidences of skin, lung, and bladder cancer as well as its overall contribution to chronic toxicity (see the section on arsenic on page 14 for more information). Intentionally added to poultry feed to increase poultry size, it also accumulates in the edible broiler meat, as well as in the litter left behind.[924]

Roxarsone and arsenic in chickens

For many decades, chicken producers had been adding roxarsone, sold under the trade name 3-Nitro and composed of 4-hydroxy-3-nitrobenzenearsonic acid, to the majority of conventionally raised meat-producing birds consumed in the United States. Ostensibly used to treat intestinal parasites, roxarsone was put into the chicken feed of more than 90 percent of starter broilers and grower broilers and about 75 percent of withdrawal broilers (chickens whose feed is restricted in preparation for slaughter) up until the mid-1990s to promote fast growth in these meat-producing birds.[925]

Use declined somewhat until sales of the arsenic growth promoter were voluntarily suspended in the United States by Pfizer-subsidiary Alpharma in 2007, after studies showed its occurrence was widespread in U.S. chicken samples. Levels were more than twice as high in conventional poultry than in antibiotic-free or organic chicken.[926] Presumably, roxarsone continues to be sold in other markets around the world (except where it has been banned).

Roxarsone was also given to pigs and turkeys to boost sale weight.[927] The arsenic content had been accepted because it was an organic species thought to be harmless; however, tests found inorganic arsenic in the livers of chickens, confirming that chickens were converting the organic arsenic to the inorganic type, known for its deadly toxicity. Further, inorganic arsenic turned up in chicken litter,[928] which was in turn disposed of in the land's neighboring poultry farms, leached into water supplies,[929] and used as cheap fertilizer.[930,931]

Authorities were unable to test for arsenic accumulated in the breast meat of chicken, although arsenic compounds have been known to bioaccumulate. Tests show that roxarsone introduced into the soil by litter was capable of uptake by rice crops and other vegetation, where it could re-enter the human food chain.[932]

While roxarsone was withdrawn from use in the United States, it was replaced in many cases by another arsenic-based compound, nitarsone (4-nitrophenylarsonic acid), another additive put in poultry feed to increase weight and prevent disease—in this case, blackhead disease.[933] Studies found that nitarsone was toxic in developing turkeys, killing all of the subjects at 0.08 percent in feed, and half (LD 50) at 0.05 percent over the course of twenty-eight days.[934]

In December 2015, the FDA withdrew its approval of nitarsone for use in chicken feed, stating, "Following this action, there are no FDA-approved, arsenic-based drugs for use in food-producing animals."[935]

However, it is important to note that because a very large supply of arsenic-based feed additives still exist in the supply chain, they will likely continue to be used by unscrupulous poultry producers for many years to come.

Fish farming

Farm-raised carnivorous fish such as salmon require large amounts of fish meal and fish oil feed. Three pounds of other types of fish are necessary to raise one pound of marketable salmon.[936] Because of the high feed requirements, hundreds of thousands of tons of uneaten, wasted salmon feed laced with chemicals is discharged by industrial salmon farms back into coastal ocean waters each year, according to some estimates. Studies have also shown high concentrations of pollutants in the salmon feed that contaminate the areas surrounding these farms, and the fish themselves are laden with concentrated environmental toxins, including poisonous heavy metals such as mercury.[937]

Due to expenses and the toll these operations take on the environment, the industry has been turning to cheaper alternatives for fish feed. Just as with cattle, which were never meant to eat grain, scientists have developed new farm fish feeds made of corn, wheat, and soy—turning meat-eating fish into vegetarians in the name of sustainability and profits.[938] As the Pure Salmon Campaign notes, "Industrialized salmon farming relies on a deeply flawed assumption that agricultural practices for animals can be applied to carnivorous fish."[939] These issues are complicated by the advent of genetically modified, super-aggressive, high-yield salmon currently in the process of being commercialized and approved by several biotech firms.[940] The FDA approved genetically modified salmon for human consumption in November 2015.[941]

Worse, the use of hormones to manipulate sex ratios and promote maturation has become dominant in aquatic farming for food production. Researchers say that nearly all fish bred in captivity have reproductive dysfunction, prompting those raising fish to administer reproductive hormones, with various possible effects not only on the fish, but on those who consume it.[942] Control of water temperature and the administration of hormone therapy are often necessary to achieve commercially desirable reproduction. Luteinizing hormones (LH) and gonadotropin-release hormones (GnRHa) are often administered, while interrupting normal spawning behavior under captive aquaculture conditions may also require artificial forms of fertilization.[943] In fish farming for many species, including trout and tilapia, sex-inversed males, altered via hormone treatments, are frequently used to fertilize female stocks for consumption.[944] Studies show that some of these hormones, which can disrupt the endocrine system, end up in wastewater, groundwater, and even drinking water supplies.

Waste from concentrated animal feeding operations (CAFO)

The greatest concern with concentrated livestock operations is not only the low-quality food that is produced by raising livestock animals under stressed conditions to maximize space while providing inadequate nutritional inputs and unnatural dietary choices, but also, of course, the biomagnification of toxins.

Concentrated heavy metals, antibiotics, and other pharmaceutical regiments as well as hormonal treatments pose risks for environmental systems as well as humans and other animals.

Sewage runoff from factory farms has proven to be a lasting concern. Endocrine disruptors, growth hormones, and toxic chemicals have all been found in drinking water supplies.

A 2004 Conference on Environmental Health Impacts of Concentrated Animal Feeding Operations: Anticipating Hazards—Searching for Solutions held in Iowa acknowledged significant issues involving the contamination of groundwater and urban water supplies caused by a number of variables including long-term, low-level exposure to antibiotics, veterinary pharmaceuticals, and endocrine-disrupting chemicals, which can all have serious impacts on human health and the environment.[945,946]

In 2004, civil and environmental engineering professor Edward P. Kolodziej and his colleagues participated in a number of environmental monitoring programs near cattle and fishery operations, finding elevated levels of androgens, estrogens, and progestins from steroid hormones in the discharged water.[947] Even background levels of 1ng/L are enough to pose risks to fish and amphibian life, and the endocrine-disrupting compounds have been known to change the sexual behavior and physiology of aquatic wildlife, posing population risks.

Kolodziej and his colleagues found a dairy-waste lagoon with especially high levels of 650 ng/L for hormones such as 17β-estradiol and estrone, as well as the androgens testosterone and androstenedione; these impact groundwater consumed by wildlife and humans. In humans, endocrine disruption can cause infertility, reproductive blowback, and certain types of cancer, including breast cancer.

More recently in 2013, Kolodziej and other colleagues published findings that the synthetic anabolic steroid trenbolone acetate (TBA), which is widely used in an estimated 20 million beef cattle to promote growth, was entering water near feedlots via runoff as the metabolized 17α-trenbolone, a strong endocrine disruptor.[948] While scientists knew this was impacting aquatic life and the environment more broadly, they discovered that the 2013 study underestimated the 17α-trenbolone levels: Even though 17α-trenbolone and other similar compounds were found to break down by day in the light, it was discovered they regenerate at night under shifting pH conditions.[949]

Food contamination scandals and failed safety regulations

A number of high-profile contamination scandals have drawn attention to factory-farming practices and prompted reforms in the food industry.

In 2007, more than 850,000 frozen beef hamburger patties produced at a Cargill meat packing plant and sold at Walmart and Sam's Club stores were recalled over E. coli-contamination concerns. Several lawsuits, including one filed by a woman left paralyzed by the E. coli outbreak and a ten-year-old girl who required a kidney transplant as a result of eating the tainted burgers, were spotlighted in the media.[950,951,952] *The New York Times* exposed the lack of true oversight in the meat-packing business. The guise of food safety, it reported, was an illusion, as real testing was discouraged within the industry, and ground beef burgers often came from grinded parts from different cuts and different slaughterhouses, including lower-grade parts more likely to have been contaminated by bacteria.

The 2011 Food Safety Modernization Act nominally addressed many of these issues, though watchdogs and organic producers have heavily critiqued it for burdening small producers while failing to address the systematic issues in concentrated animal feeding operations and confined pens in densely packed poultry houses and pig factories.

The much-hailed reforms of the U.S. Food Safety Modernization Act— sold to the public as a solution to E. coli outbreaks in produce and salmonella outbreaks in egg production—have in many ways only made it more difficult for the little guy to compete.[953]

The most recent regulations for egg production and leafy vegetable farming—imposed by the FDA and USDA under the guise of food safety— have created extensive requirements for inspections, tracking, and standards compliance that significantly increase the time, manpower, and costs necessary for small- and medium-size farms to remain in operation, while significant loopholes exist for major factory farms that not only exempt them from the same standards but ensure that smaller operations cannot compete on a cost basis.[954]

Finding alternatives to factory-farmed meats

Though you'll pay a premium for them in grocery stores, purchasing USDA-certified organic meats are the best way to avoid the issues discussed here. They come from livestock raised without the use of antibiotics or synthetic growth hormones that are emblematic of concentrated animal feeding operations;[955] moreover, they must have access to pastures during the grazing season.[956] USDA certification for grass-fed beef requires that cattle have year-round access to grass and other forage crops, and that they cannot be fed any grain (from corn, soy, and so on), although they may eat these crops in the vegetative state.[957]

Free-range poultry and egg production often go hand in hand with organically raised livestock, though not always. In the best operations, offering shelter to chickens is balanced with true and free access to the outdoors without crowding.[958] New regulations under the Food Safety Modernization Act have watered down the requirements for "organic" egg production, allowing many factory farms to more easily pass for organic.[959] Look into the source of your eggs for better information; the Cornucopia Institute has an organic egg scorecard to evaluate how producers really stack up.[960]

In many cases, local producers, including those represented at farmer's markets, have the best commercially available meats and eggs. Even without USDA-certification for organic foods, which can be quite costly for smaller producers, local family farms offer better-sourced foods. As common sense dictates, seek out farmers you know or are familiar with, and who support transparency in the food supply by answering questions and opening up to scrutiny.

PART 2

THE HEALTH RANGER'S GUIDE TO

NATURAL DETOXIFICATION

The cost of allowing your body and brain to be systematically poisoned is enormous: increased risks of disease, deteriorating quality of life, increased medical costs, loss of cognitive function, and much more. That's why it makes such good sense to invest in the knowledge and behavioral strategies needed to eliminate and avoid the toxins that cause these costly problems.

However, there's a lot of misinformation about so-called "detox" in the natural products marketplace. One of the most unfortunate detox memes that has emerged over the past few decades is the idea that detox must be "heroic" to be effective. This idea says, essentially, that an effective detox can only be achieved through tremendous suffering and pain.

Usually, this is the narrative offered by companies selling aggressive detox products made with ingredients like cascara sagrada, which can cause extreme digestive irritation and diarrhea. While a little cascara can be quite useful in a holistic formula that's also made with supportive, healing herbs, some companies go too far and formulate their products to cause what they label a "healing crisis," a process in which the body attempts to eliminate toxins faster that it can dispose of, resulting in a temporary increase in symptoms.

The term "healing crisis" is often used to explain away the fact that many detox products are too aggressive and too difficult for consumers to endure. The idea that "you have to get worse before you can get better" is similar to the iron-pumping slogan "No pain, no gain."

Both ideas are blatantly false.

In fact, detoxification is an everyday biological process that your body has been automatically achieving from the moment you were conceived. Every cell in your body has a detox process, and your body as a holistic system is also quite adept at detoxification. You have entire organs dedicated to it: the liver, kidneys, and bladder, to name three. Lungs are also involved in detoxification, as is your body's largest organ, your skin.

Realizing this allows us to step back, calm down, avoid the cascara sagrada overdose, and think about how to truly support our body's own natural ability to remove toxins.

Avoidance of Toxins

Because your body is "detoxing" every second, the best strategy starts with avoidance of toxins. Although this position won't make me popular among detox supplement companies, I believe that the first and foremost strategy for detoxification must be *avoidance*.

"Avoidance" means avoiding continued exposure to toxic chemicals and heavy metals in your food, water, medicine, personal care products, home-cleaning products, lawn-care products, pet-care products, and even the air you breathe. Everything you touch, inhale, or consume becomes a part of your physical body. So removing toxic chemicals and poisons from your system must begin with eliminating new exposures to those very poisons.

In fact, the most sound method for what I call "lifetime detox" works like this:

1. **Identify** the sources of your exposure to toxic substances.
2. **Avoid** those substances by reforming your behavior and consumption choices to break the cycle of repeated exposure to those toxins.
3. **Support** your body's natural detoxification and elimination abilities.

Let's explore each of these:

Identify: Identifying means educating yourself about all the sources of exposure to toxic chemicals and heavy metals that are negatively impacting your health and brain function right now. This step is especially difficult because all the corporations that are poisoning you with their toxic medications, toxic food additives, toxic herbicides, and toxic home-care products are precisely the same corporations who insist all of their products are safe and even "green"!

To really get up to speed on this step, you'll need to learn to read and understand food labels, educate yourself about hidden toxic chemicals in personal care products, learn about the toxic additives hidden in vaccines and medications, and awaken to the dishonest marketing strategies invoked by corporations to deceive consumers about the supposed safety of their products.

Avoid: Once you identify the sources of these toxins, you must then pursue aggressive reforms in your own behavior and consumption patterns to avoid purchasing or consuming these products.

To begin, this step will require you to go through your entire home—medicine cabinets, pantry, refrigerator, garage, laundry room, and so on—and throw out the products containing toxic chemicals. Yes, this includes your toxic laundry detergent and dryer sheets, your toxic cough medicine, your toxic weed killers and bug sprays, and even your toxic air fresheners (which actually pollute your air rather than clean it).

This step can sometimes stress marriages, so here's a tip: If you're a woman reading this, and you want to get your husband on board with the household detox program, just remind him that these toxic chemicals reduce sperm count, diminish male virility, and interfere with male hormones (which is true). Tell him that endocrine disruptors really do promote the "feminization" of males, and pesticides like atrazine really do cause amphibians to develop dual sets of sex organs (both male and female). This knowledge might open your husband's eyes to a legitimate reason to reduce your household chemical exposure.

If you're a man who wants to convince your wife to get on board with the program, explain to her that corporations have been violating our bodies with toxic chemicals that aggressively promote cancers, including breast cancer and ovarian cancer (this is also true). Suggest that one of the best ways to reduce cancer risk is to get the toxic chemicals out of your lives, which will also reduce future hospitalization expenses and healthcare costs.

An important strategy here centers around the *replacement* of toxic products with non-toxic products. For example, after you throw out your toxic laundry detergent, you'll need to discover non-toxic brands that still clean clothes without saturating them in synthetic chemicals. (Ecos is my favorite brand of natural laundry detergent.)

Nearly all the toxic soap products in your home can be replaced with Dr. Bronner's Magic Soaps. Toxic air fresheners can be replaced with essential oil diffusers. Toxic personal care products can be replaced with organic products made with non-toxic ingredients. For every toxic product you used to buy, there's a non-toxic alternative available today that's safer, cleaner, and more environmentally friendly.

The Environmental Working Group, a consumer advocacy organization that specializes in identifying toxic chemicals, launched an online database called Skin Deep that provides safety profiles for cosmetics and personal care products, as well as offers tips on how to find the best products. EWG scientists compare the ingredients of personal care product labels to information in nearly sixty toxicity and regulatory databases, giving consumers the power to protect their health by choosing clean, non-toxic personal care products and cosmetics. You can access the profiles and tips at ewg.org/skindeep.

Support: Your body won't eliminate water-soluble toxins very well if you're dehydrated, so drinking plenty of water is the first and simplest way to support your everyday detox strategy. Physical movement is also key, both for circulating lymph, which allows your body to expel metabolic garbage more proficiently, and producing sweat, which eliminates toxins through your skin. However, when you sweat, many of these toxins then get absorbed by your clothing. This is why changing your clothes after a sweaty workout is very important.

It's also smart to support the healthy function of your detox organs (liver, kidneys, and digestive system) with medicinal herbs, superfoods, and other dietary strategies. Herbs such as yarrow, dandelion, and yellow dock are well known to support liver function. Superfoods like chlorella help support the body's natural elimination of toxic elements through its ability to remove chemicals and heavy metals. Even everyday foods like beets and white carrots are "liver cleansers" that give the liver added support to carry out its job.

Supporting your body's detox of toxic elements also requires boosting your body's intake of nutritive elements. Nearly all people living in Western society today are chronically deficient in vitamin D, zinc, selenium, magnesium, and many other nutrients. Nutritional supplements that help restore the balance of those vitamins and minerals in the body are remarkably safe, effective, and affordable. But the best method for acquiring minerals is to grow them into your own food, and then eat that food.

To help support this effort, I launched www.foodrising.org, which features a revolutionary nonelectric hydroponics grow technology you can easily build yourself. Using mineral-enhanced plant-food formulas I developed in my lab (using an Agilent ICP-MS instrument, a tool capable of detecting metals and nonmetals at very low levels), you can grow your own lettuce, strawberries, tomatoes, and other garden vegetables that are exceptionally high in selenium, zinc, and other trace minerals. By eating or juicing these foods, you are "supplementing" your body with potent sources of truly organic (plant-based) forms of these nutritive minerals.

Eliminating toxins in household water

Hydrating our bodies with clean, non-toxic water is equally as important as avoiding toxins found in food. Unfortunately, tap water contains many impurities, highlighting the need for an optimal water filter.

I strongly recommend every household be equipped with some sort of water-filtration device due to the very poor quality of municipal water. Never drink unfiltered tap water! It contains many toxic chemicals, including some that were not added by the water treatment plant but rather formed in the pipes as the water made its way to your home. One example would be chloramines, carcinogenic chemical compounds that form when ammonia and chlorine are added to treated tap water.

As part of my laboratory investigations into healthy living, I've conducted detailed testing of the ability of commercial water filters to remove toxic heavy metals.

I tested all the popular gravity water filters, including the Big Berkey, ProPur, Zen Water Systems, and other brands. All the results are published on www.waterfilterlabs.com.

The bottom line from the results? The Big Berkey (with fluoride filters) and Zen Water Systems filters removed the most toxic heavy metals in our tests. For countertop water filters, the Zerowater brand did the best job by far. In fact, many cheaper brands of countertop water filters barely removed any toxic heavy metals at all. Reverse osmosis (RO) water systems do an excellent job of removing toxic heavy metals.

I am also the inventor of 3D-printable water-filtration devices that you can download free and print on most 3D printers. You'll find those downloadable water-filter files at www.foodrising.org.

Use plants and air filters to clean your indoor air

In addition to filtering your water, you can also remove toxic elements from your indoor air. Indoor air quality is atrociously bad in newer U.S. homes due to the "green building" phenomenon. Although it seems counter-intuitive, "green" construction techniques actually make homes more airtight (and thus, because of reduced air loss, less expensive to heat and air condition, which is what most of the "green" claims are based on), not less toxic.

By making your home more airtight, this construction method also means toxic chemicals remain at a higher concentration inside your home. With less outside air mixing with indoor air, there's nowhere for toxic off-gassing chemicals to go. That's why so-called "green homes" can actually be extremely toxic environments, filled with VOCs from glues, formaldehyde, plastics, wood varnishes, home-cleaning chemicals, molds, fungi spores, and more.

There are essentially two ways to eliminate these: living plants and high-efficiency particulate arresting (HEPA) air filtration.

Living plants are living air filters. The more plants you have thriving in your home, the cleaner and more refreshing your indoor air will be. So make it a goal to invite more living plants into your home and enjoy their many benefits on your physical and even emotional health (plants make humans happy!).

For serious HEPA air filtration, the best unit on the market today, to my knowledge, is the IQAir system. It's European-made and very expensive, but it works flawlessly, moving a tremendous volume of air through its HEPA filtration modules.

Personally, I use HEPA filtration in my lab but living plants in my home.

Avoiding toxins where you live

Another key strategy in avoiding toxins is choosing to live far from congested cities. It's no exaggeration to say that if you're living in Los Angeles, you're inhaling tens of millions of microscopic pollution particles each day. If you live in Mexico City, it's far worse.

Living in cities subjects you to pollution sources you may have never considered. For example, vehicle brake pads often contain cadmium. Walking along the sidewalk of a busy street quite literally results in you inhaling trace levels of cadmium with every breath.

Lead is still used in the aviation fuel used by most small aircraft. (The fuel is called 100LL, in which "LL" stands for "Low Lead.") If you live underneath a common approach or departure path of small aircraft, you are literally having microscopic lead particles dropped on your yard and house any time aircraft are directly overhead.

Chemically ignorant municipal decision makers may also subject you to toxic chemical fumigation by spraying your entire neighborhood with malathion or other mosquito-deterrent chemicals. City workers are also known to spray toxic glyphosate (Roundup) to kill weeds on sidewalks, curbs, and city streets, subjecting you to an extremely toxic substance that has been linked to cancer at parts per billion concentrations.

The only reliable way to control your environment and dramatically reduce your exposure to toxic chemicals is to get out of the city. Move out to the country and buy as much land as you can to have buffer space between yourself and your chemically ignorant neighbors who still probably douse their weeds with cancer-causing chemicals. Buying a buffer zone is really the best strategy, because you can only truly control the land that you own.

Defensive Eating

The avoidance of dietary toxins can be significantly enhanced through the defensive eating strategies covered in this book. By consuming chlorella (an extremely nutritious green algae), strawberries, fruit fibers, or even activated charcoal at the same time you are eating a meal with a high likelihood of containing toxic substances, you can "bind up" those toxins, push them through your digestive tract, and avoid absorbing them into your blood and tissues.

I won't eat at a restaurant without bringing a bottle of chlorella with me. I never eat meat unless it's accompanied by a fresh salad or whole fruit. Interestingly, the Southern tradition of eating barbecued meat with coleslaw coincides with this same advice: The cabbage in the slaw helps block the carcinogenic effects of the charred meats. Vitamin C found in fresh fruits creates yet more of a protective effect during digestion. That's why people who say things like, "It's unhealthy to eat barbecued meats" really only grasp half the picture. In truth, it's only unhealthy to eat such meats *if they are consumed alone*, without the protective benefits of fresh fruits and vegetables.

One of the simplest things you can do when eating questionable meals is to take extra fiber supplements. The best choice for this, in terms of ability to bind with toxic elements, is apple fiber. As it turns out, apple fiber supplements are very inexpensive and readily available.

Detoxing from bad fats

How do you eliminate the toxic effects of bad fats you've consumed in the past? The simplest answer is also the most scientifically sound: Start eating good fats, and they will eventually replace the bad fats.

It really is that simple. Every cell in your body consists of a membrane made with fat molecules. But those molecules aren't stuck there for life; they are eliminated and rebuilt quite frequently, using whatever fat molecules are found in your blood at the time. If your blood is circulating fats from cheese, onion rings, and genetically modified soybean oil, then all your body's cells will have membranes made from the same garbage. But if your blood is circulating fats from avocados, chia seeds, clean fish oils, and high omega-3 oils, then all your body's cells get a boost in health as a result.

If you want to change your body, I've always said, change your blood first. And if you want to change what's in your blood, *change what you eat*.

Detoxing from heavy metals

Eliminating heavy metals from your body can be significantly more challenging than a chemical detox. There's evidence that mercury, for example, can almost never be fully eliminated from brain tissue. This is a good reason, by the way, to completely avoid all vaccines that contain thimerosal as a preservative. It's made from mercury, and it's still used in vaccines sold in the United States (despite the myth being circulated that falsely claims mercury was removed from all vaccines).

Heavy metals are difficult to remove once they become embedded in tissues. Just as mercury clings to neural tissue, lead gets "stuck" in skeletal tissue (your bones). Once lead is integrated into your bone tissue, it's almost impossible to eliminate it unless you undergo hormonal changes that lead to an erosion of bone mass. This is why many elderly women can experience lead poisoning even when they have had no recent exposures to lead: The heavy metal is literally coming out of their bones.

The difficulty of eliminating heavy metals from body tissues and organs underscores the importance of avoidance. Your first defense against heavy metals is to avoid exposure. Your second defense is to block them during digestion with formulas such as the one I invented and patented called Heavy Metals Defense (www.HeavyMetalsDefense.com). It's made of an ion-exchange material, derived from selected seaweeds, which efficiently binds with lead, cadmium, mercury, and other toxic heavy metals. When taken during a meal, it can bind with those substances during digestion, supporting your body's natural elimination process that shuttles these substances out of your body through digestive elimination.

If you don't have specialized formulas like Heavy Metals Defense, eating more whole fruits and vegetables with every meal can go a long way toward blocking the absorption of heavy metals.

Health foods that are high in heavy metals

Beware of the hidden sources of heavy metals in many foods and beverages that are considered "healthy." Many tea products are shockingly high in lead and fluoride. Tea is generally viewed as a healthy beverage, which may lead many people to mistakenly believe all teas are "clean" when it comes to heavy metals. They aren't.

As I also documented, many organic foods and superfoods can be surprisingly high in toxic heavy metals. As a general rule, anything imported from China, India, or Thailand has a much higher risk of being contaminated than foods from North America, Europe, New Zealand, or many South

American countries like Bolivia. In my lab, I've seen alarmingly high levels of lead in mangosteen powder, and almost all spices from India are consistently contaminated with lead.

China's agricultural sector is widely contaminated with heavy metals, so almost anything you get from China is immediately suspect.

On the good news side, however, fresh foods grown in the United States and sold at local grocery stores were found to be quite clean. Local farmers' markets in North America or Europe are likely to offer foods with very low levels of toxic heavy metals. Purchasing and eating more of these foods is a smart way to avoid toxic heavy metals in your diet.

Infrared saunas and sweat lodges

One other option for healthy detox is infrared saunas. These saunas can be extraordinarily useful but only if these two conditions are met:

1. The sauna must cause your body to produce sweat. If there's no sweat, there isn't much health benefit; the toxins aren't exiting your body.
2. You must wash off the sweat immediately after exiting the sauna. (Otherwise, the toxins can be reabsorbed through your skin.)

Some people are fans of sweat lodges, which sometimes combine fasting with heat-induced delirium that's marketed as a spiritual experience. After the death of two sweat lodge customers a few years ago in a "spiritual journey" retreat, more people have come to realize just how dangerous it can be to torment your body in the hopes of detoxing it. The key to staying safe is to limit your time in the heat to fifteen minutes or less per session.

The bottom-line truth I hope you remember in all of this is: Detoxification doesn't have to hurt. It should never involve diarrhea, hallucinations, or intense cramping. You don't have to suffer to heal. In fact, the best way to detox your body is to pursue cleansing diligently but patiently, allowing time to work in your favor as your body does what it already knows how to do: heal itself from within.

PART 3

THE DATA

What follows here are the analysis data derived from testing popular foods and beverages for specific elements, including heavy metals such as lead. Dark gray cells indicate numbers of particular concern due to their high concentrations.

Here are some important notes about the data:

- All concentration numbers are listed as parts per billion; 1,000 parts per billion (ppb) equals 1 part per million (ppm).
- All testing was conducted on an Agilent 7700x instrument with the addition of a Niagara Plus sample injection system from Glass Expansion (which improves stability and accuracy).
- Analysis runs were all initiated with a four-point calibration curve using external multi-element standards from Inorganic Ventures. In addition, midrun calibration checks are performed after every tenth unknown sample.
- For each product tested, three samples were run and the results were averaged, then rounded to the nearest ppb.
- These data show a "snapshot" of that one product sample that was tested. It does not mean that these concentrations of elements will be the same across different production run lots of the same product. Because the origins of ingredients often vary widely from one lot to

the next, you cannot assume that these numbers represent all lots from the same manufacturer.

- Brand names are not shown for many products for the simple reason that a book takes so long to publish, print, and distribute that many of the products tested here may have shifted in composition over that time. As online publishing encourages more up-to-date data, you can find more recent testing results for specific brands at http://labs.naturalnews.com.

- Not all elements should cause the same degree of concern at the same levels. Mercury, for example, becomes worrisome at relatively low levels in foods (50 ppb, for example), while cadmium at 50 ppb is generally not much of a concern. But at 500 ppb, cadmium may begin to warrant attention, and if cadmium is present at 5,000 ppb, then it's clearly a concern.

What concentrations are "acceptable" in your food? Opinions vary widely on this subject, but I've put together a guide at lowheavymetalsverified.org that describes the rating system we use on the Natural News website and the Natural News Store (store.NaturalNews.com), where every product that's offered to the public is tested for heavy metals. All products must meet "A" or better to remain in our store.

Verified A+++
Lead < 0.025 ppm
Cadmium < 0.1 ppm
Arsenic < 0.62 ppm
Mercury < 0.006 ppm

Verified A++
Lead < 0.05 ppm
Cadmium < 0.25 ppm
Arsenic < 1.25 ppm
Mercury < 0.012 ppm

Verified A+
Lead < 0.12 ppm
Cadmium < 0.5 ppm
Arsenic < 2.5 ppm
Mercury < 0.025 ppm

Verified A
Lead < 0.25 ppm
Cadmium < 1.0 ppm
Arsenic < 5.0 ppm
Mercury < 0.050 ppm

Verified B
Lead < 0.5 ppm
Cadmium < 2.0 ppm
Arsenic < 10.0 ppm
Mercury < 0.1 ppm

Verified C
Lead < 1.0 ppm
Cadmium < 4.0 ppm
Arsenic < 20.0 ppm
Mercury < 0.2 ppm

Verified D
Lead < 2.0 ppm
Cadmium < 8.0 ppm
Arsenic < 40.0 ppm
Mercury < 0.4 ppm

Verified F
Anything worse than "D"

GROCERIES

Organic Produce

Element	Mg	Al	Cu	Zn	As
Organic Cucumber	106,703	438	648	2,326	42
Organic Kale Greens	401,460	1,872	1,148	3,693	2
Organic Cilantro	471,021	12,798	6,840	2,659	38
Organic Italian Parsley	453,554	26,161	12,412	4,337	27
Organic Romaine Lettuce	133,591	20,082	516	2,605	12
Organic Celery	159,039	5,970	327	1,040	10
Organic Avocado	1,381,515	1,516	14,779	18,768	11
Organic Tomato	38,001	99	1,689	1,723	0
Organic Strawberries	177,434	2,089	670	1,271	3
Organic Broccoli	711,276	1,604	1,384	13,469	4
Organic Raspberries	211,097	1,460	638	2,765	2
Organic Carrots	117,277	112	832	2,994	2
Organic Potato	274,585	26,915	883	1,868	13
Organic Plum	44,615	229	404	200	2
Organic Blueberries	58,551	3,377	704	1,780	11
Organic Orange	93,629	0	436	329	0
AVG PPM	302.08	6.55	2.77	3.86	0.01

Sr	Cd	Cs	Hg	Pb	U
2,857	0	1	0	2	0
21,948	52	8	1	4	1
9,929	42	3	0	26	11
18,555	6	9	2	32	6
1,650	33	4	0	26	2
10,890	2	1	0	4	2
7,091	22	110	1	27	2
227	10	7	0	2	1
151	2	1	0	6	0
4,087	64	0	0	2	1
2,204	8	1	0	2	1
9,689	184	1	0	1	5
2,367	19	7	0	40	33
58	0	1	0	0	0
124	1	29	0	10	0
1,328	0	0	0	1	0
5.82	0.03	0.01	0.00	0.01	0.00

Conventional Produce

Element	Mg	Al	Cu	Zn	As
Cucumber	138,028	97	312	2,235	39
Kale Greens	698,227	7,762	1,687	4,765	26
Fuji Apple Flesh	41,447	0	305	172	0
Fuji Apple Skin	263,153	4,130	1,476	1,392	4
Tomato	128,673	66	370	1,275	0
Blueberries	59,371	3,771	216	553	20
Potato	253,623	48,175	1,192	2,413	33
Lettuce	425,484	4,362	1,053	4,109	10
Orange	126,722	429	317	689	0
Grapes	127,778	2,996	2,732	1,197	0
Avocado	250,115	932	2,979	6,691	2
Broccoli	395,535	3,903	971	6,275	4
Carrots	96,080	495	768	3,484	9
Banana	356,667	119	1,767	2,844	0
AVG PPM	**240.06**	**5.52**	**1.15**	**2.72**	**0.01**

Breakfast Cereals

Element	Mg	Al	Cu	Zn	As
Cap'n Crunch Oops! All Berries	2,596,460	5,414	5,301	18,519	70
Puffins Peanut Butter	420,846	941	1,194	7,675	14
Froot Loops Marshmallow	12,266	517	34	227	0
Flax Plus Multibran Flakes	1,709,020	5,683	4,881	34,263	24
Honey Bunches of Oats	67,435	340	262	1,073	5
Uncle Sam	1,269,693	750	4,067	24,519	6
Corn Pops	66,812	1,220	363	142,589	10

Sr	Cd	Cs	Hg	Pb	U
3,198	3	1	0	1	0
50,300	145	6	4	14	3
374	0	1	0	0	1
946	1	1	0	16	1
769	11	1	0	1	0
322	0	6	0	5	0
2,308	26	9	0	54	7
1,628	68	5	0	8	6
1,586	0	2	0	1	0
853	0	6	0	1	1
62	17	37	0	0	0
3,238	40	3	0	5	4
8,193	2	1	0	3	1
628	0	1	0	0	0
5.31	**0.02**	**0.01**	**0.00**	**0.01**	**0.00**

Sr	Cd	Cs	Hg	Pb	U
	6	6	0	13	21
	4	4	3	15	2
	1	0	6	0	3
	45	2	2	0	1
	1	2	1	0	0
	36	10	2	0	14
	3	2	2	0	12

Breakfast Cereals (Cont.)

Element	Mg	Al	Cu	Zn	As
Frosted Mini-Wheats Strawberry	824,198	4,711	3,241	35,301	10
Corn Flakes	104,630	1,180	540	2,654	4
Kix	673,213	3,821	1,252	184,056	17
Honey-Ful Wheat	439,802	684	1,823	10,343	11
Honey Nut Squares	138,051	2,223	420	200,224	18
Corn Chex	652,381	2,407	1,112	226,836	14
Honey Graham Oh's	440,484	1,386	965	207,290	36
Raisin Bran	1,588,877	1,049	4,462	82,626	26
Frosted Flakes	154,681	168	286	2,106	8
Total Blueberry Pomegranate	1,193,419	10,735	3,181	468,009	88
Frosted Mini-Wheats Blueberry	776,888	27,667	2,866	33,925	11
Special K	2,237,304	3,044	8,175	43,962	52
Heart to Heart	1,112,080	16,355	2,655	56,707	24
Honey Nut Chex	500,879	2,590	869	203,318	20
Whole O's	607,353	43,979	2,030	10,222	117
Coconut Granola	958,170	1,361	3,353	19,090	22
Ezekiel 4-9 Golden Flax	2,055,529	3,285	7,086	38,413	31
Crunchy Flax	1,738,601	11,945	5,353	31,190	26
Crunchy Rice	930,803	5,156	2,659	16,025	188
Cheerios	915,980	1,648	2,580	131,350	56
Veganic Sprouted Brown Rice Crisps	1,251,279	5,883	2,439	14,336	964
Veganic Sprouted Brown Rice Cacao Crisps	1,019,337	8,394	3,649	12,500	717
Veganic Sprouted Ancient Maize Flakes	950,322	7,599	1,721	18,551	2
AVG PPM	913.56	6.07	2.63	75.93	0.09

Sr	Cd	Cs	Hg	Pb	U
	38	1	1	0	0
	2	3	2	0	0
	10	7	2	4	16
	74	0	1	0	0
	1	4	1	3	25
	7	6	1	0	12
	3	4	1	0	0
	49	1	1	0	0
	2	6	6	0	11
	48	5	1	17	51
	27	1	3	0	0
	53	9	1	0	1
	17	5	1	0	2
	6	5	1	0	10
	2	5	1	1	1
	5	20	1	0	2
	71	3	1	0	1
	47	6	1	169	1
	11	10	2	0	10
1,511	18	9	6	4	15
941	10	31	0	7	1
1,357	43	89	0	8	1
1,359	4	22	0	47	2
1.29	0.02	0.01	0.00	0.01	0.01

Teas

Element	Mg	Al	Cu	Zn	As
Earl Grey	1,967,501	759,557	20,643	35,525	74
Earl Grey	1,900,852	1,368,988	24,641	36,756	212
Earl Grey	1,762,202	754,684	13,077	23,358	19
Double Bergamot Earl Grey	2,002,568	1,593,905	22,828	30,057	52
Classics Earl Grey	1,811,920	953,068	16,729	23,355	73
Earl Grey	1,992,055	811,489	16,555	27,720	97
Earl Grey	1,911,883	858,071	19,102	25,743	18
Black	19,068	4,388	161	599	1
Green	23,092	4,127	191	740	2
AVG PPM	**1,487.90**	**789.81**	**14.88**	**22.65**	**0.06**

Spices

Element	Mg	Al	Cu	Zn	As
Coriander Powder	3,414,735	76,723	14,303	38,400	52
Cinnamon Powder	545,651	36,170	2,883	260,537	23
Ground Mustard	3,331,588	33,024	3,275	46,448	21
Onion Granules	1,125,436	6,221	4,392	20,333	39
Cilantro Powder	4,743,767	134,607	12,225	36,731	163
Cilantro Leaf	3,515,451	61,400	8,095	29,512	63
Organic Balti Curry	2,336,328	40,034	8,208	30,539	88
Organic Oregano	2,055,258	102,065	6,775	23,643	243
Organic Ground Coriander	3,538,894	34,761	10,561	49,825	44
AVG PPM	**2,734.12**	**58.33**	**7.86**	**59.55**	**0.08**

Sr	Cd	Cs	Hg	Pb	U
16,003	31	96	5	690	4
18,311	69	382	5	1,650	24
33,303	36	119	0	166	9
19,883	40	1,330	2	683	6
21,101	58	323	1	783	8
21,577	35	331	2	398	16
21,502	38	77	0	372	6
34	0	5	1	6	0
28	0	18	0	4	0
16.86	**0.03**	**0.30**	**0.00**	**0.53**	**0.01**

Sr	Cd	Cs	Hg	Pb	U
37,799	48	10	0	47	0
113,000	213	306	0	0	0
	40	81	3	51	2
	16	0	0	8	3
40,819	114	35	2	272	41
31,193	51	8	8	69	16
46,179	136	40	2	119	8
22,735	16	56	11	360	10
27,761	167	10	1	37	2
45.64	**0.09**	**0.06**	**0.00**	**0.11**	**0.01**

Sodas & Juices

Element	Mg	Al	Cu	Zn	As
Mountain Dew	1,631	76	3	91	1
Coca-Cola	4,741	82	0	117	10
Carrot Orange Juice	99,178	212	392	1,059	2
Apple Cucumber Juice	117,967	635	223	584	3
AVG PPM	55.88	0.25	0.15	0.46	0.00

Sweeteners

Element	Mg	Al	Cu	Zn	As
Maple Sugar Crystals	203,981	284	77	4,586	5
Corn Syrup	1,294	248	0	135	11
Sucralose	1,541	1,942	10	947	47
Cane Sugar	2,259	224	58	304	28
AVG PPM	52.27	0.67	0.04	1.49	0.02

Sr	Cd	Cs	Hg	Pb	U
	0	0	0	2	0
	1	1	1	0	0
	58	7	1	1	1
	6	0	0	0	0
	0.02	0.00	0.00	0.00	0.00

Sr	Cd	Cs	Hg	Pb	U
	0	6	2	8	0
	0	0	1	0	0
	2	0	1	0	0
0	1	2	7	0	3
0	0.00	0.00	0.00	0.00	0.00

Grains

Element	Mg	Al	Cu	Zn	As
Brown Rice	836,716	890	1,786	12,635	147
White Rice	54,291	475	713	5,291	165
White Flour	299,291	2,321	1,527	7,459	0
Wheat Flour	1,571,989	3,809	5,093	40,619	16
California Sushi Rice	254,501	2,498	2,004	9,979	59
California White Jasmine Rice	391,653	1,107	2,040	15,047	75
California Brown Jasmine Rice	1,256,658	1,046	3,069	20,956	207
Thai Hom Mali Rice	1,155,111	1,542	2,250	23,983	234
Brown Short Grain Rice	1,175,728	1,089	2,390	14,746	247
AVG PPM	**777.33**	**1.64**	**2.32**	**16.75**	**0.13**

Candy Bars

Element	Mg	Al	Cu	Zn	As
Butterfinger	519,510	3,065	1,864	7,592	6
3 Musketeers	367,552	2,673	2,753	6,452	14
Crunch	577,976	3,906	3,655	10,546	14
AVG PPM	**488.35**	**3.21**	**2.76**	**8.20**	**0.01**

Sr	Cd	Cs	Hg	Pb	U
	0	1	2	9	0
	2	35	3	0	3
	19	0	5	0	0
	29	1	2	79	0
147	9	0	0	3	1
148	6	0	0	3	1
451	12	0	1	2	1
297	72	115	0	12	2
452	3	1	1	5	1
0.30	0.02	0.02	0.00	0.01	0.00

Sr	Cd	Cs	Hg	Pb	U
906	21	13	4	1	4
1,438	12	9	8	1	3
2,150	29	17	7	7	8
1.50	0.02	0.01	0.01	0.00	0.01

Pastries and Cookies

Element	Mg	Al	Cu	Zn	As
Donut Sticks	56,211	6,799	256	1,872	14
Oreo	119,564	1,279	1,034	3,290	0
Granola Bars	788,519	1,268	2,794	13,561	24
Swiss Rolls	193,754	169,038	1,332	3,631	21
Jet-Puffed Marshmallows	22,669	1,400	21	672	5
AVG PPM	236.14	35.96	1.09	4.61	0.01

Snack Chips

Element	Mg	Al	Cu	Zn	As
Cheetos	210,534	1,051	452	4,649	15
Doritos	808,533	1,506	887	15,994	37
Nut-Thins	572,466	3,382	2,294	12,737	162
Pringles Sour Cream & Onion	414,957	4,614	1,431	6,250	5
Goldfish	317,119	2,115	1,503	15,412	22
Milano Minis	323,432	15,730	2,322	6,210	11
Lays Classic	605,699	2,202	3,613	11,163	11
Crackers	154,613	1,446	1,003	5,260	30
Wheat Thins	841,379	1,447	2,923	18,070	29
Potato Chips Nori Seaweed	905,019	18,278	1,708	9,240	84
AVG PPM	515.38	5.18	1.81	10.50	0.04

Soups

Element	Mg	Al	Cu	Zn	As
SpaghettiOs	164,916	837	668	5,233	0
Chicken Noodle Soup	40,148	2,273	164	2,062	0
Mini Ravioli	136,837	1,000	805	3,431	0
AVG PPM	113.97	1.37	0.55	3.58	0

Sr	Cd	Cs	Hg	Pb	U
808	9	26	10	10	86
412	10	9	12	7	25
1,283	12	7	0	0	7
694	18	3	15	1	3
122	2	3	0	1	1
0.66	**0.01**	**0.01**	**0.01**	**0.00**	**0.02**

Sr	Cd	Cs	Hg	Pb	U
	0	6	1	8	11
	9	1	1	0	0
	8	3	2	0	0
1,289	77	3	0	0	0
1,782	35	2	0	0	0
1,217	31	9	25	2	2
2,883	35	3	0	0	0
1,005	37	1	0	0	8
1,427	51	0	6	0	3
2,588	87	2	3	0	4
1.74	**0.04**	**0.00**	**0.00**	**0.00**	**0.00**

Sr	Cd	Cs	Hg	Pb	U
396	18	4	0	0	22
230	5	0	0	0	17
535	15	1	0	0	10
0.39	**0.01**	**0.00**	**0**	**0**	**0.02**

Seafood and Fish

Element	Mg	Al	Cu	Zn	As
Saki Ika Squid	553,102	1,452	1,676	31,701	1,764
Dried Shrimp	2,648,471	7,365	11,478	21,080	3,844
Dried Shaved Bonito	773,548	593	4,172	20,509	5,024
Powdered Krill	8,130,211	99,506	52,786	61,492	2,975
Bento	335,491	2,344	1,504	8,568	1,106
Dried Shrimp	1,411,838	4,028	9,677	34,662	14,161
Dried White Shrimp	717,064	7,173	596	28,000	1,553
Dried Shrimp	1,383,777	7,044	6,776	36,466	2,687
Dried Whole Anchovy	2,682,815	65,015	2,373	46,763	3,068
Dried Shaved Bonito	910,362	316	1,739	14,888	5,030
Dried Shaved Bonito	1,134,391	402	4,696	24,790	6,343
Shaved Bonito Flakes	948,742	817	7,159	21,999	7,387
Dried Shaved Bonito	817,762	156	3,647	18,663	5,906
Prepared Shredded Squid	348,873	170	1,170	30,077	1,322
Dried Bonito Shavings	1,142,991	0	6,901	26,936	9,217
Dried Sliced Pollack	1,407,259	75	1,215	24,136	9,052
Chopped Clams	307,175	6,156	1,803	20,461	674
AVG PPM	1,509.05	11.92	7.02	27.72	4.77

Sr	Cd	Cs	Hg	Pb	U
	71	5	39	51	12
	62	9	5	91	8
	30	70	167	16	3
582,235	323	18	6	23	102
	203	21	45	43	14
53,325	22	13	9	29	19
22,453	65	10	9	82	20
47,919	48	7	14	8	7
101,484	274	48	15	247	83
498	35	82	119	0	0
780	70	94	213	6	15
899	56	70	96	8	11
894	31	86	136	0	0
1,583	126	0	153	1	0
2,368	52	78	110	0	0
8,125	15	97	62	2	8
	141	2	15	237	28
68.55	0.10	0.04	0.07	0.05	0.02

Indian Spices

Element	Mg	Al	Cu	Zn	As
Coriander Powder	3,211,418	46,556	15,729	33,788	48
Turmeric Powder	2,964,728	131,210	5,724	20,781	25
Curry Powder Hot	2,535,155	45,569	8,929	23,432	23
Turmeric Powder	3,650,387	147,928	5,310	41,296	131
Turmeric Powder	2,709,861	151,380	4,753	11,373	21
Turmeric Powder	2,579,554	82,512	4,184	11,168	25
Garam Masala	2,848,422	369,542	15,744	29,578	153
Chili Powder	2,163,319	178,136	15,672	20,782	26
Cinnamon	2,190	963	14	65	0
Cinnamon	71,377	7,434	436	819	4
Cinnamon	19,050	2,186	120	223	1
AVG PPM	2,068.68	105.77	6.97	17.57	0.04

Sunflower Seeds

Element	Mg	Al	Cu	Zn	As
Sunflower Seeds	3,704,645	1,392	20,655	67,808	0
Sunflower Seeds Unsalted	3,205,342	612	12,205	48,595	0
Sunflower Seeds Unsalted	3,264,960	6,739	20,941	71,302	0
Honey Roasted Sunflower Kernels	2,828,970	1,369	19,876	65,228	0
Seaweed Flavored Sunflower Seeds	1,390,213	67,528	8,640	29,795	37
Sunflower Seeds	1,761,850	4,139	16,214	51,649	2
Sunflower Seeds	3,201,503	1,877	21,024	72,007	0
Sunflower Seeds	3,364,601	2,540	17,100	60,052	0
Sunflower Seeds	3,867,853	836	19,282	59,056	0
Sunflower Seeds	2,232,347	9,594	14,848	37,737	16
Sunflower Seeds	3,812,895	3,702	22,684	56,098	0
Sunflower Seeds	2,804,716	813	22,856	69,829	0

Sr	Cd	Cs	Hg	Pb	U
47,630	105	25	4	485	2
13,354	22	3	2	65	9
24,142	51	14	6	174	4
18,201	66	11	2	517	25
9,218	9	3	2	33	6
8,193	25	1	2	46	4
46,371	73	114	59	674	31
9,590	42	14	2	101	5
360	1	1	0	2	0
1,279	2	2	1	13	2
143	0	1	0	4	0
16.23	0.04	0.02	0.01	0.19	0.01

Sr	Cd	Cs	Hg	Pb	U
4,368	1,155	1	0	9	7
2,629	707	4	0	0	0
5,160	56	11	0	7	0
5,934	74	586	0	0	0
5,755	73	12	0	8	1
15,952	196	15	0	9	0
3,953	361	0	0	0	0
7,147	94	10	0	27	1
1,865	1,350	4	0	6	0
14,518	31	13	0	29	17
9,442	101	2	0	12	1
2,858	1,197	0	0	0	2

Sunflower Seeds (Cont.)

Element	Mg	Al	Cu	Zn	As
Pumpkin Seeds	5,926,038	2,081	16,614	69,857	2
Sunflower Seed Spread	2,309,836	784	15,811	43,928	0
Sunflower Seed Spread	2,841,759	656	19,046	56,329	0
Sunflower Seed Butter	2,450,390	542	15,732	46,890	0
Sunflower Seed Butter	2,893,619	575	18,017	48,196	0
AVG PPM	3,050.68	6.22	17.74	56.14	0.00

Baking Powders

Element	Mg	Al	Cu	Zn	As
Baking Powder	292,221	31,975	855	1,819	442
Baking Powder	1,375,733	60,892	404	1,564	562
Baking Powder	200,293	26,641,036	498	2,030	120
Baking Powder	84,253	223,299	419	3,805	229
AVG PPM	488.13	6,739.30	0.54	2.30	0.34

Sr	Cd	Cs	Hg	Pb	U
2,146	3	5	0	8	2
1,426	430	7	0	0	1
1,780	749	13	0	0	1
1,771	484	7	0	0	0
1,882	430	4	0	0	1
5.21	0.44	0.04	0.00	0.01	0.00

Sr	Cd	Cs	Hg	Pb	U
25,341	775	10	9	25	169
26,658	83	7	5	24	280
10,340	82	0	336	21	95
5,864	96	0	1	21	250
17.05	0.26	0.00	0.09	0.02	0.20

SUPERFOODS

Sea Vegetables

Element	Mg	Al	Cu	Zn	As
Modifilan	5,724,376	35,423	991	22,529	39,197
Kelp Granules	8,104,823	916,992	2,497	8,542	82,512
Agar Agar	1,425,730	128,713	499	2,444	210
Wakame Flakes	2,164,829	2,354	376	17,840	33,800
Seaweed Ca	23,132,805	102,165	1,162	1,978	1,202
Wakame Flakes	7,313,252	66,135	946	17,082	53,042
Kelp Help	5,411,955	44,287	81,545	129,277	11,523
Sea Lettuce Flakes	14,913,864	512,414	5,861	10,310	4,385
Kombu Flakes	9,705,962	39,855	1,444	29,732	102,058
Kelp Powder	8,922,247	115,374	1,247	32,003	30,159
Dulse	2,203,750	101,470	3,275	21,100	14,699
New Zealand Wakame	570,534	2,968	57	753	1,999
Fueru Wakame Dried Seaweed	8,085,079	37,238	1,209	57,283	34,845
Wakame	8,309,616	39,836	1,582	36,536	30,238
Pacific Wakame	10,633,995	44,838	1,390	34,800	37,568
Seasoned & Roasted Seaweed	1,950,866	50,009	3,725	20,316	11,324
Roasted Seaweed	1,517,760	25,677	3,029	14,729	11,063
Roasted Seaweed Snack	1,650,407	11,139	2,885	13,173	9,712
Roasted Seaweed	1,920,403	29,649	2,677	12,800	10,860
Roasted Seaweed	1,909,139	108,024	2,446	14,502	10,343
Roasted Seasoned Seaweed	1,789,767	13,575	3,598	25,448	12,805
Sushi Nori	2,832,595	2,828	6,859	48,000	23,517
AVG PPM	5,917.90	110.50	5.88	25.96	25.78

Sr	Cd	Cs	Hg	Pb	U
600,389	948	7	12	297	419
962,547	771	24	5	0	156
	402	21	5	169	15
	1,369	1	7	116	381
	553	3	2	14	3,754
1,087,944	1,979	26	6	83	299
238,126	255	18	6	30	155
44,612	143	77	5	429	38
	233	82	20	252	434
641,174	381	22	34	259	289
15,373	428	97	0	139	50
51,661	69	9	0	0	32
666,758	1,959	3	13	851	383
1,113,319	1,946	4	24	250	949
754,251	1,247	2	13	631	453
11,931	696	8	3	81	118
34,596	560	2	1	67	237
16,547	486	7	1	81	1,347
12,690	271	1	3	66	200
41,466	357	14	1	130	244
25,488	1,418	11	1	74	314
26,403	64	10	0	112	352
352.52	**0.75**	**0.02**	**0.01**	**0.19**	**0.48**

Fruit Powders & Freeze-Dried Fruits

Element	Mg	Al	Cu	Zn	As
Strawberry Powder	1,431,346	40,840	3,527	10,045	107
Mango Powder	731,612	23,219	5,160	4,795	60
Maqui Powder	558,737	19,946	3,384	7,621	18
Acai Powder	964,367	2,139	10,711	15,213	51
Camu Camu Powder	523,182	2,118	2,970	11,959	43
Blackberry Halves	1,345,558	4,578	8,510	13,930	17
Rooibos Powder	5,638,136	31,828	1,565	25,860	199
Whole Blueberries	322,612	7,850	6,832	3,966	15
Mangosteen Powder	704,415	6,444	7,557	12,096	33
Mango Pieces	689,836	291	3,171	2,745	18
Strawberry Pieces	1,257,559	17,205	3,275	7,974	106
Whole Raspberries	1,324,365	1,406	5,874	22,113	26
Blueberry	449,203	44,357	3,913	8,321	38
Acai	1,469,172	4,015	26,461	40,743	10
Maqui Powder	799,022	177,284	4,891	9,126	11
Camu Powder	614,751	1,828	7,193	17,806	112
Pomegranate Powder	176,318	7,158	1,282	3,550	9
Apple Fiber Powder	457,342	3,034	5,665	6,091	15
Grapefruit Pectin Powder	728,888	1,859	2,883	5,908	0
Goji Berries	865,637	39,913	5,847	8,286	54
Organic Goji Berries—Saint World Group	1,098,088	12,056	8,364	11,043	21
AVG PPM	1,054.77	21.40	6.14	11.87	0.05

Sr	Cd	Cs	Hg	Pb	U
9,272	114	0	0	0	26
5,272	0	0	0	0	0
	8	28	1	67	3
	199	146	6	25	3
	7	16	0	38	4
	57	3	4	22	3
	53	117	9	111	9
	7	6	1	15	14
	105	729	5	174	5
	11	40	0	11	5
	24	64	1	55	7
	329	6	2	10	2
2,911	15	18	0	82	6
12,655	133	152	0	52	1
27,368	2	30	0	39	23
2,899	0	14	0	7	0
1,583	3	35	0	65	8
3,233	2	7	0	62	1
59,757	7	10	0	8	1
3,606	40	19	0	101	7
8,934	43	3	0	33	4
12.50	**0.06**	**0.07**	**0.00**	**0.05**	**0.01**

Exotic Superfood Powders

Element	Mg	Al	Cu	Zn	As
100% Organic Maca Root Powder	993,805	84,279	2,792	35,631	244
Mesquite Powder	966,088	7,442	4,923	12,811	13
Mesquite Powder	1,063,160	21,318	8,838	16,854	2
Beet Powder	401,771	13,906	3,417	11,728	33
AVG PPM	856.21	31.74	4.99	19.26	0.07

Grass Powders

Element	Mg	Al	Cu	Zn	As
Barley Grass Powder	5,865,061	74,596	13,921	32,521	2,883
Alfalfa Powder	4,286,643	232,594	8,725	28,514	109
Wheatgrass	3,561,101	294,655	9,755	23,357	715
Shavegrass Powder	5,211,869	473,355	20,970	25,418	207
Barley Grass	4,215,879	44,505	9,374	20,547	841
Alfalfa Grass Powder	4,662,295	199,201	11,606	45,156	1,897
Wheat Grass Powder	3,861,893	164,920	8,117	25,571	1,899
Just Barley	2,752,029	64,996	10,760	27,253	985
Green Magma	1,527,519	55,759	5,716	14,194	769
Warrior Food	2,753,051	52,761	31,442	157,570	119
Alfalfa Leaf Powder	5,288,886	252,697	13,090	42,496	163
AVG PPM	3,998.75	173.64	13.04	40.24	0.96

Sr	Cd	Cs	Hg	Pb	U
	614	54	1	361	13
	7	26	2	22	5
25,090	4	20	0	171	0
750	57	13	0	36	1
12.92	**0.17**	**0.03**	**0.00**	**0.15**	**0.00**

Sr	Cd	Cs	Hg	Pb	U
	249	193	13	143	32
653,692	18	0	22	219	56
39,787	61	67	4	153	31
50,885	59	226	16	155	0
23,985	187	158	0	0	7
117,846	87	43	11	308	47
43,569	54	56	4	135	51
34,729	42	21	1	50	11
24,610	40	25	3	87	12
9,795	1,329	7	34	253	5
1,092,198	31	17	22	541	121
209.11	**0.20**	**0.07**	**0.01**	**0.19**	**0.03**

Chia Seeds

Element	Mg	Al	Cu	Zn	As
White Chia Seeds	3,515,192	3,280	19,561	62,307	8
Black Chia Seeds	2,750,920	5,223	19,817	65,771	21
AVG PPM	3,133.06	4.25	19.69	64.04	0.01

Chlorella

Element	Mg	Al	Cu	Zn	As
Chlorella	2,528,947	47,736	2,691	16,039	782
Chlorella	2,655,482	19,801	3,179	20,130	823
Chlorella	2,896,765	2,457	5,793	18,480	17
Chlorella	2,078,317	8,045	4,871	11,438	42
Chlorella	3,303,590	13,842	3,485	20,994	300
Chlorella	2,640,708	20,328	1,926	9,995	53
Chlorella	2,845,314	13,393	9,116	73,963	25
Chlorella	3,450,038	15,713	10,954	22,444	56
Chlorella	3,739,117	19,484	2,000	10,643	504
Chlorella	2,204,012	18,864	3,346	18,623	927
Chlorella	3,307,107	8,869	6,607	43,086	56
Chlorella	2,978,301	6,785	4,930	20,742	1,204
Chlorella	2,535,224	4,810	5,892	12,178	125
Chlorella	2,671,924	178,756	3,500	11,860	1,048
Chlorella	3,748,179	1,704	4,770	24,440	58
Chlorella	3,798,197	1,149	3,222	9,644	52
Chlorella	2,829,608	64,652	2,059	31,874	491
Chlorella	2,546,824	51,763	1,854	31,013	723
Chlorella	2,552,197	59,577	2,044	32,473	447
AVG PPM	2,911.04	29.35	4.33	23.16	0.41

Sr	Cd	Cs	Hg	Pb	U
29,149	3	4	0	0	1
34,455	27	1	1	16	1
31.80	**0.02**	**0.00**	**0.00**	**0.01**	**0.00**

Sr	Cd	Cs	Hg	Pb	U
5,903	10	11	0	153	17
7,624	13	5	0	272	6
4,696	0	0	0	0	17
9,902	3	0	0	0	8
11,852	4	0	0	23	0
34,118	237	0	0	0	295
35,445	4	0	0	117	0
57,355	7	0	0	125	0
49,047	8	0	3	116	22
	48	8	3	952	3
	12	0	3	262	12
	4	4	2	97	7
	8	0	2	79	25
25,385	11	44	0	552	44
34	9	0	0	0	0
0	0	0	0	0	0
14,469	16	0	1	129	42
13,839	15	14	1	252	58
14,904	25	19	1	288	63
18.97	**0.02**	**0.01**	**0.00**	**0.18**	**0.03**

Nut & Seed Oils & Spreads

Element	Mg	Al	Cu	Zn	As
Organic Coconut Oil—Extra Virgin	3,943	548	4	184	0
Sunflower Seed Spread	2,457,875	1,001	17,131	48,750	0
Coconut Butter	555,186	1,184	5,105	9,662	0
Hazelnut Butter	1,485,490	5,990	13,871	19,179	0
Almond Butter	2,482,271	4,652	9,099	32,217	8
Coconut Butter	106,179	267	994	2,296	0
AVG PPM	1,181.82	2.27	7.70	18.71	0.00

Rice Protein Powders

Element	Mg	Al	Cu	Zn	As
Rice Protein	1,842,642	45,371	17,408	97,924	90
Rice Protein	2,825,046	32,787	21,296	139,964	72
Rice Protein	1,527,659	45,303	31,067	153,914	154
Rice Protein	1,176,557	25,842	25,839	130,438	89
Rice Protein	1,382,376	58,241	22,981	119,730	117
Rice Protein	1,027,569	47,821	20,085	96,876	109
Rice Protein	1,744,047	89,242	21,476	111,168	775
Rice Protein	1,697,844	30,024	26,402	136,952	517
Rice Protein	1,521,875	31,150	25,577	130,264	521
Rice Protein	1,296,835	37,000	21,595	111,053	265
Rice Protein	372,666	21,133	8,321	72,462	165
Rice Protein	322,334	9,830	13,559	59,696	68
Rice Protein	756,257	15,652	14,329	59,533	61
Rice Protein	1,836,481	59,102	23,842	114,555	104
Rice Protein	4,033,836	53,129	27,954	153,306	125
Rice Protein	1,062,196	29,826	25,317	139,163	118
Rice Protein	1,987,490	24,498	28,075	139,981	178
Rice Protein	2,099,013	55,340	21,434	111,682	64

Sr	Cd	Cs	Hg	Pb	U
12	0	0	0	0	0
1,664	452	8	0	3	3
255	20	66	0	39	1
9,022	129	120	0	15	2
19,609	15	16	0	27	1
40	3	3	0	4	0
5.10	**0.10**	**0.04**	**0**	**0.01**	**0.00**

Sr	Cd	Cs	Hg	Pb	U
	1,365	14	23	371	7
	1,540	14	37	253	7
4,900	2,407	20	38	833	16
2,872	1,637	10	27	194	8
4,334	1,874	18	33	377	12
4,270	1,699	7	26	533	12
3,999	1,716	13	34	535	14
4,714	2,533	9	30	312	4
5,116	1,370	13	27	292	3
5,816	1,296	22	14	384	14
11,071	46	6	0	38	130
10,880	30	0	0	14	163
11,483	73	0	0	20	125
6,725	1,042	10	25	260	8
7,722	1,889	18	30	313	8
3,901	1,372	5	46	324	9
4,336	1,895	28	26	302	8
4,238	1,449	4	23	235	2

Rice Protein Powders (Cont.)

Element	Mg	Al	Cu	Zn	As
Rice Protein	2,756,332	26,812	11,598	48,115	30
Rice Protein	638,864	23,888	22,726	140,969	48
Rice Protein	959,196	27,513	16,513	77,949	163
Rice Protein	1,814,902	62,154	28,167	149,103	123
Rice Protein	1,521,178	39,635	27,664	139,360	107
Rice Protein	1,508,410	26,165	28,560	157,462	281
AVG PPM	1,571.32	38.23	22.16	116.32	0.18

Protein Powders

Element	Mg	Al	Cu	Zn	As
Protein Powder	4,824,636	38,862	17,976	100,457	82
Protein Powder	9,334,813	2,047	2,047	165,195	37
Protein Powder	1,787,996	21,704	12,976	116,280	58
Protein Powder	356,566	21,052	4,111	61,640	18
Protein Powder	10,816,773	5,654	33,770	171,519	33
Protein Powder	7,125,498	15,337	20,776	104,921	280
Protein Powder	9,114,413	56,860	30,431	148,085	57
Protein Powder	632,324	7,449	8,440	119,909	32
Protein Powder	558,389	4,042	3,596	7,740	18
Protein Powder	209,948	955	356	1,982	11
Protein Powder	214,025	3,096	509	3,385	19
Protein Powder	2,871,890	15,383	7,799	67,430	360
Protein Powder	699,270	24,565	10,719	96,766	15
Protein Powder	1,374,000	16,005	10,492	74,573	15
Protein Powder	1,066,753	8,705	7,050	79,044	10
Protein Powder	4,153,022	18,928	15,162	531,078	23
Protein Powder	687,860	4,690	3,457	6,895	7
Protein Powder	623,377	15,401	14,142	67,927	10
Protein Powder	1,168,619	5,531	6,008	12,380	6
Protein Powder	943,579	15,117	9,084	83,868	15

Sr	Cd	Cs	Hg	Pb	U
6,527	292	21	2	204	2
3,176	700	3	18	323	4
6,752	391	21	10	169	12
	1,912	22	43	353	14
6,022	1,600	8	29	343	10
2,800	1,681	8	29	273	3
5.79	**1.33**	**0.01**	**0.02**	**0.30**	**0.02**

Sr	Cd	Cs	Hg	Pb	U
	54	40	8	55	8
	25	14	7	9	2
	90	8	10	43	39
1,367	19	1	0	17	14
	13	14	12	22	4
	31	55	8	67	25
	58	20	8	53	6
3,854	53	23	2	140	70
2,703	11	19	9	0	19
1,324	0	11	5	0	20
1,441	0	7	4	0	17
15,446	85	14	15	40	67
4,849	50	0	1	40	14
29,952	113	15	4	52	90
29,796	66	1	1	22	77
7,506	34	6	0	16	66
3,877	17	10	0	9	8
1,575	27	1	2	46	2
3,116	78	10	0	12	4
4,550	54	0	2	19	32

Protein Powders (Cont.)

Element	Mg	Al	Cu	Zn	As
Protein Powder	658,782	2,735	593	3,494	5
Protein Powder	750,983	5,298	2,338	5,044	1
Protein Powder	316,425	1,475	677	11,977	0
Protein Powder	4,591,118	17,941	22,958	71,910	42
Protein Powder	4,787,781	17,416	24,386	79,204	56
Protein Powder	7,728,351	3,221	20,227	111,817	13
AVG PPM	2,976.82	13.44	11.16	88.64	0.05

Cacao

Element	Mg	Al	Cu	Zn	As
Cacao	2,639,292	633	20,517	38,583	23
Cacao	5,670,241	21,976	40,293	83,320	101
Cacao	6,327,818	39,225	43,157	96,838	79
Cacao	4,769,525	16,600	42,232	74,285	30
Cacao	2,684,050	1,411	11,841	37,970	8
Cacao	4,486,781	44,617	38,306	71,825	26
Cacao	2,289,344	581	20,330	41,969	14
Cacao	5,370,389	108,178	35,346	89,691	33
Cacao	5,624,522	54,372	38,986	91,965	62
Cacao	4,876,786	43,064	37,494	79,400	72
Cacao	4,957,378	39,082	39,162	72,276	16
Cacao	2,979,522	9,020	21,567	39,117	18
Cacao	3,019,270	3,612	20,402	43,326	10
Cacao	2,266,803	7,943	16,896	40,250	7
AVG PPM	4,140.12	27.88	30.47	64.34	0.04

Sr	Cd	Cs	Hg	Pb	U
2,557	0	9	0	0	1
2,167	18	9	0	0	1
1,163	0	3	0	0	2
189,494	89	262	0	96	2
214,602	136	280	0	114	2
4,811	3	36	0	10	3
26.31	**0.04**	**0.03**	**0.00**	**0.03**	**0.02**

Sr	Cd	Cs	Hg	Pb	U
	652	12	0	13	6
10,834	2,318	86	1	44	5
21,039	1,826	84	2	66	4
5,840	1,324	17	12	154	5
9,837	162	75	0	0	78
21,622	1,738	147	3	95	6
7,049	184	4	1	2	0
21,647	1,083	78	6	257	5
25,086	2,138	46	3	78	3
18,920	2,448	42	0	67	1
25,181	196	46	0	51	1
9,193	326	9	2	30	0
5,360	258	17	0	14	1
4,880	298	7	1	9	1
14.35	**1.07**	**0.05**	**0.00**	**0.06**	**0.01**

Greens Blends Powders

Element	Mg	Al	Cu	Zn	As
Greens Blends	4,187,833	50,169	9,537	39,985	296
Greens Blends	1,551,786	87,661	4,700	15,873	313
Greens Blends	1,990,810	423,899	4,760	23,719	994
Greens Blends	2,335,527	91,354	7,823	18,891	1,055
Greens Blends	3,126,985	232,570	6,709	24,311	1,794
Greens Blends	2,851,849	141,969	6,151	34,089	3,369
Greens Blends	3,891,676	135,467	10,706	32,365	580
Greens Blends	2,702,532	95,343	7,936	38,184	500
Greens Blends	5,780,336	19,024	25,610	436,516	327
Greens Blends	1,928,171	58,744	20,042	90,758	90
Greens Blends	1,797,902	44,008	24,197	152,709	45
Greens Blends	3,730,573	54,245	21,648	132,666	49
Greens Blends	3,335,790	133,677	6,117	31,602	3,781
Greens Blends	2,648,296	95,326	7,957	29,560	470
Greens Blends	5,890,633	48,844	25,036	379,012	430
Greens Blends	1,449,698	120,466	4,062	17,510	281
Greens Blends	1,277,516	89,353	3,571	14,042	403
Greens Blends	4,596,059	141,357	8,956	25,240	476
Greens Blends	2,620,035	146,957	4,893	22,140	1,804
AVG PPM	**3,036.53**	**116.34**	**11.07**	**82.06**	**0.90**

Spirulina

Element	Mg	Al	Cu	Zn	As
Spirulina	5,419,993	24,693	677	110,941	135
Spirulina	2,599,188	348,196	6,615	18,422	569
Spirulina	2,365,873	248,619	6,428	16,510	458
Spirulina	3,160,249	33,814	668	6,343	506
Spirulina	3,492,239	26,535	3,446	17,799	48
Spirulina	4,110,488	264,945	5,081	25,784	320
Spirulina	2,467,549	47,464	3,829	13,718	180

Sr	Cd	Cs	Hg	Pb	U
22,562	65	5	1	99	16
17,205	35	17	0	261	11
31,786	111	55	3	446	89
19,123	105	63	3	182	27
80,085	70	68	3	1,046	50
78,985	162	76	2	364	89
40,606	171	60	0	164	35
29,351	130	208	0	134	26
15,580	347	20	6	133	31
18,825	460	29	0	176	12
6,024	170	3	0	99	9
13,467	140	8	0	90	8
89,407	136	53	0	261	52
28,609	136	207	1	134	27
19,218	382	44	5	172	60
30,101	69	36	1	413	41
15,140	51	39	3	418	53
88,333	91	47	4	276	44
52,508	76	58	0	290	42
36.68	**0.15**	**0.06**	**0.00**	**0.27**	**0.04**

Sr	Cd	Cs	Hg	Pb	U
44,237	67	6	8	83	27
26,580	85	71	11	1,059	118
22,320	81	49	13	1,040	106
27,518	15	18	2	173	131
6,428	14	12	3	64	20
10,975	23	35	22	50	80
39,052	3	8	1	34	42

Spirulina (Cont.)

Element	Mg	Al	Cu	Zn	As
Spirulina	5,551,238	199,278	9,576	20,483	330
Spirulina	3,977,964	131,650	7,792	15,794	276
Spirulina	5,247,732	205,319	6,419	31,550	261
Spirulina	2,944,323	6,912	2,816	13,814	70
AVG PPM	3,757.89	139.77	4.85	26.47	0.29

Organic Mushrooms from USA (not China)

Element	Mg	Al	Cu	Zn	As
Organic Cordyceps	1,187,969	4,045	2,439	20,086	270
Organic Reishi	1,333,792	7,299	2,352	22,158	298
Organic Turkey Tail	1,089,297	1,600	2,881	13,853	144
Organic Lions Mane	1,001,816	1,559	2,846	12,653	101
Organic Maitake	1,255,611	5,278	2,454	20,887	281
Organic Chaga	3,236,283	13,193	5,728	55,691	765
Organic Shiitake	1,551,840	3,012	2,694	25,501	888
AVG PPM	1,522.37	5.14	3.06	24.40	0.39

Sr	Cd	Cs	Hg	Pb	U
	51	44	15	93	80
	36	37	11	73	75
12,714	36	35	18	75	105
7,086	5	6	3	33	15
21.88	**0.04**	**0.03**	**0.01**		**0.07**

Sr	Cd	Cs	Hg	Pb	U
633	6	3	2	0	1
436	8	3	0	0	1
566	6	0	0	0	0
540	5	0	0	0	0
604	7	1	0	0	0
1,224	15	6	0	0	1
523	11	4	0	0	0
0.65	**0.01**	**0.00**	**0.00**	**O**	**0.00**

SUPPLEMENTS & VITAMINS

Iodine

Element	Mg	Al	Cu	Zn	As
Lugol's Iodine		7,419	150	39,000	213
Iodine for Life		2,944	127	39,417	75
Magnascent Nascent Iodine		1,676	89	107,115	58
J Crow's Iodine		2,091	51	27,866	36
Atomic Iodine		1,318	63	3,239	10
Nascent Iodine		1,461	50	2,080	0
Detoxidine		1,460	56	8,800	34
Atomidine		1,184	135	9,156	36
Liquid Iodine Forte		1,009	102	5,204	0
Original Nascent Iodine		864	22	1,156	0
AVG PPM		2.14	0.08	24.30	0.05

Ginkgo

Element	Mg	Al	Cu	Zn
Ginkgo Biloba 120mg	2,342,057	912,292	7,082	34,306
Ginkgo Biloba 120mg	3,469,127	22,699	3,733	28,381
Ginkgo Biloba 120mg	930,165	13,384	2,283	7,428
Ginkgo Biloba 24%	2,807,023	11,280	356	20,561
Ginkgo Standardized	3,671,831	822,954	9,879	84,246
Ginkgo Leaf	445,803	467	706	1,163
Ginkgo Smart	944,452	130,365	6,403	8,338
Extra Strength Ginkgo	157,404	80,943	1,694	6,184
Ginkgo Leaf	3,906,120	84,917	2,298	8,336
Ginkgo Biloba	683,337	3,133	2,670	14,100
AVG PPM	1,935.73	208.24	3.71	21.30

Sr	Cd	Cs	Hg	Pb	U
0	4	18	26	11	18
0	16	18	7	0	27
0	0	8	8	0	15
0	4	7	5	0	11
0	7	23	9	7	39
0	10	11	7	5	22
0	7	16	5	29	18
0	16	130	7	14	176
0	5	10	6	63	13
0	4	6	10	9	17
0	0.01	0.02	0.01	0.01	0.04

As	Sr	Cd	Cs	Hg	Pb	U
836	30,138	432	713	3	1,711	148
603	5,800	14	19	1	74	25
255	2,570	14	3	4	207	34
296	19,647	16	11	2	250	10
189	81,960	307	215	8	949	26
59	127	1	8	0	7	0
274	15,369	30	30	13	980	10
278	1,044	4	3	0	365	18
221	35,734	65	45	99	2,402	10
340	358	18	0	2	42	1
0.34	19.27	0.09	0.10	0.01	0.70	0.03

Liquid Minerals

Element	Mg	Al	Cu	Zn	As
Fulvic Mineral Complex	999	55	3	15	2
Optimal Nutrition	35,800	23	8,799	93,224	1
Vitality Boost HA	1,054	25,341	74	91	2
Colloidal Minerals	11,045	39,392	2	208	1
Liquid Light Fulvic Acid	7,758	13,744	6	366	4
Plant Derived Minerals	73,316	286,377	3	1,269	2
Organic Life Vitamins	218,240	196	20	107,232	7
Colloidal Minerals	50,827	6,904	217	176	3
Trace Mineral Drops	13,397,090	78	5	110	416
ColloidaLife	7,053	2	1	93	2
Sea Minerals	9,381,937	59	1	81	13
Multiple Mineral	26,794	397	113	1,913	2
Blood Sugar Support	14,314	11	2	784,394	1
Ionic Trace Minerals	4,688,362	21	3	135	109
Super Ionic Concentrated Fulvic	5,925,285	42,360	28	891	84
Oceans Alive! Marine Phytoplankton	8,271,764	217	16	831	15
Ionic Minerals Concentrated X350	249,849	763,172	200	23,111	1,858
Raw Unheated Ocean Minerals	67,895	45	2	44	5
Sea Minerals	9,275,842	65	2	146	12
Sea Minerals	9,076,029	250	56	214	13
Sea Minerals	7,162,844	17	14	83	12

Sr	Cd	Cs	Hg	Pb	U
47	0	0	0	0	0
26	0	0	0	0	0
115	0	0	0	37	9
82	1	0	0	0	0
213	0	0	0	0	1
507	5	3	0	0	1
143	0	0	1	1	0
367	1	0	0	13	2
30	0	7	1	3	7
50	0	0	0	0	0
32	0	1	0	1	4
46	0	0	0	0	0
0	2	0	0	1	0
13	0	18	0	0	7
5,990	10	1	0	10	20
67	0	0	0	1	2
791	160	0	0	0	298
386	0	0	0	0	0
31	0	0	0	1	3
61	0	0	0	3	3
105	0	1	0	2	2

Liquid Minerals (Cont.)

Element	Mg	Al	Cu	Zn	As
Optimal Liposomal Magnesium	18,377,115	6,992	185	1,459	13
True Colloidal Silver	9,096	1,917	24	990	2
Sovereign Silver	182	139	0	109	0
AVG PPM	3,597.10	49.49	0.41	42.38	0.11

Zeolites

Element	Mg	Al	Cu	Zn	As
Zeolites	4,769,997	37,670,948	2,036	38,608	35,247
Zeolite Capsules	2,866,082	31,327,069	2,772	35,851	3,791
ZeoForce	1,143,945	24,702,813	2,799	40,594	5,449
AVG PPM	2,926.67	31,233.61	2.54	38.35	14.83

Children's Vitamins

Element	Mg	Al	Cu	Zn	As
Children's Multivitamin	8,992	17,725	72	499,785	2
Multivitamin Gummies	13,000	1,216	113	543,730	0
Spongebob Gummies Multivitamins	8,693	14,731	85	620,297	4
Avengers Multivitamin Gummies	512,211	3,793	176	469,324	5
Gummy Vites	8,342	2,722	43	530,089	1
FlintStones Complete	1,188,826	2,527,622	1,277,443	8,728,236	111

Sr	Cd	Cs	Hg	Pb	U
754	1	1	0	8	16
22	0	0	0	0	0
2	0	0	0	0	0
0.41	0.01	0.00	0.00	0.00	0.02

Sr	Cd	Cs	Hg	Pb	U
	278	18,447	14	66,453	6,468
	214	3,056	5	27,198	3,625
	223	3,041	5	26,846	6,209
	0.24	8.18	0.01	40.17	5.43

Sr	Cd	Cs	Hg	Pb	U
506	1	0	0	9	3
747	1	0	1	16	2
648	0	0	0	6	4
662	1	0	0	10	1
741	0	0	1	6	3
82,807	11	0	2	59	45

Children's Vitamins (Cont.)

Element	Mg	Al	Cu	Zn	As
Disney Multivitamin	326,143	2,760	381	314,354	0
Almased	935,098	8,985	8,880	71,439	0
Gummy Vitamins	21,495	29,978	2,091	571,574	41
AVG PPM	335.87	289.95	143.25	1,372.09	0.02

Multivitamins

Element	Mg	Al	Cu	Zn	As
Raw Prenatal	23,995,482	54,479	354,332	5,546,687	314
Raw One Multivitamin	37,575,719	61,823	2,828,770	20,291,082	249
Whole Foods Multivitamin Men	8,425,503	25,276	719,542	7,310,991	190
Whole Foods Multi-vitamin Women	9,389,918	29,810	705,276	4,771,354	197
Cal-Mag	64,738,293	57,636	281	5,464	502
High Potency Multi	17,509,648	15,548	584,005	6,200,634	99
Whole Foods Daily Without Iron	44,224,365	29,294	718,538	12,918,207	273
High Proency Multi	18,075,034	15,239	457,418	4,250,634	98
AVG PPM	27,991.75	36.14	796.02	7,661.88	0.24

Liquid Vitamins

Element	Mg	Al	Cu	Zn	As
Optimal Liposomal Vitamin C	243,649	385	142	438	1
Optimal Liposomal Vitamin C	224,289	584	150	556	3
Optimal Liposomal Active B12	323,694	321	172	2,205	20
AVG PPM	263.88	0.43	0.15	1.07	0.01

Sr	Cd	Cs	Hg	Pb	U
243	1	0	0	6	0
3,189	28	1	1	12	4
1,101	40	34	0	50	13
10.07	**0.01**	**0.00**	**0.00**	**0.02**	**0.01**

Sr	Cd	Cs	Hg	Pb	U
428,163	205	13	8	137	630
192,796	110	23	5	157	267
66,908	69	64	1	49	109
75,409	71	89	2	59	126
55,696	166	4	1	162	510
32,865	22	0	0	47	158
54,533	96	8	0	43	240
25,952	27	1	3	43	186
116.54	**0.10**	**0.03**	**0.00**	**0.09**	**0.28**

Sr	Cd	Cs	Hg	Pb	U
316	1	0	0	0	0
334	1	1	0	1	0
492	2	1	0	3	0
0.38	**0.00**	**0.00**	**0.00**	**0.00**	**0**

Calcium

Element	Mg	Al	Cu	Zn	As
Raw Calcium	138,366,005	1,722,935	1,568	13,060	10,057
Strontium Carbonate	1,854	5,717	468	1,148	0
Gypsum	990,307	118,712	2,121	1,365	76
Coral Calcium	1,719,588	54,704	1,641	1,666	240
Calcium Carbonate	10,425,502	277,707	325	2,835	284
Calcium	630,119	7,057	424	1,840	135
Calcium	1,656,335	59,673	698	9,344	700
Calcium & Magnesium	100,200,081	63,387	4,545	5,309	381
Liquid Calcium	51,069	0	62	664	53
Calcium + Vitamin D3	313,844	16,796	186	1,691	53
Calcium	385,788	16,001	193	1,100	69
Calcium	2,477,299	296,441	2,493	14,752	871
Calcium	9,329,871	153,438	251,141	1,881,202	202
Calcium	669,248	30,351	330	2,832	281
Calcium	1,929,744	82,132	969	4,192	658
Calcium	1,205,691	15,511	664	2,659	249
Calcium	88,847,644	73,746	604	3,094	419
Calcium Pyruvate	495,879	4,204	1,463	2,514	508
Calcium Citrate	1,203,566	203,797	248	11,462	165
Calcium Ascorbate	4,691	529	99	551	31
Calcium Carbonate Oyster Shell	1,477,195	176,018	554	3,051	225
AVG PPM	17,256.25	160.90	12.90	93.63	0.75

Sr	Cd	Cs	Hg	Pb	U
2,020,966	81	72	9	2,352	1,631
OUT OF RANGE	1	2	0	28	0
871,433	5	16	1	373	48
319,691	461	4	3	286	1,126
5,128,936	4	10	0	52	1,818
53,625	477	0	1	32	703
84,661	351	12	2	194	774
93,350	160	2	0	201	549
1,971,899	0	210	0	1	17
22,329	9	1	0	24	47
25,157	12	3	0	25	59
1,104,351	54	57	8	756	281
34,864	90	1	0	65	129
127,736	149	7	0	118	1,134
224,218	7	5	0	474	1,194
89,011	666	3	1	42	1,123
116,153	75	23	0	305	295
48,219	211	2	22	8,304	23
110,707	153	3	0	65	741
37,685	4	1	0	18	0
339,260	21	4	0	411	65
641.21	**0.14**	**0.02**	**0.00**	**0.67**	**0.56**

Plant Extract Supplements

Element	Mg	Al	Cu	Zn	As
Quercetin	857,341	5,821	768	669	0
Amla Gold	1,201,014	101,343	713	27,966	78
Green Tea Extract	663,507	291,671	1,950	3,617	40
Citrus Bioflavonoids	763,179	18,223	3,001	7,998	18
Quercetin	1,621,746	2,411	426	1,060	5
Organic Turmeric	2,087,926	261,281	6,367	20,716	26
AVG PPM	**1,199.12**	**113.46**	**2.20**	**10.34**	**0.03**

Vitamin Mineral Powders

Element	Mg	Al	Cu	Zn	As
Mineral Powder	13,692,891	12,647	17,790	995,814	. 102
Multi-Nutrient Formula	29,977,817	20,106	111,288	3,666,833	50
Multivitamin Minerals	2,735,803	15,861	83,635	635,254	10
Mineral Booster	508,913	2,241	5,851,429	55,580,885	11
Trace Minerals	376,724	30,506	4,284,499	67,778,284	25
Raw Shilajit	7,477,248	461,684	7,312	49,267	468
Trace Minerals	409,087	44,792	383	77,388,168	32
Source Minerals	61,424,031	15,884,806	1,705	224,580	324
Mineral Complex	2,915,032	56,772	125,909	205,317	395
AVG PPM	**13,279.73**	**1,836.60**	**1,164.88**	**22,947.16**	**0.16**

Sr	Cd	Cs	Hg	Pb	U
415	2	0	18	6	10
25,407	19	35	0	241	34
5,185	11	160	0	195	5
48,468	4	14	20	145	50
8,605	7	1	7	16	11
18,895	110	13	13	301	11
17.83	**0.03**	**0.04**	**0.01**	**0.15**	**0.02**

Sr	Cd	Cs	Hg	Pb	U
22,359	92	0	6	68	165
15,939	38	3	3	53	39
2,808	8	0	3	18	19
432	381	0	0	0	0
12,398	398	1	0	0	0
118,953	57	44	1	371	355
14,735	292	0	0	655	104
72,546	1,739	102	2	47	142
68,092	73	1	2	132	32
36.47	**0.34**	**0.02**	**0.00**	**0.15**	**0.10**

Magnesium Liquids

Element	Mg	Al	Cu	Zn	As
Magnesium Chloride Liquid	25,801,871	255	37	514	0
Magnesium Chloride Brine	85,536,454	182	62	969	2
Prehistoric Magnesium Oil	66,699,274	1	41	5,431	6
Pure Magnesium Oil	71,968,482	0	24	305	4
Biogenics Magnesium Lotion	29,237,542	0	80	1,327	0
Magnesium Oil	76,415,449	0	19	332	4
Topical Magnesium Spray	13,464,836	0	17	481	514
Magnesium Oil	19,836,699	0	11	181	0
Magnesium Oil	40,102,323	32	34	456	528
Magnesium Bath Flakes	116,500,981	0	64	2,409	5
AVG PPM	54,556.39	0.05	0.04	1.24	0.11

PERSONAL CARE

Skin Whitening

Element	Mg	Al	Cu	Zn	As
Deep-Whitening Facial Foam	87,120	1,332	54	1,276	34
Brightening Gel	448	619	47	573	7
Skin Whitener	224	610	20	594	2
Crema AclaRante Natural	288,594	24,758	766	99,999,999	79
Skin-Lightening Cream	0	605	11	14,424	10
AVG PPM	75.28	5.58	0.18	20,003.37	0.03

Sr	Cd	Cs	Hg	Pb	U
44,328	0	11	0	1	1
5,524	0	12	0	32	4
256	0	0	0	175	0
4,483	0	8	0	26	0
166	0	1	0	3	0
4,782	0	8	0	28	2
4,625	0	54	1	0	20
28	0	0	0	0	0
154	4	65	0	7	1
6,463	0	15	0	41	0
7.08	**0.00**	**0.02**	**0.00**	**0.03**	**0.00**

Sr	Cd	Cs	Hg	Pb	U
115	2	23	1	8	2,214,783
6	1	3	0	3	2,435,220
25	1	0	0	3	1
20,902	142	0	0	1,893	43
19	2	0	0	4	2
4.21	**0.03**	**0.01**	**0.00**	**0.38**	**930.01**

Mascara

Element	Mg	Al	Cu	Zn	As
Washable Mascara	69,469	1,254	123	4,556	41
Clump Crusher Mascara	3,643,805	151,821	382	9,502	105
Waterproof Mascara	6,257,488	363,442	405	21,801	148
Mascara	69,882	260,161	200	5,078	58
100% Pure Fruit Pigmented Mascara	10,574	280,069	304	5,119	85
Mascara	9,899	60,179	261	5,371	56
Mascara	23,897	379,498	342	6,120	49
Doll Eye Mascara	59,194	10,162	333	24,715	50
Mascara	131,842	2,570	220	7,106	23
AVG PPM	1,141.78	167.68	0.29	9.93	0.07

Tattoo Ink

Element	Mg	Al	Cu	Zn	As
Tattoo Ink Orange	60,389	307,171	1,093	1,645	76
Tattoo Ink Rose	17,804	4,179	225	913	1
Tattoo Ink Gray	67,673	1,957	12,847,139	2,455	77
Tattoo Ink Mid Yellow	66,743	2,311	3,449	4,607	8
Tattoo Ink Grass Green	44,173	329,530	2,661,422	1,535	6
AVG PPM	51.36	129.03	3,102.67	2.23	0.03

Sr	Cd	Cs	Hg	Pb	U
78	3	1	0	14	8
7,986	6	216	3	115	15
12,753	9	403	10	242	33
629	41	0	0	42	20
48	11	5	0	77	19
122	10	7	2	77	37
250	7	2	0	53	12
3,770	5	29	1	62	17
1,488	0	3	3	13	9
3.01	**0.01**	**0.07**	**0.00**	**0.08**	**0.02**

Sr	Cd	Cs	Hg	Pb	U
1,423	0	81	3	743	32
1,199	0	0	0	26	44
1,995	18	0	6,078	147	4
19,754	1	0	45	23	35
9,930	0	2	11	55	30
6.86	**0.00**	**0.02**	**1.23**	**0.20**	**0.03**

PETS & PLANTS

Plant Fertilizers

Element	Mg	Al	Cu	Zn	As
Volcanic Mineral	1,511,099	18,392,579	1,781	24,990	2,140
Azomite	1,529,070	3,507,080	1,449	5,896	1,082
Liquid Kelp	39,454	200,623	20,673	17,563	192
Sea Minerals	1,073,328	13,554	110	428	3
Humic Acid	37,248	269,770	1,041	1,080	672
Liquid Seaweed	234,560	2,710	200	3,409	3,560
Iorn Chelate	10,077	15,810	177	227	0
Fruit & Citrus Fertilizer	4,618,486	4,592,898	45,962	279,917	2,288
Glacial Rock Dust	9,521,521	16,651,020	113,971	171,572	11,648
Elemite	1,235,561	8,997,425	4,330	32,255	2,091
Potassium Nitrate 13.7-0-46.3	18,284	40,109	617	2,791	2
Mono Potassium Posphate 0-52-34	7,703	6,846	36	2,082	602
AVG PPM	1,653.03	4,390.87	15.86	45.18	2.02

Pet Treats

Element	Mg	Al	Cu	Zn	As
Gourmet Munchy Rawhide	1,869,052	575,096	2,616	20,992	606
Jerky Naturals	650,141	16,660	5,188	32,129	15
Rawhide	152,989	26,714	767	8,294	29
Beef Jerky Treats	634,354	6,080	4,143	89,253	18
Natural Rawhide Rings	1,685,162	363,786	4,320	17,363	387
Premium Duck Filet	735,848	2,125	8,441	48,834	0
Herring Strips for Dogs	422,120	1,267	3,674	28,147	7,074

Sr	Cd	Cs	Hg	Pb	U
79,975	67	2,410	1	14,487	1,184
45,707	34	1,694	0	6,145	448
2,520	3	14	1	164	57
13,311	2	11	0	60	5
4,179	6	26	6	144	48
14,685	15	4	1	19	46
1,084	3	0	3	25	141
158,149	238	394	12	7,154	4,594
49,507	372	1,006	25	22,056	493
101,921	79	4,657	3	36,167	1,449
812	1	2	0	18	3
2,221	1	0	0	3	33
39.51	**0.07**	**0.85**	**0.00**	**7.20**	**0.71**

Sr	Cd	Cs	Hg	Pb	U
54,603	47	74	136	1,854	341
3,660	52	18	3	58	16
12,248	9	3	0	350	27
8,589	14	8	0	40	5
43,045	41	26	1	1,422	134
242	1	30	2	20	2
2,839	34	4	172	27	8

Pet Treats (Cont.)

Element	Mg	Al	Cu	Zn	As
Organic Blueberry Dog Treats	2,367,632	35,634	5,748	23,625	82
Munchy Stix	1,373,993	103,590	1,546	14,906	311
Sweet Potato Dog Chewz	1,150,498	35,041	9,059	35,947	43
Premium Dog Food	2,001,562	29,091	20,360	219,845	81
Bite Size Dog Food	2,608,516	17,034	14,000	246,602	34
Friskies	155,814	1,125	1,350	20,990	0
Dog Meals	228,276	3,077	5,721	56,559	12
Red Shrimp	3,506,174	93,807	37,582	49,414	45,137
Dried Fish For Cats	2,304,386	5,362	2,172	52,697	5,964
Freeze Dried Ocean Whitefish Cat Treat	3,774,984	1,095	2,648	14,804	3,060
AVG PPM	1,507.15	77.45	7.61	57.67	3.70

Cigarettes

Element	Mg	Al	Cu	Zn	As
Cigarettes / Mellow	3,715,358	125,717	6,867	23,683	226
Cigarettes / Full Bodied Taste	6,401,988	246,938	12,476	42,593	337
Cigarettes / Filtered	5,464,621	257,907	11,804	41,353	294
Cigarettes / Blue	5,225,485	210,734	10,501	43,514	249
Cigarettes / Red	5,226,049	249,931	11,432	33,061	339
Cigarettes / Blue	5,013,729	228,245	10,646	32,283	222
Cigarettes / Filtered	4,825,585	227,508	9,766	35,222	321
Cigarettes / Gold	5,017,451	230,400	9,914	35,867	329

Sr	Cd	Cs	Hg	Pb	U
2,185	33	52	1	95	5
23,992	25	10	2	695	109
5,773	49	12	0	97	6
21,063	34	21	1	102	18
12,034	31	16	1	102	54
1,446	3	6	0	13	1
2,865	25	3	1	12	4
658,885	5,930	55	36	303	36
50,294	911	22	20	89	12
19,180	8	131	506	0	14
54.29	**0.43**	**0.03**	**0.05**	**0.31**	**0.05**

Sr	Cd	Cs	Hg	Pb	U
57,131	1,086	43	16	308	53
70,273	1,432	60	22	466	84
112,501	1,152	106	14	606	43
102,002	998	68	15	415	35
84,328	917	95	13	511	31
122,544	927	76	12	526	34
65,402	935	79	15	496	39
60,727	1,041	85	14	483	35

Cigarettes (Cont.)

Element	Mg	Al	Cu	Zn	As
Cigarettes	5,194,016	225,380	7,440	36,485	238
Cigarettes / Gold	4,859,093	186,413	9,597	38,139	174
Cigarettes / Red	5,989,114	244,375	10,845	41,291	216
Bob Marley Hemp Rolling Papers	351,383	68,870	1,552	2,477	0
300's Raw Hemp Rolling Paper	292,229	14,384	4,063	2,234	15
Top Rolling Papers	559,411	37,676	402	1,218	43
AVG PPM	4,152.54	182.46	8.38	29.24	0.21

FAST FOOD RESTAURANTS

McDonald's

Element	Mg	Al	Cu	Zn	As
Big Mac Bun	529,883	2,361	2,549	13,282	9
Big Mac Lettuce	84,756	1,057	445	2,904	2
Big Mac Patty	228,815	2,255	769	50,373	4
Big Mac Pickles	98,188	60,513	318	2,284	5
Big Mac Sauce	100,874	1,173	550	2,819	3
Big Mac Cheese	235,704	236	236	22,048	2
Fish Filet Bun	214,510	2,271	1,185	6,919	4
Fish Filet Tartar Sauce	52,310	316	452	3,367	21
Fish Filet Breading	186,312	16,452	414	2,342	779
Fish Filet Cheese	255,540	577	257	23,986	16
Fish Filet Patty	354,118	730	131	3,526	1,156
French Fries	312,585	642	1,853	6,510	7
Chicken Nugget Breading	150,028	261,654	1,095	4,063	5
Chicken Nugget Meat	196,320	1,296	265	5,937	0

Sr	Cd	Cs	Hg	Pb	U
84,015	1,032	96	14	439	61
87,423	966	99	12	421	27
92,119	1,089	66	15	472	34
6,651	4	6	7	189	17
11,193	14	2	0	142	36
48,371	22	0	0	76	59
71.76	**0.83**	**0.06**	**0.01**	**0.40**	**0.04**

Sr	Cd	Cs	Hg	Pb	U
9,083	37	7	0	0	1
2,114	12	3	0	0	1
2,164	0	12	0	0	1
19,394	2	7	0	35	1
3,749	9	0	0	6	3
7,058	3	2	0	1	0
7,786	21	4	0	0	1
982	2	0	0	11	0
731	8	10	0	2	0
7,972	0	0	0	1	0
953	0	19	7	0	0
1,559	60	1	0	9	3
589	8	0	0	2	1
136	1	6	0	0	1

McDonald's (Cont.)

Element	Mg	Al	Cu	Zn	As
Apple Pie Crust	121,512	2,304	999	4,907	4
Apple Pie Filling	50,416	1,841	392	863	3
Sweet 'n' Sour Sauce	62,308	1,851	226	801	0
BBQ Ranch Burger Bun	214,105	2,660	1,157	7,950	4
BBQ Ranch Burger Chips	496,532	615	771	10,014	1
BBQ Ranch Burger Cheese	276,893	3,384	3,384	27,336	5
BBQ Ranch Burger Patty	226,835	1,631	735	56,587	7
McDouble Bun	163,973	1,581	870	5,646	3
McDouble Patty	183,885	5,906	729	43,472	7
McDouble Cheese	309,761	3,998	265	29,066	4
McDouble Pickles	182,551	41,585	391	6,427	2
McChicken Bun	178,897	2,004	975	5,817	2
McChicken Chicken Breading	170,554	3,704	780	5,089	0
McChicken Chicken Meat	197,280	0	504	16,276	0
McChicken Mayo	66,953	0	123	1,818	0
McChicken Lettuce	200,627	425	797	4,677	6
AVG PPM	**203.43**	**14.17**	**0.79**	**12.57**	**0.07**

Burger King

Element	Mg	Al	Cu	Zn	As
Whopper Jr. Bun	190,106	2,367	1,047	7,214	8
Whopper Jr. Mayo	11,794	259	39	1,158	0
Whopper Jr. Patty	328,922	0	1,028	79,978	9
Whopper Jr. Onion Ring Breading	150,485	96,106	643	5,275	8

Sr	Cd	Cs	Hg	Pb	U
833	11	4	0	29	0
771	1	2	0	3	1
743	0	0	0	2	2
8,119	20	5	0	0	1
2,914	2	0	0	1	0
7,361	0	3	0	12	1
931	0	9	0	2	1
5,896	16	6	0	0	1
1,901	11	8	0	5	2
9,486	0	5	0	19	1
13,523	12	12	0	40	1
6,467	17	4	0	0	0
966	6	2	0	4	0
161	1	4	0	0	0
425	3	0	0	0	0
4,589	30	5	0	0	0
4.31	**0.01**	**0.00**	**0.00**	**0.01**	**0.00**

Sr	Cd	Cs	Hg	Pb	U
1,717	43	16	0	0	0
142	0	0	0	0	0
313	0	40	0	0	1
1,939	11	0	0	5	2

Burger King (Cont.)

Element	Mg	Al	Cu	Zn	As
Whopper Jr. Onion Ring Onion	204,528	19,787	481	8,771	12
French Fries	410,761	1,436	1,128	6,280	9
Onion Ring Breading	215,450	99,117	910	7,007	14
Onion Ring Onion	227,034	37,884	670	4,127	22
Chicken Nugget Breading	176,770	224,171	610	3,869	3
Chicken Nugget Meat	244,567	17,908	409	9,956	0
Sweet & Sour Sauce	115,711	1,093	351	552	0
Original Chicken Sandwich Mayo	40,440	0	186	1,930	0
Original Chicken Sandwich Bun	276,156	3,053	1,596	9,013	17
Original Chicken Sandwich Breading	168,944	8,124	753	3,687	2
Original Chicken Sandwich Meat	296,422	1,658	540	11,711	4
Whopper Bun	664,762	1,844	3,023	18,297	10
Whopper Tomato	227,593	6,835	801	5,585	0
Whopper Onion	196,026	16,188	353	5,996	0
AVG PPM	230.36	29.88	0.81	10.58	0.01

Jack in the Box

Element	Mg	Al	Cu	Zn	As
Chicken Nugget Breading	235,904	10,748	859	5,294	5
Chicken Nugget Meat	539,721	4,213	1,238	13,148	12
Taco Shell	744,810	1,188	970	16,133	3
Taco Meat	275,865	1,288	1,123	14,298	0

Sr	Cd	Cs	Hg	Pb	U
4,282	12	0	0	12	1
978	200	0	0	0	1
3,394	15	0	0	9	2
4,797	16	0	0	17	2
524	5	0	0	0	4
371	1	5	0	0	2
724	4	5	0	4	0
129	10	0	0	0	1
2,223	19	0	0	1	2
646	5	2	0	3	0
553	4	3	0	2	0
3,226	145	0	0	0	0
1,378	20	7	0	0	0
3,387	62	8	0	2	0
1.71	0.03	0.00	-	0.00	0.00

Sr	Cd	Cs	Hg	Pb	U
743	14	7	0	2	2
1,205	12	8	0	3	4
2,190	0	1	0	5	4
1,680	6	2	0	3	0

Jack in the Box (Cont.)

Element	Mg	Al	Cu	Zn	As
Big Cheeseburger Bun	202,502	3,529	1,010	5,900	6
Big Cheeseburger Patty	272,589	397	929	66,427	5
Chicken Sandwich Bun	158,858	2,993	843	5,209	4
Chicken Sandwich Breading	154,588	3,699	631	4,226	0
Chicken Sandwich Meat	297,132	1,247	494	9,748	4
Jumbo Jack Bun	313,393	3,467	1,605	8,182	3
Jumbo Jack Patty	223,048	379	818	54,435	3
AVG PPM	310.76	3.01	0.96	18.45	0.00

Wendy's

Element	Mg	Al	Cu	Zn	As
French Fries	445,907	2,373	2,511	6,729	8
Chicken Nugget Breading	274,183	141,857	832	6,826	2
Chicken Nugget Meat	288,458	16,260	702	11,666	3
BBQ Sauce	204,027	2,318	962	6,159	0
Chicken Sandwich Breading	138,860	38,925	620	2,163	0
Chicken Sandwich Meat	306,341	46,264	673	7,800	1
Chicken Sandwich Bun	265,154	2,615	1,531	11,307	2
Hamburger Bun	200,011	2,185	1,231	11,071	0
Hamburger Patty	247,256	3,799	1,180	55,504	5
Chicken Wrap Tortilla	206,502	1,911	1,277	6,311	11
Chicken Wrap Cheese	222,384	1,923	443	33,207	1

Sr	Cd	Cs	Hg	Pb	U
7,566	42	2	0	2	2
3,279	4	4	0	0	0
5,536	31	2	0	1	1
697	9	3	0	0	1
436	1	5	0	1	10
6,177	33	2	0	2	1
762	1	2	0	0	1
2.75	0.01	0.00	0	0.00	0.00

Sr	Cd	Cs	Hg	Pb	U
1,514	27	12	0	0	1
831	8	9	0	1	0
214	1	8	0	2	0
2,515	17	8	0	17	0
446	3	6	0	0	1
550	4	6	0	0	3
1,604	17	0	0	0	1
1,329	13	4	0	1	1
918	11	3	0	0	0
1,514	31	2	0	3	5
3,754	4	0	0	1	0

Wendy's (Cont.)

Element	Mg	Al	Cu	Zn	As
Chicken Wrap Breading	195,372	1,478	1,255	4,832	0
Chicken Wrap Meat	252,635	1,168	834	4,708	0
Chili	161,251	2,525	1,187	12,087	0
Baked Potato	329,090	488	393	4,360	0
Baked Potato Sour Cream	86,051	247	54	4,632	0
AVG PPM	238.97	16.65	0.98	11.84	0.00

Dairy Queen

Element	Mg	Al	Cu	Zn	As
Patty Melt Bun	160,082	1,014	857	7,875	0
Patty Melt Cheese	319,294	2,574	386	25,616	0
Chicken Crisp Sandwich Bun	212,114	3,750	1,141	8,703	1
Chicken Crisp Sandwich Breading	182,075	40,509	689	4,072	0
Chicken Crisp Sandwich Meat	204,905	26,487	835	8,972	0
Burger Bun	199,920	1,403	922	7,160	0
Burger Patty	221,337	444	833	60,044	1
Burger Cheese	297,788	1,153	464	22,134	1
AVG PPM	224.69	9.67	0.77	18.07	0.00

Domino's

Element	Mg	Al	Cu	Zn	As
Pizza Cheese	234,963	816	483	30,710	1
Pizza Sausage	173,694	1,255	1,008	35,440	0
Pizza Olive	354,286	3,309	3,262	4,526	0
Pizza Pepperoni	347,491	1,573	1,795	46,696	0

Sr	Cd	Cs	Hg	Pb	U
1,081	7	7	0	2	4
946	7	9	0	2	6
2,947	7	2	0	4	0
1,514	44	2	0	0	0
256	0	0	0	0	0
1.37	**0.01**	**0.00**	**0**	**0.00**	**0.00**

Sr	Cd	Cs	Hg	Pb	U
1,984	30	1	0	0	2
8,339	15	3	0	6	0
3,875	18	2	2	5	5
395	12	4	0	1	1
383	40	3	0	1	0
2,383	36	2	0	1	3
526	0	2	0	0	1
7,643	19	2	0	6	2
3.19	**0.02**	**0.00**	**0.00**	**0.00**	**0.00**

Sr	Cd	Cs	Hg	Pb	U
11,886	1	2	0	7	2
772	3	4	0	6	1
32,256	24	3	0	36	3
1,900	4	19	0	5	1

Domino's (Cont.)

Element	Mg	Al	Cu	Zn	As
Pizza Mushroom	328,002	17,046	10,530	20,322	34
Pizza Green Pepper	154,269	1,363	1,014	2,073	0
Pizza Onion	124,761	1,767	585	4,485	0
Pizza Ham	166,241	1,438	966	21,088	0
Pizza Crust	246,146	3,891	1,335	7,043	0
Pizza Sauce	276,444	4,668	1,660	7,723	0
Pizza Parmesan Cheese	62,480	169,159	173	6,552	0
Pizza Crushed Red Pepper	1,371,363	9,091	6,859	14,673	0
AVG PPM	320.01	17.95	2.47	16.78	0.00

KFC

Element	Mg	Al	Cu	Zn	As
Gravy	113,035	1,530	586	6,379	1
Mashed Potatoes	128,524	1,505	737	1,766	0
Chicken Breading	205,994	2,999	737	3,676	0
Chicken	215,755	410	491	11,106	3
Chicken Sandwich Bun	211,418	2,336	1,217	9,862	1
Chicken Sandwich Cheese	196,528	3,884	165	23,735	0
Chicken Sandwich Bacon	215,824	6,086	1,307	47,298	0
Chicken Sandwich Breading	133,362	1,995	601	2,038	0
Chicken Sandwich Meat	285,056	691	600	13,213	0
AVG PPM	189.50	2.38	0.72	13.23	0.00

Sr	Cd	Cs	Hg	Pb	U
1,089	14	6	2	12	0
653	24	5	0	0	0
914	12	6	0	6	0
1,762	16	13	0	12	5
721	15	1	0	1	0
3,210	22	26	0	15	0
1,685	6	0	0	25	1
3,493	29	34	0	38	0
5.03	**0.01**	**0.01**	**0.00**	**0.01**	**0.00**

Sr	Cd	Cs	Hg	Pb	U
486	4	1	0	1	2
954	18	1	0	1	2
1,145	6	15	0	4	4
207	0	21	0	1	4
1,037	15	2	0	2	3
2,613	0	4	0	2	1
398	3	5	0	3	1
428	0	6	0	1	1
343	12	8	0	2	2
0.85	**0.01**	**0.01**	**0**	**0.00**	**0.00**

Taco Bell

Element	Mg	Al	Cu	Zn	As
Taco Shell	459,104	1,213	993	12,815	0
Taco Meat	297,035	1,956	867	29,735	7
Beefy Five Layer Burrito Tortilla	167,612	2,277	762	4,813	0
Beefy Five Layer Burrito Beans	396,201	994	2,725	10,778	0
Chicken Burrito Rice	133,708	3,183	635	2,383	25
Chicken Burrito Tortilla	177,245	2,479	895	4,912	13
Chicken Burrito Meat	232,666	1,317	1,187	13,284	28
Soft Taco Tortilla	173,934	9,269	809	5,884	3
Soft Taco Meat	241,457	3,971	875	23,317	12
Doritos Taco Shell	584,135	435,036	986	18,899	0
AVG PPM	228.92	16.85	0.97	12.82	0.02

Sr	Cd	Cs	Hg	Pb	U
1,561	15	4	0	1	0
1,291	8	6	0	9	7
1,275	19	2	0	2	2
1,960	6	2	0	2	1
788	8	7	0	5	5
1,564	22	3	0	3	2
915	20	3	0	2	1
1,696	28	3	0	4	2
1,367	34	5	0	5	6
2,284	15	3	0	20	1
2.65	0.01	0.00	0.00	0.00	0.00

ABOUT THE AUTHOR

Mike Adams—a.k.a., the "Health Ranger"—gained fame as an outspoken clean-food advocate and critic of the over-drugging of America with toxic pharmaceuticals. As an award-winning investigative journalist, he pursued a path of discovery into food ingredients, composition, and contamination, ultimately transitioning to a food scientist with a world-class analytical laboratory he built from the ground up.

Today, Adams is a member of the Association of Analytical Communities and is the lab science director of an ISO-accredited analytical laboratory conducting commercial food testing for heavy metals, pesticides, herbicides, and other chemical contaminants. In pursuing the science of food safety, Adams has become one of the leading food research scientists in the industry, helping validate scientific methodologies for glyphosate detection while revealing startling details about food contamination sources such as dental offices (for mercury), "biosolids" human waste compost, and industrial heavy metals pollution that now contaminates organic food products grown in China.

Over the last three years, Adams has become a recognized expert in running ICP-MS instrumentation and routinely publishes online videos to help other scientists troubleshoot problems and obstacles with ICP-MS methodologies. Adams is also fully versed in LC/MS instrumentation and currently runs a Time-of-Flight mass spec instrument made by Agilent.

Adams is also the inventor of the Food Rising Mini-Farm Grow System (FoodRising.org) and holds two pending patents on dietary formulations for eliminating heavy metals and radioactive Cesium-137. He has conducted extensive research into dietary defense strategies to help protect humanity from radiation in water, lead in water, and mercury in foods. All of his research has been privately funded, using no funds whatsoever from the NIH, government grants, or academic sources.

His commercial laboratory is described at CWClabs.com. Adams also serves as the editor of NaturalNews.com, a popular natural health website that Adams has taken in the direction of food science and clean-food activism. Learn more about Adams at HealthRanger.com.

ACKNOWLEDGMENTS

Producing a book like this one is no small feat, and it's never really the effort of just one person. In one way or another, over a dozen people took part in this project. Those who deserve special recognition include Reno, Leah, Brad, Julie, Aaron, Michael, Heather, Glenn, and all the good people at BenBella who had the patience to give me the time to complete this book nearly two years later than originally planned.

Also deserving credit is Maryfrances, who first taught me how to write, and both of my parents, who taught me how to love nature while thinking critically about the world around me. Without their nurturing and support, I never would have pursued this path of discovery about food, science, and nature. This book is dedicated to my father, who passed away during its writing. He was the person who put me on the technology path and had me writing original computer code in 1979 on the very first Apple II personal computer (today I run an analytical laboratory full of multimillion-dollar Agilent instrumentation).

Finally, I acknowledge the mysterious, divine forces that brought matter, energy, life, and consciousness into existence in our universe, creating an exciting and rich adventure of discovery for us all. Even with our most advanced scientific instruments and super colliders, we have only barely begun to unravel the most mundane secrets of nature and the world around us. Perhaps in another thousand years—if we don't destroy ourselves first—we might actually be worthy of calling ourselves an "intelligent" species.

ENDNOTES

Introduction

1. U.S. Environmental Protection Agency. Drinking water contaminants—standards and regulations. Updated January 6, 2016. http://water.epa.gov/drink/contaminants/basicinformation/mercury .cfm; World Health Organization. Mercury and health. Updated January 2016. http://www.who .int/mediacentre/factsheets/fs361/en

2. Knott L. Tungsten poisoning. Patient. 2014. http://patient.info/doctor/tungsten-poisoning; Jaslow R. Exposure to tungsten metals found in phones, computers linked to strokes. CBS News. November 13, 2013. http://www.cbsnews.com/news/tungsten-metals-found-in-phones-computers-linked -to-strokes

3. Peeples L. Lead poisoning threshold lowered by CDC, five times more children now considered at risk. The Huffington Post. May 16, 2012. http://www.huffingtonpost.com/2012/05/16/lead -poisoning-cdc_n_1522448.html

4. World Health Organization. Arsenic. December 2012. http://www.who.int/mediacentre/ factsheets/fs372/en/; U.S. Environmental Protection Agency. Chemical contaminant rules. November 9, 2015. http://water.epa.gov/lawsregs/rulesregs/sdwa/arsenic/index.cfm

5. U.S. Environmental Protection Agency. Cadmium compounds (A). September 9, 2015. http:// www.epa.gov/ttnatw01/hlthef/cadmium.html; Agency for Toxic Substances and Disease Registry. Cadmium toxicity: what health effects are associated with acute high-dose cadmium exposure? May 12, 2008. http://www.atsdr.cdc.gov/csem/csem.asp?csem=6&po=11

6. Wong E. China to reward cities and regions making progress on air pollution. The New York Times. February 13, 2014. http://www.nytimes.com/2014/02/14/world/asia/china-to-reward -localities-for-improving-air-quality.html

7. Julshamn K, Maage A, Noril HS, et. al. Determination of arsenic, cadmium, mercury, and lead in foods by pressure digestion and inductively coupled plasma/mass spectrometry: first action 2013.06. 2013. National Center for Biotechnology Information. http://www.ncbi.nlm.nih.gov/ pubmed/24282954

Part 1: Everything You Need to Know About Toxic Elements

8. U.S. Center for Disease Control and Prevention. Fourth national report on human exposure to environmental chemicals. 2009. Updated 2013. http://www.cdc.gov/exposurereport/pdf/ FourthReport_UpdatedTables_Sep2013.pdf#page=1&zoom=auto,0,800

9. Salnikow K & Zhitkovich A. Genetic and epigenetic mechanisms in metal carcinogenesis and co-carcinogenesis: nickel, arsenic and chromium. Chemical Research in Toxicology. 2008. 21(1):28–44. PMCID:PMC2602826. PubMed. http://www.ncbi.nlm.nih.gov/pmc/articles/PMC2602826/

10. Wright RO, Schwartz J, Bacharelli A, et al. Biomarkers of lead exposure and DNA methyl-

ation within retrotransposons. Environmental Health Perspectives. 2010. 118(6):790–795. PMID:20064768. PubMed. http://www.ncbi.nlm.nih.gov/pubmed/20064768

11. Pilsner JR, Hu H, Ettinger A, et al. Influence of prenatal lead exposure on genomic methylation of cord blood DNA. Environmental Health Perspectives. 2009. 117(9):1466–71. PMID:19750115. PubMed. http://www.ncbi.nlm.nih.gov/pubmed/19750115

12. Kile ML, Baccarelli A, Hoffman E, et al. Prenatal arsenic exposure and DNA methylation in maternal and umbilical cord blood leukocytes. Environmental Health Perspectives. 2012. 120(7):1061–6. PMID:22466225. PubMed. http://www.ncbi.nlm.nih.gov/pubmed/22466225

13. Reichard J, Schnekenburger M, & Puga A. Long-term low-dose arsenic exposure induces loss of DNA methylation. Biochemical and Biophysical Research Communications. 2007. 352(1):188–92. PMID: 17107663. PubMed. http://www.ncbi.nlm.nih.gov/pubmed/17107663

14. Ibid.

15. Kalia K & Flora SJ. Strategies for safe and effective therapeutic measures for chronic arsenic and lead poisoning. Journal of Occupational Health. 2005. 47(1):1–21. PMID: 15703449. PubMed. http://www.ncbi.nlm.nih.gov/pubmed/15703449

16. George GN, Prince RC, Gailer J, et al. Mercury binding to the chelation therapy agents DMSA and DMPS and the rational design of custom chelators for mercury. Chemical Research in Toxicology. 2004. 17(8):999–1006. PMID:15310232. PubMed. http://www.ncbi.nlm.nih.gov/pubmed/15310232

17. Mercola J & Klinghardt D. Mercury toxicity and systemic elimination agents. Journal of Nutritional and Environmental Medicine. 2001. 11(1):53–62. doi:10.1080/13590840020030267

18. Kaplan D, Christiaen D, & Arad S. Chelating properties of extracellular polysaccharides from chlorella spp. Applied and Environmental Microbiology. 1987. 53(12):2953–2956. PMCID: PMC204228. PubMed. http://www.ncbi.nlm.nih.gov/pmc/articles/PMC204228

19. Arslanoglu H, Altundogan HS, & Tumen F. Heavy metals binding properties of esterified lemon. Journal of Hazardous Materials. 2009. 164(2–3):1406–13. PMID:18980807. PubMed. http://www.ncbi.nlm.nih.gov/pubmed/18980807

20. Xavier LL, Murphy R, Thandapilly SJ, et al. Garlic extracts prevent oxidative stress, hypertrophy and apoptosis in cardiomyocytes: a role for nitric oxide and hydrogen sulfide. BMC Complementary and Alternative Medicine. 2012. 12:140. PMCID: PMC3519616. PubMed. http://www.ncbi.nlm.nih.gov/pmc/articles/PMC3519616

21. Gadd GM & Griffiths AJ. Microorganisms and heavy metal toxicity. Microbial Ecology. 1977. 4(4):303–317. Springer Link. http://link.springer.com/article/10.1007%2FBF02013274

22. Thirumavalavan M, Lai YL, Lin LC, et al. Cellulose-based native and surface modified fruit peels for the adsorption of heavy metal ions from aqueous solution: langmuir adsorption isotherms. Journal of Chemical & Engineering Data. 2010. 55(3):1186–1192. doi:10.1021/je900585t. ACS Publications. http://pubs.acs.org/doi/abs/10.1021/je900585t

23. Sears ME, Kerr KJ, & Bray RI. Arsenic, cadmium, lead, and mercury in sweat: a systematic review. Journal of Environmental and Public Health. 2012. Article ID:184745. doi:10.1155/2012/184745. Hindawi. http://dx.doi.org/10.1155/2012/184745

24. Genuis S, Birkholz D, Rodushkin I, et al. Blood, urine, and sweat (BUS) study: monitoring and elimination of bioaccumulated toxic elements. Archives of Environmental Contamination and Toxicology. 2011. 61(2):344–357. doi:10.1007/s00244-010-9611-5. Springer Link. http://link.springer.com/article/10.1007/s00244-010-9611-5

25. University of South Carolina. Obese Americans get less than one minute of vigorous activity per day, research shows. Newswise. February 12, 2014. http://www.newswise.com/articles/obese-americans-get-less-than-one-minute-of-vigorous-activity-per-day-research-shows

Heavy Metals: Arsenic
26. Schuhmacher-Wolz U, Dieter HH, Klein D, et al. Oral exposure to inorganic arsenic: evaluation of its carcinogenic and non-carcinogenic effects. Critical Reviews in Toxicology. 2009. PMID:19235533. Science.NaturalNews. http://science.naturalnews.com/pubmed/19235533.html

27. Arsenic and inorganic arsenic compounds CAS No. 7440-38-2 (Arsenic). Report on Carcinogens, 13th edition. 2011. National Toxicology Program, Department of Health and Human Services. PubMed. http://ntp.niehs.nih.gov/ntp/roc/content/profiles/arsenic.pdf

28. Ravenscroft P. Predicting the global distribution of arsenic pollution in groundwater. Royal Geographical Society Annual International Conference. 2007. Department of Geography, University of Cambridge. http://www.geog.cam.ac.uk/research/projects/arsenic/symposium/S1.2_P_ Ravenscroft.pdf

29. Chowdhury UK, Biswas BK, Chowdhury TR, et al. Groundwater arsenic contamination in Bangladesh and West Bengal, India. Environmental Health Perspectives. 2000. 108(5):393–397. PMCID:PMC1638054. PubMed. http://www.ncbi.nlm.nih.gov/pmc/articles/PMC1638054

30. Khan MMH, Sakauchi F, Sonoda T, et al. Magnitude of arsenic toxicity in tube-well drinking water in Bangladesh and its adverse effects on human health including cancer: evidence from a review of the literature. Asian Pacific Journal of Cancer Prevention. 2003. 4(1):7–14. PMID:12718695. PubMed. http://www.ncbi.nlm.nih.gov/pubmed/12718695

31. Heck J, Chen Y, Grann V, et al. Arsenic exposure and anemia in Bangladesh: a population-based study. Journal of Occupational and Environmental Medicine. 2008. PMID:18188085. Science. NaturalNews. http://science.naturalnews.com/pubmed/18188085.html

32. Brammer H & Ravenscroft P. Arsenic in groundwater: a threat to sustainable agriculture in south and south-east Asia. Environment International. 2009. PMID:19110310. Science.NaturalNews. http://science.naturalnews.com/pubmed/19110310.html

33. Reichard J, Schnekenburger M, & Puga A. Long term low-dose arsenic exposure induces loss of DNA methylation. Biochemical and Biophysical Research Communications. 2007. 352(1):188–92. PMID:17107663. PubMed. http://www.ncbi.nlm.nih.gov/pubmed/17107663

34. Hood E. The apple bites back: claiming old orchards for residential development. Environmental Health Perspectives. 2006. 114(8): A470–A476. PMCID:PMC1551991. PubMed. http://www .ncbi.nlm.nih.gov/pmc/articles/PMC1551991

35. Tracy D & Baker B. Heavy metals in fertilizers used in organic production. Organic Materials Review Institute. 2005. http://web.archive.org/web/20090307203634/http://www.omri.org/ AdvisoryCouncil/Metals_in_Fertilizers-b6-2005-02-14.pdf

36. U.S. Environmental Protection Agency. Arsenical pesticides, man and the environment. Office of Pesticides Programs. 1972. http://nepis.epa.gov/Exe/ZyPURL.cgi?Dockey=9100CC80.txt

37. U.S. Environmental Protection Agency. EPA proposing to revoke tolerances for pesticides calcium arsenate and lead arsenate. Pesticide Management Education Program, Cornell University. 1988. http://pmep.cce.cornell.edu/profiles/insect-mite/fenitrothion-methylpara/lead-arsenate/prof -calcium-ars-revoke.html

38. U.S. Environmental Protection Agency. Arsenal pesticides, man and the environment. Office of pesticides programs. 1972. http://nepis.epa.gov/Exe/ZyPURL.cgi?Dockey=9100CC80.txt

39. Qi Y & Donahoe R. The environmental fate of arsenic in surface soil contaminated by historical herbicide application. Science of the Total Environment. 2008. PMID:18706676. Science.NaturalNews. http://science.naturalnews.com/pubmed/18706676.html

40. Peryea F. Historical use of lead arsenate insecticides, resulting soil contamination and implications for soil remediation. Presented at the 16th World Congress of Soil Science in Montpellier, France. August 20–26, 1998. http://soils.tfrec.wsu.edu/leadhistory.htm

41. Hood E. The apple bites back: claiming old orchards for residential development. Environmental Health Perspectives. 2006. 114(8): A470–A476. PMCID:PMC1551991. PubMed. http://www .ncbi.nlm.nih.gov/pmc/articles/PMC1551991

42. Newton K, Amarasiriwardena D, & Xing B. Distribution of soil arsenic species, lead and arsenic bound to humic acid molar mass fractions in a contaminated apple orchard. Environmental Pollution. 2006. PMID:16480799. PubMed. http://science.naturalnews.com/pubmed/16480799.html

43. Su C, Lin Y, & Chang T. Incidence of oral cancer in relation to nickel and arsenic concentrations in farm soils of patients' residential areas in Taiwan. BMC Public Health. 2010. PMID:20152030. Science.NaturalNews. http://science.naturalnews.com/pubmed/20152030.html

44. Zagury G, Dobran S, Estrela S, et al. Inorganic arsenic speciation in soil and groundwater near in service chromated copper arsenate treated wood poles. Environmental Toxicology and Chemistry. 2008. PMID:18333683. Science.NaturalNews. http://science.naturalnews.com/pubmed/18333683.html

45. U.S. Environmental Protection Agency. Chromated copper arsenate (CCA): evaluating the wood preservative chromated copper arsenate (CCA). Pesticides: regulating pesticides. 2002. http://web.archive.org/web/20150905235743/http://www.epa.gov/oppad001/reregistration/cca/cca_evaluating.htm

46. Zagury G, Dobran S, Estrela S, et al. Inorganic arsenic speciation in soil and groundwater near in-service chromated copper arsenate treated wood poles. Environmental Toxicology and Chemistry. 2008. PMID:18333683. Science.NaturalNews. http://science.naturalnews.com/pubmed/18333683.html

47. Balasubramaniam CS. Product recall: the concept, procedures and emerging challenges. Abhinav: International Monthly Refereed Journal of Research in Management & Technology, volume II. 2013. ISSN 2320-0073.

48. U.S. Center for Disease Control and Prevention. Arsenic toxicity: what is the biologic fate of arsenic in the body? Agency for Toxic Substances and Disease Registry. 2009. http://www.atsdr.cdc.gov/csem/csem.asp?csem=1&po=9

49. The Dr. Oz Show. Arsenic in apple juice. September 9, 2011. http://www.doctoroz.com/videos/arsenic-apple-juice

50. Consumer Reports. Debate grows over arsenic in apple juice. September 14, 2011. http://www.consumerreports.org/cro/news/2011/09/debate-grows-over-arsenic-in-apple-juice/index.htm

51. U.S. Food and Drug Administration. Questions & answers: apple juice and arsenic. July 2013. http://www.fda.gov/Food/ResourcesForYou/Consumers/ucm271595.htm

52. European Food Safety Authority Panel on Contaminants in the Food Chain (CONTAM). Scientific opinion on arsenic in food. EFSA Journal. 2009. 7(10):1351. doi:10.2903/j.efsa.2009.1351

53. Nachman KE, Baron PA, Raber G, et al. Arsenic levels in chicken: Nachman et al. respond. Environmental Health Perspectives. 2013. 121(9):a267–a268. PMCID:PMC3764093. PubMed. http://www.ncbi.nlm.nih.gov/pmc/articles/PMC3764093

54. European Food Safety Authority Panel on Contaminants in the Food Chain (CONTAM). Scientific opinion on arsenic in food. EFSA Journal. 2009. 7(10):1351. doi:10.2903/j.efsa.2009.1351

55. Lasky T, Sun W, Kadry A, et al. Mean total arsenic concentrations in chicken 1989–2000 and estimated exposures for consumers of chicken. Environmental Health Perspectives. 2004. 112(1):18–21. PMCID:PMC1241791. PubMed. http://www.ncbi.nlm.nih.gov/pmc/articles/PMC1241791/pdf/ehp0112-000018.pdf

56. U.S. Food and Drug Administration. Questions and answers regarding 3-nitro (roxarsone). Updated April 1, 2015. http://www.fda.gov/AnimalVeterinary/SafetyHealth/ProductSafetyInformation/ucm258313.htm

57. Lasky T, Sun W, Kadry A, et al. Mean total arsenic concentrations in chicken 1989–2000 and estimated exposures for consumers of chicken. Environmental Health Perspectives. 2004. 112(1):18–21. PMCID:PMC1241791. PubMed. http://www.ncbi.nlm.nih.gov/pmc/articles/PMC1241791/pdf/ehp0112-000018.pdf

58. U.S. Food and Drug Administration. Questions and answers regarding 3-Nitro (roxarsone). April 1, 2015. http://www.fda.gov/AnimalVeterinary/SafetyHealth/ProductSafetyInformation/ucm258313.htm#uses_of_arsenic-based_animal_drugs

59. Nachman KE, Baron PA, Raber G, et al. Roxarsone, inorganic arsenic, and other arsenic species in chicken: a U.S.-based market basket sample. Environmental Health Perspectives. 2013. Johns Hopkins Center for a Livable Future. http://www.jhsph.edu/research/centers-and-institutes/johns-hopkins-center-for-a-livable-future/research/clf_publications/pub_rep_desc/arsenic_chicken.html

60. Nachman KE, Baron PA, Raber G, et al. Arsenic levels in chicken: Nachman et al. respond. Environmental Health Perspectives. 2013. 121(9):a267–a268. PMCID:PMC3764093. PubMed. http://www.ncbi.nlm.nih.gov/pmc/articles/PMC3764093

61. U.S. Food and Drug Administration. Questions and answers regarding 3-nitro (roxarsone). Updated April 1, 2015. http://www.fda.gov/AnimalVeterinary/SafetyHealth/ProductSafetyInformation/ucm258313.htm

62. Li D, An D, Zhou Y, et al. Current status and prevention strategy for coal arsenic poisoning in Guizhou, China. Journal of Health, Population and Nutrition. September 2006. Science.NaturalNews. http://science.naturalnews.com/pubmed/17366768.html

63. U.S. Environmental Protection Agency. Regulatory actions. Final mercury and air toxics standards (MATS) for power plants. December 2, 2015. http://www.epa.gov/mats/actions.html

64. Hughes MF. Arsenic toxicity and potential mechanisms of action. Toxicology Letters. 2002. 133(1):1–16. PMID:12076506. PubMed. http://www.ncbi.nlm.nih.gov/pubmed/12076506

65. Schiller CM, Fowler BA, & Woods JS. Effects of arsenic on pyruvate dehydrogenase activation. Environmental Health Perspectives. 1977. 19:205–207. PMCID:PMC1637432. PubMed. http://www.ncbi.nlm.nih.gov/pmc/articles/PMC1637432

66. U.S. National Library of Medicine. Pyruvate dehydrogenase deficiency. Genetics Home Reference. Reviewed July 2012. http://ghr.nlm.nih.gov/condition/pyruvate-dehydrogenase-deficiency

67. Flora SJ, Bhadauria S, Kannan GM, et al. Arsenic induced oxidative stress and the role of antioxidant supplementation during chelation: a review. Journal of Environmental Biology. 2007. 28(2 Suppl):333–47. PMID:17929749. PubMed. http://www.ncbi.nlm.nih.gov/pubmed/17929749

68. Levander OA. Metabolic interrelationships between arsenic and selenium. Environmental Health Perspectives. 1977. 19:159–164. PMCID:PMC1637401. PubMed. http://www.ncbi.nlm.nih.gov/pmc/articles/PMC1637401/

69. Spallholz, JE, La Porte PF, & Ahmed S. Selenium in the treatment of arsenic toxicity and cancers. ClinicalTrials.gov Identifier: NCT01442727. http://clinicaltrials.gov/show/NCT01442727

70. Choudhury AR, Das T, Sharma A, et al. Dietary garlic extract in modifying clastogenic effects of inorganic arsenic in mice: two-generation studies. Mutation Research/Environmental Mutagenesis and Related Subjects. 1996. 359(3):165–170. Science.NaturalNews. http://science.naturalnews.com/pubmed/8618548.html

71. Choudhury AR, Das T, & Sharma A. Mustard oil and garlic extract as inhibitors of sodium arsenite-induced chromosomal breaks in vivo. Cancer Letters. 1997. 121(1):45–52. Science Direct. http://www.sciencedirect.com/science/article/pii/S0304383597003340

72. Flora SJS, Mehta A, & Gupta R. Prevention of arsenic-induced hepatic apoptosis by concomitant administration of garlic extracts in mice. Chemico-Biological Interactions. 2009. 177(3):227–233. Science Direct. http://www.sciencedirect.com/science/article/pii/S0009279708004791

Heavy Metals: Mercury

73. Moskowitz C. The secret tomb of China's 1st emperor: will we ever see inside? August 17, 2012. Live Science. http://www.livescience.com/22454-ancient-chinese-tomb-terracotta-warriors.html

74. Norn S, Permin H, Kruse E, et al. Mercury—a major agent in the history of medicine and alchemy. Dansk Medicinhistorisk Arbog. 2008. 36:21–40. PMID:19831290. PubMed. http://www.ncbi.nlm.nih.gov/pubmed/19831290

75. Kitzmiller KJ. Science connections: the not-so-mad hatter: occupational hazards of mercury. American Chemical Society. Chemical Abstracts Service. July 21, 2010. http://www.cas.org/news/insights/science-connections/mad-hatter

76. Kark RAP, Poskanzer DC, Bullock JD, et al. Mercury poisoning and its treatment with N-Acetyl-D,L-Penicillamine. The New England Journal of Medicine. 1971. 285:10–16 doi:10.1056/NEJM197107012850102. http://www.nejm.org/doi/pdf/10.1056/NEJM197107012850102

77. Stemp-Morlock G. Mercury: cleanup for broken CFLs. Environmental Health Perspectives. 2008. 116(9): A378. PMCID:PMC2535642. PubMed. http://www.ncbi.nlm.nih.gov/pmc/articles/PMC2535642/

78. Mironava T, Hadjiargyrou M, Simon M, et al. The effects of UV emission from compact fluorescent light exposure on human dermal fibroblasts and keratinocytes in vitro. Photochemistry and Photobiology. 2012. 88(6):1497–1506. doi:10.1111/j.1751–1097.2012.01192.x. Wiley Online Library. http://onlinelibrary.wiley.com/doi/10.1111/j.1751-1097.2012.01192.x/abstract

79. Dooley D. Letter to state of California health and human services agency. October 9, 2015. http://www.cdph.ca.gov/programs/immunize/Documents/Secretary%20Dooley%20Notification%20Letter%20to%20Legislature.pdf

80. Adams, Mike. California begins injecting children with mercury…flu shot 'shortage' cited as bizarre justification…state safety laws nullified to push vaccines. Natural News, November 2015. http://www.naturalnews.com/052142_flu_shots_mercury_in_vaccines_California_child_safety_laws.html#

81. Natural Resources Defense Council. Mercury contamination in fish: a guide to staying healthy and fighting back. Updated 2014. http://www.nrdc.org/health/effects/mercury/sources.asp

82. U.S. Environmental Protection Agency. 40 CFR parts 60 and 63: national emission standards for hazardous air pollutants from coal- and oil-fired electric utility steam generating units and standards of performance for fossil-fuel-fired electric utility, industrial-commercial-institutional, and small industrial-commercial-institutional steam generating units; final rule. February 16, 2012. U.S. Government Printing Office. http://www.gpo.gov/fdsys/pkg/FR-2012-02-16/pdf/2012-806.pdf

83. University of Michigan. Fires fuel mercury emissions, University of Michigan study finds. January 10, 2007. http://ns.umich.edu/new/releases/3089

84. Great Lakes Protection Fund, Great Lakes Pollution Prevention Centre & Western Lake Superior Sanitary District. Blueprint for mercury elimination: mercury reduction project guidance for wastewater treatment plants. Revised January 2002. http://wlssd.com/wp-content/uploads/2014/12/Revised-Blueprint-for-Mercu.pdf

85. U.S. Environmental Protection Agency. Summary review of health effects associated with mercuric chloride: health issue assessment. EPA/600/R-92/199. Office of Health and Environmental Assessment, Washington, DC. 1994.

86. Ibid.

87. Consumer Wellness Center Forensic Food Lab. Proteins-rice heavy metals rating according to lowheavymetalsverified.org. Accessed February 22, 2015. http://labs.naturalnews.com/heavy-metals-chart-Proteins-rice.html

88. RIA Novosti. 200 tons of banned pesticides containing mercury found near central Russian village. May 31, 2011. Sputnik News. http://en.ria.ru/russia/20110531/164348627.html

89. U.S. Environmental Protection Agency. Summary review of health effects associated with mercuric chloride: health issue assessment. EPA/600/R-92/199. Office of Health and Environmental Assessment, Washington, DC. 1994.

90. U.S. Food and Drug Administration. Appendix I: summary of changes to the classification of dental amalgam and mercury. 2009. http://www.fda.gov/MedicalDevices/ProductsandMedicalProcedures/DentalProducts/DentalAmalgam/ucm171120.htm

91. American Dental Association. Statement on dental amalgam. ADA Council on Scientific Affairs. Updated August 25, 2009. http://www.ada.org/1741.aspx

92. Ngim CH, Foo SC, Boey KW, et al. Chronic neurobehavioural effects of elemental mercury in dentists. British Journal of Industrial Medicine. 1992. 49(11):782–790. PMCID:PMC1039326. PubMed. http://www.ncbi.nlm.nih.gov/pmc/articles/PMC1039326

93. Sikorski R, Juszkiewicz T, Paszkowski T, et al. Women in dental surgeries: reproductive hazards in occupational exposure to metallic mercury. International Archives of Occupational and Environmental Health. 1987. 59(6):551–7. PMID: 3679554. PubMed. http://www.ncbi.nlm.nih.gov/pubmed/3679554

94. Stoner GE, Senti SE, & Gileadi E. Effect of sodium fluoride and stannous fluoride on the rate of corrosion of dental amalgams. Journal of Dental Research. 1979. 58:576–583. Sage Journals. http://jdr.sagepub.com/content/50/6/1647.short

95. Geier DA, Kern JK, & Geier MR. A prospective study of prenatal mercury exposure from maternal dental amalgams and autism severity. Acta Neurobiologiae Experimentalis. 2009. 69(2):189–97. PMID:19593333. Science.NaturalNews. http://science.naturalnews.com/pubmed/19593333.html

96. Mortazavi SM, Daiee E, Yazdi A, et al. Mercury release from dental amalgam restorations after magnetic resonance imaging and following mobile phone use. Pakistan Journal of Biological

Sciences. 2008. 11(8):1142–6. PMID:18819554. Science.NaturalNews. http://science.naturalnews .com/pubmed/18819554.html

97. Yoshida M, Kishimoto T, Yamamura Y, Tabuse M, Akama Y, & Satoh H. Amount of mercury from dental amalgam filling released into the atmosphere by cremation. Japanese Journal of Public Health. 1994. 41(7):618–24. PMID: 7919469. PubMed. http://www.ncbi.nlm.nih.gov/pubmed/7919469

98. European Union. Consolidated version of cosmetics directive 76/768/EEC. Revised 2010. European Union Health and Consumers. http://eur-lex.europa.eu/LexUriServ/LexUriServ.do?uri= CONSLEG:1976L0768:20100301:en:PDF

99. RIA Novosti. EU bans mercury exports from 2011. September 26, 2008. Sputnik News. http:// en.ria.ru/world/20080926/117137589.html

100. U.S. Food and Drug Administration. Prohibited & restricted ingredients. Updated January 26, 2015. http://www.fda.gov/cosmetics/guidanceregulation/ucm127406.htm

101. The Associated Press. Mercury in mascara? Minn. law bans it. 2007. http://web.archive.org/ web/20071223182712/http://news.yahoo.com/s/ap/20071214/ap_on_he_me/mercury_in _mascara

102. Agency for Toxic Substances and Disease Registry (ATSDR). Toxicological profile for mercury: health effects. Public Health Service, U.S. Department of Health and Human Services. 1999. http://www.atsdr.cdc.gov/toxprofiles/tp46-c2.pdf

103. U.S. Environmental Protection Agency. Mercury study report to Congress, vol. 7. Environmental Protection Agency, Washington D.C., USA, 1997. http://www3.epa.gov/airtoxics/112nmerc /volume7.pdf

104. Alissa EM & Ferns GA. Heavy metal poisoning and cardiovascular disease. Journal of Toxicology. 2011. Article ID 870125. Hindawi. http://dx.doi.org/10.1155/2011/870125

105. Burros M. High mercury levels are found in tuna sushi. The New York Times. January 23, 2008. http://www.nytimes.com/2008/01/23/dining/23sushi.html

106. Dorea JG, Farina M, & Rocha JB. Toxicity of ethylmercury (and thimerosal): a comparison with methylmercury. Journal of Applied Toxicology. 2013. 33(8):700–11. doi:10.1002/jat.2855. http:// www.ncbi.nlm.nih.gov/pubmed/23401210

107. Sharpe M, Livingston A, & Baskin D. Thimerosal-derived ethylmercury is a mitochondrial toxin in human astrocytes: possible role of Fenton chemistry in the oxidation and breakage of mtDNA. Journal of Toxicology. 2012. ID: 373678. 12 pages. doi:10.1155/2012/373678. http://www .hindawi.com/journals/jt/2012/373678

108. Waste & Materials Management Program. Managing excess vaccines. Publication WA 841. Revised March 2014. Wisconsin Department of Natural Resources. http://dnr.wi.gov/files/PDF/pubs/wa/ wa841.pdf

109. EWG's Skin Deep Cosmetics Database. Thimerosal. Environmental Working Group. Accessed February 22, 2016. http://www.ewg.org/skindeep/ingredient/706528/THIMEROSAL/#

110. Magos L, Brown AW, Sparrow S, et al. The comparative toxicology of ethyl- and methylmercury. Archives of Toxicology. 1985. 57(4):260–7. PMID:4091651. PubMed. http://www.ncbi.nlm.nih .gov/pubmed/4091651

111. Barrett J. Thimerosal and animal brains: new data for assessing human ethylmercury risk. Environmental Health Perspectives. 2005. 113(8):A543–A544. PMCID:PMC1280369. PubMed. http:// www.ncbi.nlm.nih.gov/pmc/articles/PMC1280369

112. Bernard S, Enayati A, Redwood L, et al. Autism: a novel form of mercury poisoning. Medical Hypotheses. 2001. 56(4):462–471. PMID:11339848. PubMed. http://www.ncbi.nlm.nih.gov/ pubmed/11339848

113. Muth MK, Karns SA, Nielsen SJ, et al. Consumer-level food loss estimates and their use in the ERS loss-adjusted food availability data. Technical bulletin no. (TB-1927). January 2011. U.S. Department of Agriculture. http://www.ers.usda.gov/Publications/TB1927/TB1927.pdf

114. Dufault R, LeBlanc B, Schnoll R, et al. Mercury from chlor-alkali plants: measured concentrations in food product sugar. Environmental Health. 2009. 8:2 doi:10.1186/1476-069X-8-2. Science. NaturalNews. http://science.naturalnews.com/pubmed/19171026.html

115. Sorensen J, Mottl P, & Yablon B. Not so sweet: missing mercury and high fructose corn syrup. 2009. Institute for Agriculture and Trade Policy. http://www.iatp.org/files/421_2_105026.pdf

116. The FDA sat on evidence of mercury-tainted high-fructose corn syrup. January 27, 2009. Grist. http://grist.org/article/some-heavy-metal-with-that-sweet-roll

117. Stanhope KL & Havel PJ. Endocrine and metabolic effects of consuming beverages sweetened with fructose, glucose, sucrose, or high fructose corn syrup. American Journal of Clinical Nutrition. December 2008. 88(6):1733S–1737S. PMCID:3037017. PubMed. http://www.ncbi.nlm.nih.gov/pmc/articles/PMC3037017

118. Warner MA. A sweetener with a bad rap. The New York Times. July 2, 2006. http://www.nytimes.com/2006/07/02/business/yourmoney/02syrup.html

119. Fetzer WR, Crosby EK, Engel CE, et al. Effect of acid and heat on dextrose and dextrose polymers. Industrial & Engineering Chemistry. May 1953. 45(5):1075–1083. doi:10.1021/ie50521a056. American Chemical Society. http://pubs.acs.org/doi/abs/10.1021/ie50521a056?journalCode=iechad

120. White JS. Straight talk about high-fructose corn syrup: what it is and what it ain't. American Journal of Clinical Nutrition. 2008. 88:1716S–1721S. http://ajcn.nutrition.org/content/88/6/1716S.full.pdf

121. Mitchell DO. Sugar policies: opportunity for change. The World Bank: Development Prospects Group. World Bank Policy Research working paper 3222. February 2004.

122. Leu GJM, Schmitz A, & Knutson RD. Gains and losses of sugar program policy options. American Journal of Agricultural Economics. 1987. 69(3):591–602. doi:10.2307/1241694. Oxford Journals. http://ajae.oxfordjournals.org/content/69/3/591.short

123. The Associated Press. High fructose corn syrup sales down 11%. June 2, 2010. Chicago Breaking Business. http://archive.chicagobreakingbusiness.com/2010/06/high-fructose-corn-syrup-sales-down-11.html

124. Vuilleumier S. Worldwide production of high-fructose syrup and crystalline fructose. American Journal of Clinical Nutrition. November 1993. 58(5):733S–736S. http://ajcn.nutrition.org/content/58/5/733S.short

125. Buck AW. "High fructose corn syrup." Alternative Sweeteners, third edition, revised and expanded. 2001. Marcel Dekker: New York, NY.

126. Williams O & Bessler DA. Cointegration: implications for the market efficiencies of the high fructose corn syrup and refined sugar markets. Applied Economics. 1997. 29(2):225–232. doi:10.1080/000368497327281. Michigan State University. https://www.msu.edu/course/aec/845/READINGS/WilliamsBessler1997.pdf

127. Dufault R, LeBlanc B, Schnoll R, et al. Mercury from chlor-alkali plants: measured concentrations in food product sugar. Environmental Health. 2009. 8:2. doi:10.1186/1476-069X-8-2. http://www.ehjournal.net/content/8/1/2

128. McLaughlin, L. Is high-fructose corn syrup really good for you? TIME. September 17, 2008. http://content.time.com/time/health/article/0,8599,1841910,00.html

129. Institute for Agriculture and Trade Policy. Much high fructose corn syrup contaminated with mercury, new study finds. January 25, 2009. http://www.iatp.org/documents/much-high-fructose-corn-syrup-contaminated-with-mercury-new-study-finds

130. Duffey KJ & Popkin BM. High-fructose corn syrup: is this what's for dinner? American Journal of Clinical Nutrition. December 2008. 88(6):1722S–1732S. doi:10.3945/ajcn.2008.25825C. http://ajcn.nutrition.org/content/88/6/1722S.short

131. Bray GA, Nielsen SJ, & Popkin BM. Consumption of high-fructose corn syrup in beverages may play a role in the epidemic of obesity. American Journal of Clinical Nutrition. April 2004. 79(4):537–543. http://ajcn.nutrition.org/content/79/4/537.short

132. Ferder L, Ferder MD, & Inserra F. The role of high-fructose corn syrup in metabolic syndrome and hypertension. Current Hypertension Reports. April 2010. 12(2):105–112. doi:10.1007/s11906-010-0097-3. Springer Link. http://link.springer.com/article/10.1007/s11906-010-0097-3

133. Angelopoulos TJ, Lowndes J, Zukley L, et al. The effect of high-fructose corn syrup consump-

tion on triglycerides and uric acid. The Journal of Nutrition. April 2009. 139(6):1242S–1245S. doi:10.3945/jn.108.098194. http://jn.nutrition.org/content/139/6/1242S.short

134. Stanhope KL & Havel PJ. Endocrine and metabolic effects of consuming beverages sweetened with fructose, glucose, sucrose, or high-fructose corn syrup. American Journal of Clinical Nutrition. December 2008. 88(6):1733S–1737S. doi:10.3945/ajcn.2008.25825D. http://ajcn.nutrition.org/content/88/6/1733S.short

135. Ouyang X, Cirillo P, Sautin Y, et al. Fructose consumption as a risk factor for non-alcoholic fatty liver disease. Journal of Hepatology. June 2008. 48(6):993–999. PMCID:2423467. PubMed. http://www.ncbi.nlm.nih.gov/pmc/articles/PMC2423467/

136. Duke University Medical Center. High fructose corn syrup linked to liver scarring, research suggests. March 23, 2010. Science Daily. http://www.sciencedaily.com/releases/2010/03/100322204628.htm

137. Sanchez-Lozada LG, Mu W, Roncal C, et al. Comparison of free fructose and glucose to sucrose in the ability to cause fatty liver. European Journal of Nutrition. February 2010. 49(1):1–9. doi:10.1007/s00394-009-0042-x. Springer Link. http://link.springer.com/article/10.1007%2Fs00394-009-0042-x#

138. Soenen S & Westerterp-Plantenga MS. No differences in satiety or energy intake after high-fructose corn syrup, sucrose, or milk preloads. American Journal of Clinical Nutrition. December 2007. 86(6):1586–1594. http://ajcn.nutrition.org/content/86/6/1586.short

139. Bocarsly ME, Powell ES, Avena NM, et al. High-fructose corn syrup causes characteristics of obesity in rats: increased body weight, body fat and triglyceride levels. Pharmacology, Biochemistry and Behavior. November 2010. 97(1):101–6. PMID:20219526. PubMed. http://www.ncbi.nlm.nih.gov/pubmed/20219526

140. Bray GA. Soft drink consumption and obesity: it is all about fructose. Current Opinion in Lipidology. February 2010. 21(1):51–7. PMID:19956074. PubMed. http://www.ncbi.nlm.nih.gov/pubmed/19956074

141. Stanhope KL, Griffen SC, Bair BR, et al. Twenty-four-hour endocrine and metabolic profiles following consumption of high-fructose corn syrup-, sucrose-, fructose-, and glucose-sweetened beverages with meals. American Journal of Clinical Nutrition. May 2008. 87(5):1194–1203. http://ajcn.nutrition.org/content/87/5/1194.short

142. Stanhope KL & Havel PJ. Endocrine and metabolic effects of consuming beverages sweetened with fructose, glucose, sucrose, or high fructose corn syrup. American Journal of Clinical Nutrition. December 2008. 88(6):1733S–1737S. PMCID:3037017. PubMed. http://www.ncbi.nlm.nih.gov/pmc/articles/PMC3037017/

143. Cha SH, Wolfgang M, Tokutake Y, et al. Differential effects of central fructose and glucose on hypothalamic malonyl-CoA and food intake. Proceedings of the National Academy of Sciences of the United States of America. November 2008.105(44):16871–5. PMID:18971329. PubMed. http://www.ncbi.nlm.nih.gov/pubmed/18971329

144. Duffey KJ & Popkin BM. High-fructose corn syrup: is this what's for dinner? American Journal of Clinical Nutrition. December 2008. 88(6):1722S–1732S. doi:10.3945/ajcn.2008.25825C. http://ajcn.nutrition.org/content/88/6/1722S.short

145. Wallinga D, Sorenson J, Mottl P, et al. Not so sweet: missing mercury and high fructose corn syrup. 2009. Institute for Agriculture and Trade Policy. http://www.iatp.org/files/421_2_105026.pdf

146. Dean S. Make your own high-fructose corn syrup with artist Maya Weinstein. Bon Appétit. May 30, 2013. http://www.bonappetit.com/trends/article/make-your-own-high-fructose-corn-syrup-with-artist-maya-weinstein

147. Dufault R, LeBlanc B, Schnoll R, et al. Mercury from chlor-alkali plants: measured concentrations in food product sugar. Environmental Health. January 2009. 8(2). PMCID:2637263. PubMed. http://www.ncbi.nlm.nih.gov/pmc/articles/PMC2637263

148. Institute for Agriculture and Trade Policy. Much high fructose corn syrup contaminated with mercury, new study finds. January 25, 2009. http://www.iatp.org/documents/much-high-fructose-corn-syrup-contaminated-with-mercury-new-study-finds

149. Dufault R, LeBlanc B, Schnoll R, et al. Mercury from chlor-alkali plants: measured concentrations in food product sugar. Environmental Health. January 2009. 8(2). PMCID:2637263. PubMed. http://www.ncbi.nlm.nih.gov/pmc/articles/PMC2637263

150. Wenner M. Corn syrup's mercury surprise. Mother Jones. July/August 2009. http://www.motherjones.com/environment/2009/07/corn-syrups-mercury-surprise

151. Wallinga D, Sorenson J, Mottl P, et al. Not so sweet: missing mercury and high fructose corn syrup. 2009. Institute for Agriculture and Trade Policy. http://www.iatp.org/files/421_2_105026.pdf

152. Jacobson MF. Liquid candy: how soft drinks are harming American's health. Center for Science in the Public Interest. June 2005. http://www.cspinet.org/new/pdf/liquid_candy_final_w_new_supplement.pdf

153. Associated Press. High fructose corn syrup sales down 11%. June 2, 2010. Chicago Breaking Business. http://archive.chicagobreakingbusiness.com/2010/06/high-fructose-corn-syrup-sales-down-11.html

154. SweetSurprise.com. The facts about high fructose corn syrup. Corn Refiners Association. http://sweetsurprise.com

155. The Sugar Association. Corn Refiners Association lawsuit: why we filed. http://web.archive.org/web/20141213083137/http://sugar.org/cra-lawsuit/why-we-filed

156. Watkins, T. "Corn sugar" is false advertising, FDA warns. Associated Press. September 15, 2011. NBC News. http://www.nbcnews.com/id/44543271

157. Corn Refiners Association. Statement of Audrae Erickson, president, Corn Refiners Association on the Food & Drug Administration denial of petition. Press release. May 30, 2012. http://www.corn.org/press/newsroom/fda-petition-denial-statement

158. Adams M. Health ranger research breakthrough: how to block nearly all the mercury in your diet using common, everyday foods. March 18, 2014. NaturalNews. http://www.naturalnews.com/044339_dietary_mercury_heavy_metals_removal.html

159. Steijns M, Peppelenbos A, & Mars P. Mercury chemisorption by sulfur adsorbed in porous materials. Journal of colloid and interface science. 57(1):181–186. ISSN:0021-9797. University of Twente. http://doc.utwente.nl/68345

160. Shumill JA, Wetterhahn KE, & Barchowsky A. Inhibition of NF-κB binding to DNA by chromium, cadmium, mercury, zinc, and arsenite in vitro: evidence of a thiol mechanism. Archives of Biochemistry and Biophysics. 1998. 349(2):356–362. Science Direct. http://www.sciencedirect.com/science/article/pii/S0003986197904707

161. Nam KH, Gomez-Salazar S, & Tavlarides LL. Mercury(II) adsorption from wastewaters using a thiol functional adsorbent. Industrial & Engineering Chemistry Research. 2003. 42(9):1955–1964. doi: 10.1021/ie020834l. American Chemical Society. http://pubs.acs.org/doi/abs/10.1021/ie020834l

162. Quig D. Cystein metabolism and metal toxicity. Alternative Medicine Review. 1998. 3(4):262–270. http://www.altmedrev.com/publications/3/4/262.pdf

163. Drazic A, Tsoutsoulopoulos A, Peschek J, et al. Role of cysteines in the stability and DNA-binding activity of the hypochlorite-specific transcription factor HypT. PLOS ONE. 8(10):e75683. doi:10.1371/journal.pone.0075683. http://journals.plos.org/plosone/article?id=10.1371/journal.pone.0075683

164. Hyman M. Glutathione: the mother of all antioxidants. Huffington Post. June 10, 2010. http://www.huffingtonpost.com/dr-mark-hyman/glutathione-the-mother-of_b_530494.html

165. Amores-Sánchez MI & Medina MA. Glutamine, as a precursor of glutathione, and oxidative stress. Molecular Genetics and Metabolism. 1999. 67(2):100–5. PMID:10356308. PubMed. http://www.ncbi.nlm.nih.gov/pubmed/10356308

166. University of North Dakota Energy & Environmental Research Center. Selenium and mercury: fishing for answers. 2011. http://www.icmj.com/userfiles/files/Selenium-Mercury.pdf

167. Hoffman DJ & Heinz GH. Effects of mercury and selenium on glutathione metabolism and oxidative stress in mallard ducks. Environmental Toxicology and Chemistry. 1998. 17(2):161–166. doi:10.1002/etc.5620170204. Wiley Online Library. http://onlinelibrary.wiley.com/doi/10.1002/etc.5620170204/abstract

168. Rayman, MP. Food-chain selenium and human health: emphasis on intake. British Journal of Nutrition. 2008. 100:254–268. doi:10.1017/S0007114508939830. Cambridge University Press. http://journals.cambridge.org/download.php?file=%2FBJN%2FBJN100_02 %2FS0007114508939830a.pdf&code=0cba73616a3b154214e1b68b258da7cd

169. Ralston NVC, Ralston CR, Blackwell JL, et al. Dietary and tissue selenium in relation to methyl-mercury toxicity. 2008. Neurotoxicology. doi:10.1016/j.neuro.2008.07.007. University of Hawaii at Manoa. http://www.soest.hawaii.edu/oceanography/courses_html/OCN331/Mercury3.pdf #page=1&zoom=auto,0,800

170. Menzez B. Detoxing from mercury toxicity: what foods help chelate mercury? May 19, 2013. You-Tube. http://www.youtube.com/watch?v=-lHGHOMUQQQ

171. Abdalla FH, Bellé LP, De Bona KS, et al. Allium sativum L. extract prevents methyl mercury induced cytotoxicity in peripheral blood leukocytes (LS). Food and Chemical Toxicology. 2010. 48(1):417–21. PMID:19879309. Science.NaturalNews. http://science.naturalnews.com/ pubmed/19879309.html

172. Carvalho CML, Lu J, Zhang X, et al. Effects of selenite and chelating agents on mammalian thi-oredoxin reductase inhibited by mercury: implications for treatment of mercury poisoning. The FASEB Journal. 2011. 25(1):370–81. PMID:20810785. Science.NaturalNews. http://science .naturalnews.com/pubmed/20810785.html

173. Zhao J, Wen Y, Bhadauria M, et al. Protective effects of propolis on inorganic mercury induced oxidative stress in mice. Indian Journal of Experimental Biology. 2009. 47(4):264–9. PMID: 19382722. Science.NaturalNews. http://science.naturalnews.com/pubmed/19382722.html

Heavy Metals: Lead

174. Hernberg S. Lead poisoning in a historical perspective. American Journal of Industrial Medicine. 2000. 38(3). doi:10.1002/1097-0274(200009)38:3<244::AID-AJIM3>3.0.CO;2-F. http://science .naturalnews.com/pubmed/10940962.html

175. U.S. Environmental Protection Agency. Drinking water contaminants—standards and regulations. Updated January 6, 2016. http://www.epa.gov/dwstandardsregulations

176. U.S. Environmental Protection Agency (EPA). Basic information about lead in drinking water. Updated February 18, 2016. http://www.epa.gov/your-drinking-water/basic-information-about -lead-drinking-water

177. European Food Safety Authority. Lead dietary exposure in the European population. EFSA Jour-nal. 2012. 10(7):2831. http://www.efsa.europa.eu/en/efsajournal/pub/2831

178. Adams M. Whole Foods Market (WFM) continues to knowingly sell poison to its customers: Nat-ural News seeks class action law firm to pursue legal action. July 14, 2014. NaturalNews. http:// www.naturalnews.com/045999_Whole_Foods_Market_class_action_lawsuit_heavy_metals _contamination.html

179. European Food Safety Authority. Lead dietary exposure in the European population. EFSA Jour-nal. 2012. 10(7):2831. http://www.efsa.europa.eu/en/efsajournal/pub/2831

180. Wang ZW, Nan ZR, Wang SL, et al. Accumulation and distribution of cadmium and lead in wheat (Triticum aestivum L.) grown in contaminated soils from the oasis, north-west China. Journal of the Science of Food and Agriculture. 2011. 91(2):377–84. doi:10.1002/jsfa.4196.

181. Roca de Togores M, Farré R, Frigola AM. Cadmium and lead in infant cereals—electrothermal-atomic absorption spectroscopic determination. The Science of the Total Environment. 1999. 234(1–3):197–201. PMID:10507158. PubMed. http://www.ncbi.nlm.nih.gov/pubmed/10507158

182. U.S. Food and Drug Administration. FDA 101: infant formula. Updated December 7, 2013. http://www.fda.gov/ForConsumers/ConsumerUpdates/ucm048694.htm

183. Environmental Law Foundation. Notice of violation of the Safe Drinking Water and Toxic En-forcement Act of 1986 (Proposition 65), Section 25249.6 of the California Health and Safety Code, for exposing consumers of apple juice, grape juice, packaged peaches, packaged pears, and fruit cocktail to lead. 2010. http://oag.ca.gov/system/files/prop65/notices/2010-00214.pdf

184. U.S. Food and Drug Administration. Best Value, Inc., recalls PRAN bran turmeric powder due to elevated levels of lead. October 15, 2013. http://www.fda.gov/Safety/Recalls/ucm371042.htm

185. U.S. Food and Drug Administration. Bottled water everywhere: keeping it safe. Updated June 28, 2010. http://www.fda.gov/forconsumers/consumerupdates/ucm203620.htm

186. U.S. Food and Drug Administration. Lead in candy likely to be consumed frequently by small children: recommended maximum level and enforcement policy. December 2005. http://www.fda .gov/Food/GuidanceRegulation/GuidanceDocumentsRegulatoryInformation/Chemical ContaminantsMetalsNaturalToxinsPesticides/ucm077904.htm

187. U.S. Food and Drug Administration. Lead in food and color additives and GRAS ingredients; request for data. Federal Register. February 4, 1994. Government Printing Office. http://www.gpo .gov/fdsys/pkg/FR-1994-02-04/html/94-2472.htm

188. U.S. Consumer Product Safety Commission. Search results. Accessed February 23, 2016. http://cs.cpsc.gov/ConceptDemo/SearchCPSC.aspx?query=lead

189. U.S. Food and Drug Administration. FDA analyses of lead in lipsticks. 2010. http://www.fda.gov/cosmetics/productsingredients/products/ucm137224.htm#expanalyses

190. Al-Saleh I, Al-Enazi S, & Shinwari N. Assessment of lead in cosmetic products. Regulatory toxicology and pharmacology. 2009. 54(2):105–13. PMID:19250956.Science.NaturalNews. http://science.naturalnews.com/pubmed/19250956.html

191. U.S. Food and Drug Administration. FDA analyses of lead in lipsticks. 2010. http://www.fda.gov/Cosmetics/ProductsIngredients/Products/ucm137224.htm

192. Baghurst PA, McMichael AJ, Wigg NR, et al. Environmental exposure to lead and children's intelligence at the age of seven years, the Port Pirie cohort study. The New England Journal of Medicine. 1992. 327(18):1279–84. PubMed. http://www.ncbi.nlm.nih.gov/pubmed/1383818

193. Magzamen S, Imm P, Amato MS, et al. Moderate lead exposure and elementary school end-of-grade examination performance. Annals of Epidemiology. 2013. 23(11):700–7. doi:10.1016/j.annepidem.2013.08.007. PubMed. http://www.ncbi.nlm.nih.gov/pubmed/24095655

194. Xu J, Huai C, Yang B, et al. Effects of lead exposure on hippocampal metabotropic glutamate receptor subtype 3 and 7 in developmental rats. Journal of Negative Results in Biomedicine. 2009. 8:5. PMID:19374778. Science.NaturalNews. http://science.naturalnews.com/pubmed/19374778 .html

195. Rosin A. The long-term consequences of exposure to lead. The Israel Medical Association journal. 2009. 11(11):689–94. PMID:20108558. Science.NaturalNews. http://science.naturalnews.com/pubmed/20108558.html

196. Stewart WF & Schwartz BS. Effects of lead on the adult brain: a 15-year exploration. American journal of industrial medicine. 2007. 50(10):729–39. PMID:17311281. Science.NaturalNews. http://science.naturalnews.com/pubmed/17311281.html

197. World Health Organization. Fact sheet no. 317: cardiovascular diseases (CVDs). Updated January 2015. Media Centre. http://www.who.int/mediacentre/factsheets/fs317/en

198. Navas-Acien A, Guallar E, Silbergeld EK, et al. Lead exposure and cardiovascular disease—a systematic review. Environmental Health Perspectives. 2007. 115(3):472–82. PMID:17431501. Science.NaturalNews. http://science.naturalnews.com/pubmed/17431501.html

199. Alissa EM & Ferns GA. Heavy metal poisoning and cardiovascular disease. Journal of Toxicology. 2011. Article ID 870125. Hindawi. http://dx.doi.org/10.1155/2011/870125

200. Hegazy AA, Zaher MM, Abd El-Hafez MA, et al. Relation between anemia and blood levels of lead, copper, zinc and iron among children. BMC Research Notes. 2010. 3:133. PMID:20459857. Science.NaturalNews. http://science.naturalnews.com/pubmed/20459857.html

201. Lorenzo L, Silvestroni A, Martino MG, et al. Evaluation of peripheral blood neutrophil leucocytes in lead-exposed workers. International Archives of Occupational and Environmental Health. 2006. 79(6):491–498. Springer Link. http://link.springer.com/article/10.1007/s00420-005-0073-4

202. Kasperczyk A, Kasperczyk S, Horak S, et al. Assessment of semen function and lipid peroxidation among lead exposed men. Toxicology and Applied Pharmacology. 2008. 228(3):378–84. PMID:18252257. Science.NaturalNews. http://science.naturalnews.com/pubmed/18252257.html

203. Fatima P, Debnath BC, Hossain MM, et al. Relationship of blood and semen lead level with semen parameter. Mymensingh Medical Journal. 2010. 19(3):405–16. PMID:20639835. Science.Natur-alNews. http://science.naturalnews.com/pubmed/20639835.html

204. Flora SJS, Pachauri V, & Saxena G. Arsenic, cadmium and lead. Reproductive and Developmental Toxicology. 2011. 415–438. https://www.researchgate.net/publication/279429805_Arsenic _Cadmium_and_Lead

205. El-Said KF, EL-Ghamry AM, Mahdy NH, et al. Chronic occupational exposure to lead and its im-pact on oral health. The Journal of the Egyptian Public Health Association. 2008. 83(5–6):451–66. PMID:19493512. Science.NaturalNews. http://science.naturalnews.com/pubmed/19493512.html

206. Rosin A. The long-term consequences of exposure to lead. The Israel Medical Association Journal. 2009. 11(11):689–94. PMID:20108558. Science.NaturalNews. http://science.naturalnews.com/ pubmed/20108558.html

207. Agency for Toxic Substance and Disease Registry (ATSDR). Toxicological profile for lead: update. U.S. Department of Health & Human Services, Public Health Service. 2007. http://www.atsdr .cdc.gov/toxprofiles/tp13.pdf

208. Senapati S, Dey S, Dwivedi S, et al. Effect of thiamine hydrochloride on lead induced lipid peroxi-dation in rat liver and kidney. Veterinary and human toxicology. 2000. 42:236–7. PubMed. http:// www.ncbi.nlm.nih.gov/pubmed/10928693

209. Ahamed M & Siddiqui MK. Environmental lead toxicity and nutritional factors. Clinical nutrition. 2007. 26(4):400–8. PMID:17499891. PubMed. http://www.ncbi.nlm.nih.gov/ pubmed/17499891

210. Wang C, Liang J, Zhang C, et al. Effect of ascorbic acid and thiamine supplementation at different concentrations on lead toxicity in liver. The Annals of Occupational Hygiene. 2007. 51(6):563–9. PMID:17878260. PubMed. http://www.ncbi.nlm.nih.gov/pubmed/17878260

211. Sajitha GR, Jose R, Andrews A, et al. Garlic oil and vitamin E prevent the adverse effects of lead acetate and ethanol separately as well as in combination in the drinking water of rats. Indian Jour-nal of Clinical Biochemistry. 2010. 25(3):280–8. PMID:21731199. PubMed. http://www.ncbi .nlm.nih.gov/pubmed/21731199

212. Beecher GR. Review Overview of dietary flavonoids: nomenclature, occurrence and intake. The Journal of Nutrition. 2003. 133(10):3248S–3254S. PMID:14519822. PubMed. http://www.ncbi .nlm.nih.gov/pubmed/14519822

213. Bravo A & Anacona JR. Metal complexes of the flavonoid quercetin: antibacterial properties. Tran-sition Metal Chemistry. 2001. 26:20–23. Springer Link. http://link.springer.com/article /10.1023%2FA%3A1007128325639

214. Shukla PK, Khanna VK, Khan MY, et al. Protective effect of curcumin against lead neurotoxicity in rat. Human & Experimental Toxicology. 2003. 22(12):653–8. PMID:14992327. http://www .ncbi.nlm.nih.gov/pubmed/14992327

215. Daniel S. Limson JL, Dairam A, et al. Through metal binding, curcumin protects against lead- and cadmium-induced lipid peroxidation in rat brain homogenates and against lead-induced tissue damage in rat brain. Journal of Inorganic Biochemistry. 2004. 98(2):266–75. PMID:14729307. PubMed. www.ncbi.nlm.nih.gov/pubmed/14729307

216. Senapati SK, Dey S, Dwivedi SK, et al. Effect of garlic (Allium sativum L.) extract on tissue lead level in rats. Journal of Ethnopharmacology. 2001. 76(3):229–32. PMID:11448543. PubMed. http://www.ncbi.nlm.nih.gov/pubmed/11448543

217. Pourjafar M, Aghbolaghi PA, & Shakhse-Niaie M. Effect of garlic along with lead acetate admin-istration on lead burden of some tissues in mice. Pakistan Journal of Biological Sciences. 2007. 10(16):2772–4. PMID:19070102. PubMed. http://www.ncbi.nlm.nih.gov/pubmed/19070102

218. Chandrasekaran VR, Hsu DZ, & Liu MY. Beneficial effect of sesame oil on heavy metal toxicity. Journal of Parenteral and Enteral Nutrition. 2013. 38(2):179–85. PMID: 23744838. PubMed. http://www.ncbi.nlm.nih.gov/pubmed/23744838

219. Liao Y, Zhang J, Jin Y, et al. Therapeutic potentials of combined use of DMSA with calcium and ascorbic acid in the treatment of mild to moderately lead intoxicated mice. Biometals.

2008. 21(1):1–8. PMID:17287888. Science.NaturalNews. http://science.naturalnews.com/pubmed/17287888.html

220. Blaucok-Busch E, Amin OR, Dessoki HH, et al. Efficacy of DMSA therapy in a sample of Arab children with autistic spectrum disorder. Maedica. 2012. 7(3):214–21. PMID:23400264. PubMed. http://www.ncbi.nlm.nih.gov/pubmed/23400264

221. Bradberry S, Sheehan T, & Vale A. Use of oral dimercaptosuccinic acid (succimer) in adult patients with inorganic lead poisoning. Monthly Journal of the Association of Physicians. 2009. 102(10):721–32. PMID:19700440. Science.NaturalNews. http://science.naturalnews.com/pubmed/19700440.html

Heavy Metals: Cadmium

222. Mortvedt JJ. Heavy metal contaminants in inorganic and organic fertilizers. Fertilizer Research. 1996. 43:55–61. Springer Link. http://link.springer.com/article/10.1007/BF00747683#page-1

223. Dorris J, Atieh BH, & Gupta RC. Cadmium uptake by radishes from soil contaminated with nickel cadmium batteries: toxicity and safety considerations. Toxicology Mechanisms and Methods. 2002.12(4):265–76. PMID:20021168. Science.Naturalnews.com. http://science.naturalnews.com/pubmed/20021168.html

224. Mortvedt JJ. Heavy metal contaminants in inorganic and organic fertilizers. Fertilizer Research. 1996. 43:55–61. Springer Link. http://link.springer.com/article/10.1007/BF00747683#page-1

225. Frontline. Dr. Arjun Srinivasan: we've reached "the end of antibiotics, period." Public Broadcasting Station. 2013. http://www.pbs.org/wgbh/pages/frontline/health-science-technology/hunting-the-nightmare-bacteria/dr-arjun-srinivasan-weve-reached-the-end-of-antibiotics-period

226. U.S. Food and Drug Administration Center for Veterinary Medicine. 2009 summary report on antimicrobials sold or distributed for use in food-producing animals. December 9, 2010. http://www.fda.gov/downloads/ForIndustry/UserFees/AnimalDrugUserFeeActADUFA/UCM231851.pdf

227. Loglisci R. New FDA numbers reveal food animals consume lion's share of antibiotics. 2010. Johns Hopkins Center for a Livable Future. http://www.livablefutureblog.com/2010/12/new-fda-numbers-reveal-food-animals-consume-lion%E2%80%99s-share-of-antibiotics

228. Sapkota AR, Curriero FC, Gibson KE, et al. Antibiotic-resistant enterococci and fecal indicators in surface water and groundwater impacted by a concentrated swine feeding operation. Environmental Health Perspectives. 2007. 115(7):1040–1045. PMCID:PMC1913567. PubMed. http://www.ncbi.nlm.nih.gov/pmc/articles/PMC1913567

229. Wan Y, Bao Y, & Zhou Q. Simultaneous adsorption and desorption of cadmium and tetracycline on cinnamon soil. Chemosphere. 2010. 80(7):807–12. PMID:20510430. Science.Naturalnews. http://science.naturalnews.com/pubmed/20510430.html

230. Kitana N & Callard IP. Effect of cadmium on gonadal development in freshwater turtle (Trachemys scripta, Chrysemys picta) embryos. Journal of Environmental Science and Health. 2008. 43(3):262–71. PMID: 18205057. Science.NaturalNews. http://science.naturalnews.com/pubmed/18205057.html

231. Luca G, Lilli C, Bellucci C, et al. Toxicity of cadmium on Sertoli cell functional competence: an in vitro study. Journal of Biological Regulators and Homeostatic Agents. 2013. 27(3):805–816. PMID: 24152845. PubMed. http://www.ncbi.nlm.nih.gov/pubmed/24152845

232. Kazantzis G. Cadmium, osteoporosis and calcium metabolism. Biometals. 2004. 17(5):495–8. PMID:15688852. Science.NaturalNews. http://science.naturalnews.com/pubmed/15688852.html

233. Tsuchiya K. Causation of ouch-ouch disease (Itai-Itai Byō)—an introductory review. I. Nature of the disease. The Keio Journal of Medicine. 1969. 18(4):181–94. PMID: 4915215. PubMed. http://www.ncbi.nlm.nih.gov/pubmed/4915215

234. Paniagua-Castro N, Escalona-Cardoso G, Hernández-Navarro D, et al. Spirulina (arthrospira) protects against cadmium induced teratogenic damage in mice. Journal of Medicinal Food. 2011. 14(4):398–404. PMID:21254891. Science.Naturalnews. http://science.naturalnews.com/pubmed/21254891.html

Heavy Metals: Aluminum

235. Hetherington LE. World mineral production: 2001–2005. British Geological Survey. 2007. ISBN: 978-0-85272-592-4. Keyworth, Nottingham: British Geological Survey.
236. U.S. Environmental Protection Agency. Acid rain. Updated December 4, 2012. http://www.epa.gov/acidrain/effects/surface_water.html
237. Sayer J. Is eating and injecting aluminum safe as our regulators say? Green Med Info. May 27, 2012. http://www.greenmedinfo.com/blog/eating-aluminum-it-safe-our-regulators-say
238. European Food Safety Authority. Scientific opinion of the panel on food additives, flavourings, processing aids and food contact materials (AFC). The EFSA Journal. 2008. 754:2–34. http://www.efsa.europa.eu/en/efsajournal/doc/754.pdf
239. Ibid.
240. Chuchu N, Patel B, Sebastian B, et al. The aluminium content of infant formulas remains too high. BMC Pediatrics. 2013. 13:162. http://www.biomedcentral.com/content/pdf/1471-2431-13-162.pdf
241. Jefferson Lab. The element aluminum. Jefferson Lab Science Education. http://education.jlab.org/itselemental/ele013.html
242. Helmenstine, AM. Is alum safe to eat or use? Food and Cooking Chemistry FAQs. About.com. Updated February 16, 2016. http://chemistry.about.com/od/foodchemistryfaqs/f/Is-Alum-Safe.htm
243. Wise Geek. What are the pros and cons of alum for pickling? Accessed February 23, 2016. http://www.wisegeek.com/what-are-the-pros-and-cons-of-alum-for-pickling.htm
244. European Food Safety Authority. Scientific opinion of the panel on food additives, flavourings, processing aids and food contact materials (AFC). The EFSA Journal. 2008. 754:2–34. http://www.efsa.europa.eu/en/efsajournal/doc/754.pdf
245. Ibid.
246. Thompson J. Aluminum chelation. The Health Sciences Institute. HSI Online. June 25, 2003. http://hsionline.com/2003/06/25/aluminum-chelation
247. Schaeffer G, Fontès F, le Breton E, et al. The dangers of certain mineral baking-powders based on alum, when used for human nutrition. Journal of Hygiene. 1928. 28(1):92–99. PMCID:PMC2167687. PubMed. http://www.ncbi.nlm.nih.gov/pmc/articles/PMC2167687
248. Herzog P. Effect of antacids on mineral metabolism. Zeitschrift fur Gastroenterologie. 1983. Suppl:117–26. PMID: 6858403. PubMed. http://www.ncbi.nlm.nih.gov/pubmed/6858403
249. Tomljenovic L. Aluminum and Alzheimer's disease: after a century of controversy, is there a plausible link? Journal of Alzheimer's Disease. 2011. 23(4):567–98. PMID:21157018. PubMed. http://www.ncbi.nlm.nih.gov/pubmed/21157018
250. Hawkes N. Alzheimer's linked to aluminium pollution in tap water. The Times. April 20, 2006. http://www.thetimes.co.uk/tto/health/article1884130.ece
251. Sappino AP, Buser R, Lesne L, et al. Aluminium chloride promotes anchorage-independent growth in human mammary epithelial cells. Journal of Applied Toxicology. 2012. 32(3):233–43. PMID: 22223356. PubMed. http://www.ncbi.nlm.nih.gov/pubmed/22223356
252. U.S. Food and Drug Administration. Study reports aluminum in vaccines poses extremely low risk to infants. Vaccines, Blood & Biologics. Updated January 2012. http://www.fda.gov/BiologicsBloodVaccines/ScienceResearch/ucm284520.htm
253. Mitkus RJ, King DB, Hess MA, et al. Updated aluminum pharmacokinetics following infant exposures through diet and vaccination. Vaccine. 2011. 29(51):9538–43. PMID:22001122. PubMed. http://www.ncbi.nlm.nih.gov/pubmed/22001122
254. Vaccine. Elsevier. Accessed February 23, 2016. http://www.journals.elsevier.com/vaccine
255. Mayo Clinic. A physician's guide for anti-vaccine parents. 2012. http://www.mayoclinic.org/news2012-rst/6841.html
256. Tomljenovic L & Shaw CA. Aluminum vaccine adjuvants: are they safe? Current Medicinal Chemistry. 2011. 18(17):2630–7. PMID:21568886. PubMed. http://www.ncbi.nlm.nih.gov/pubmed/21568886

257. Tomljenovic L & Shaw CA. Do aluminum vaccine adjuvants contribute to the rising prevalence of autism? Journal of Inorganic Biochemistry. 2011. 105(11):1489–99. PMID:22099159. PubMed. http://www.ncbi.nlm.nih.gov/pubmed/22099159

258. Concerns About the Human Papillomavirus Vacine. American College of Pediatricians. January 2016. http://www.acpeds.org/the-college-speaks/position-statements/health-issues/new-concerns -about-the-human-papillomavirus-vaccine

259. Agency for Toxic Substances and Disease Registry. Public health statement for aluminum. CAS#: 7429-90-5. 2008. U.S. Centers for Disease Control and Prevention. http://www.atsdr.cdc.gov/phs/ phs.asp?id=1076&tid=34

260. Spencer H & Lender M. Adverse effects of aluminum-containing antacids on mineral metabolism. Gastroenterology. 1979. 76(3):603–6. PMID:428714. PubMed. http://www.ncbi.nlm.nih.gov/ pubmed/428714

261. Herzog P & Holtermüller KH. Antacid therapy—changes in mineral metabolism. Scandinavian Journal of Gastroenterology. 1982. 75:56–62. PMID:6293043. PubMed. http://www.ncbi.nlm. nih.gov/pubmed/6293043

262. Thompson J. Aluminum chelation. The Health Sciences Institute. HSI Online. June 25, 2013. http://hsionline.com/2003/06/25/aluminum-chelation

263. Wang L, Min M, Li Y, et al. Cultivation of green algae Chlorella sp. in different wastewaters from municipal wastewater treatment plant. Applied Biochemistry and Biotechnology. 2010. 162(4):1174–86. PMID:19937154. PubMed. http://www.ncbi.nlm.nih.gov/pubmed/19937154

Heavy Metals: Copper

264. Purest Colloids. A brief history of the health support uses of copper. Accessed February 23, 2016. http://www.purestcolloids.com/history-copper.php

265. Wilson L. Copper toxicity syndrome. The Center for Development. Revised 2015. Drwilson.com. http://www.drlwilson.com/articles/copper_toxicity_syndrome.htm

266. Singha I, Sagarea AP, Comaa M, et al. Low levels of copper disrupt brain amyloid-β homeostasis by altering its production and clearance. Proceedings of the National Academy of Sciences. 2013. 110(36):14505–14506. doi:10.1073/iti3613110. http://www.pnas.org/content/early/2013/08 /14/1302212110

267. Lovella MA, Robertson JD, Teesdale WJ, et al. Copper, iron and zinc in Alzheimer's disease senile plaques. Journal of the Neurological Sciences. 1998. 158(1):47–52. PMID:9667777. PubMed. http://www.ncbi.nlm.nih.gov/pubmed/9667777

268. Russo AJ. Decreased zinc and increased copper in individuals with anxiety. Nutrition and Metabolic Insights. 2011. 4:1–5. PMCID: PMC3738454. PubMed. http://www.ncbi.nlm.nih.gov/pmc /articles/PMC3738454

269. Russo AJ. Analysis of plasma zinc and copper concentration, and perceived symptoms, in individuals with depression, post zinc and anti-oxidant therapy. Nutrition and Metabolic Insights. 2011. 4:19–27. doi:10.4137/NMI.S6760. PubMed. http://www.ncbi.nlm.nih.gov/pubmed/23946658

270. Russo AJ, Bazin AP, Bigega R, et al. Plasma copper and zinc concentration in individuals with autism correlate with selected symptom severity. Nutrition and metabolic insights. 2012. 5:41–7. PMID:23882147. PubMed. http://www.ncbi.nlm.nih.gov/pubmed/23882147

271. Kovrižnych JA, Sotníková R, Zeljenková D, et al. Acute toxicity of 31 different nanoparticles to zebrafish (Danio rerio) tested in adulthood and in early life stages—comparative study. Interdisciplinary Toxicology. 2013. 6(2):67–73. PMID: 24179431. PubMed. http://www.ncbi.nlm.nih.gov/ pubmed/24179431

272. Shahabadi N, Khodaei MM, Kashanian S, et al. Interaction of a copper(II) complex containing an artificial sweetener (aspartame) with calf thymus DNA. Spectrochimica Acta Part A: Molecular and Biomolecular Spectroscopy. 2013. 120C:1–6. PMID:24177861. PubMed. http://www.ncbi.nlm .nih.gov/pubmed/24177861

273. Schilsky ML. Wilson disease: genetic basis of copper toxicity and natural history. Seminars in Liver Disease. 1996. 16(1):83–95. PMID:8723326. PubMed. http://www.ncbi.nlm.nih.gov/pubmed/8723326

274. Gupte A & Mumper RJ. Copper chelation by D-penicillamine generates reactive oxygen species that are cytotoxic to human leukemia and breast cancer cells. Free Radical Biology & Medicine. 2007. 43(9):1271–8. PMID:17893040. Science.NaturalNews. http://science.naturalnews.com/pubmed/17893040.html

275. Dietary Reference Intakes (DRIs): Elements. Food and Nutrition Board, Institute of Medicine, National Academies. http://iom.nationalacademies.org/~/media/Files/Activity%20Files/Nutrition/DRIs/New%20Material/6_%20Elements%20Summary.pdf

Heavy Metals: Tin

276. Eisler R. Tin hazards to fish, wildlife, and invertebrates: a synoptic review. Biological report. U.S. Department of the Interior Fish and Wildlife Service. 1989. 8(1.15). DTIC Online. http://oai .dtic.mil/oai/oai?verb=getRecord&metadataPrefix=html&identifier=ADA322822

277. Krigman MR & Silverman AP. General toxicology of tin and its organic compounds. Neurotoxicology. 1984. 5(2):129–39. PMID:6390260. PubMed. http://www.ncbi.nlm.nih.gov/pubmed/6390260

278. Eisler R. Tin hazards to fish, wildlife, and invertebrates: a synoptic review. Biological report. U.S. Department of the Interior Fish and Wildlife Service. 1989. 8(1.15). DTIC Online. http://oai .dtic.mil/oai/oai?verb=getRecord&metadataPrefix=html&identifier=ADA322822

279. Reicks M & Rader JI. Effects of dietary tin and copper on rat hepatocellular antioxidant protection. Proceedings of the Society for Experimental Biology and Medicine. 1990. 195(1):123–8. PMID:2399253. PubMed. http://www.ncbi.nlm.nih.gov/pubmed/2399253

280. Shargel L & Masnyj J. Effect of stannous fluoride and sodium fluoride on hepatic mixed-function oxidase activities in the rat. Toxicology and Applied Pharmacology. 1981. 59:452–456. PMID:7268769. PubMed. http://www.ncbi.nlm.nih.gov/pubmed/7268769

281. Burba JV. Inhibition of hepatic azo-reductase and aromatic hydroxylase by radiopharmaceuticals containing tin. Toxicology Letters. 1983. 18(3):269–72. PMID:6665800. PubMed. http://www .ncbi.nlm.nih.gov/pubmed/6665800

282. Dwivedi RS, Kaur G, Srivastava RC, et al. Lipid peroxidation in tin intoxicated partially hepatectomized rats. Bulletin of Environmental Contamination and Toxicology. 1984. 33(1):200–209. Springer Link. http://link.springer.com/article/10.1007%2FBF01625531

283. Food Safety Authority of Ireland. Mercury, lead, cadmium, tin and arsenic in food. Toxicology factsheet series. 2009. 1:1–13. www.fsai.ie/workarea/downloadasset.aspx?id=8412

284. Wong PTS, Chau YK, Kramar O, et al. Structure–toxicity relationship of tin compounds on algae. Canadian Journal of Fisheries and Aquatic Sciences. 1982. 39(3):483–488. doi:10.1139/f82-066. NRC Research Press. http://www.nrcresearchpress.com/doi/citedby/10.1139/f82-066# .Und0YySLrk0

285. Dehghan G. & Khoshkam Z. Chelation of toxic tin(II) by quercetin: a spectroscopic study. 2011 International Conference on Life Science and Technology. IPCBEE vol.3. 2011. IACSIT Press, Singapore. http://www.ipcbee.com/vol3/1-B050.pdf

286. Howe PD. Concise international chemical assessment document 65: tin and inorganic tin compounds. World Health Organization. 2005. ISBN: 9-24153-065-0. www.who.int/ipcs/publications/cicad/cicad_65_web_version.pdf

287. Nufarm agricultural products. Agri Tin®, EPA Reg No. 55146-72 specimen label, (RV092711). Nufarm Agricultural Products. Crop Data Management Systems, Inc. Accessed February 23, 2016. http://www.cdms.net/ldat/ld3MS005.pdf

288. Eisler R. Tin hazards to fish, wildlife, and invertebrates: a synoptic review. Biological report. U.S. Department of the Interior Fish and Wildlife Service. 1989. 8(1.15). DTIC Online. http://oai .dtic.mil/oai/oai?verb=getRecord&metadataPrefix=html&identifier=ADA322822

289. Agency for Toxic Substances and Disease Registry. Tin and tin compounds. Division of toxicology ToxFAQs™. U.S. Centers for Disease Control and Prevention. 2005. U.S. Centers for Disease Control and Prevention. http://www.atsdr.cdc.gov/toxfaqs/tfacts55.pdf

290. Morry DW, Steinmaus C, et al. Evidence on the carcinogenicity of fluoride and its salts. California

Environmental Protection Agency. 2011. http://www.fluoridealert.org/wp-content/uploads/ca
-oehha.2011.pdf

291. McFadden RD. $750,000 given in child's death in fluoride case. The New York Times. January 20,
1979. Fluoride Alert. http://fluoridealert.org/articles/kennerly

292. Dehghan G & Khoshkam Z. Chelation of toxic tin(II) by quercetin: a spectroscopic study. 2011
International Conference on Life Science and Technology. IPCBEE vol.3. 2011. IACSIT Press,
Singapore. http://www.ipcbee.com/vol3/1-B050.pdf

Chemical Contaminants: Bisphenol A

293. Bisphenol A Global Industry Group. Bisphenol A: information sheet. Discovery and use. October
2002. http://www.bisphenol-A.org/pdf/DiscoveryandUseOctober2002.pdf

294. Environment and Human Health, Inc. Bisphenol A introduction: plastics that may be harmful to
children and reproductive health. 2008. http://www.ehhi.org/reports/plastics/bpa_intro.shtml

295. Grignard E, Lapenna S, & Bremer S. Weak estrogenic transcriptional activities of Bisphenol A and
Bisphenol S. Toxicology in Vitro. 2012. 26(5):727–31. PMID:22507746. PubMed. http://www
.ncbi.nlm.nih.gov/pubmed/22507746

296. Zsarnovszky A, Le HH, Wang H, et al. Ontogeny of rapid estrogen-mediated extracellular
signal-regulated kinase signaling in the rat cerebellar cortex: potent nongenomic agonist and
Endocrine disrupting activity of the xenoestrogen Bisphenol A. Neuroendocrinology. 2005.
146(12):5388. http://endo.endojournals.org/content/146/12/5388.short

297. Breast Cancer Fund. Disrupted development: the dangers of prenatal BPA exposure. 2013. http://
www.breastcancerfund.org/assets/pdfs/publications/disrupted-development-the-dangers-of-prenatal
-bpa-exposure.pdf

298. Stahlhut RW, Welshons WV, & Swan SH. Bisphenol A data in NHANES suggest longer than expected
half-life, substantial nonfood exposure, or both. Environmental Health Perspectives. 2009. 117(5):784–
789. PMCID:2685842. PubMed. http://www.ncbi.nlm.nih.gov/pmc/articles/PMC2685842

299. Breast Cancer Fund. Disrupted development: the dangers of prenatal BPA exposure. 2013. http://
www.breastcancerfund.org/assets/pdfs/publications/disrupted-development-the-dangers-of-prenatal
-bpa-exposure.pdf

300. Josephson J. Chemical exposures: Prostate cancer and early BPA exposure. Environmental Health
Perspectives. 2006. 114(9):A520. PMCID: PMC1570083. PubMed. http://www.ncbi.nlm.nih
.gov/pmc/articles/PMC1570083

301. Konkel L. BPA as a mammary carcinogen: early findings reported in rats. Environmental Health
Perspectives. 2013. 121:A284. http://dx.doi.org/10.1289/ehp.121-A284

302. Lathi RB, Liebert CA, Brookfeild K, et al. Maternal serum biphenol-A (BPA) level is positively
associated with miscarriage risk. Fertility and Sterility. 2013. 100(3):S19. http://dx.doi
.org/10.1016/j.fertnstert.2013.07.183

303. Associated Press. Study ties chemical BPA to possible miscarriage risk. October 14, 2013. Fox
News. http://www.foxnews.com/health/2013/10/14/study-ties-chemical-bpa-to-possible
-miscarriage-risk/

304. Liao C & Kannan K. High levels of bisphenol A in paper currencies from several countries, and
implications for dermal exposure. Environmental science & technology. 2011. 45(16):6761–8.
PMID:21744851. PubMed. http://www.ncbi.nlm.nih.gov/pubmed/21744851

305. U.S. Food and Drug Administration. Bisphenol A (BPA): use in food contact application. Updated
March 2013. http://www.fda.gov/newsevents/publichealthfocus/ucm064437.htm

306. U.S. Department of Health and Human Services. NTP-CERHR monograph on the potential
human reproductive and developmental effects of bisphenol A. National Toxicology Program.
Center for the Evaluation of Risks to Human Reproduction. September 2008. NIH publication
no. 08-5994. http://ntp.niehs.nih.gov/ntp/ohat/bisphenol/bisphenol.pdf

307. Kumanyika SK & Grier S. Targeting interventions for ethnic minority and low-income populations. The future of children. 2006. 16(1):187–207. doi:10.1353/foc.2006.0005. Project MUSE.
http://muse.jhu.edu/journals/foc/summary/v016/16.1kumanyika.html

308. Treviño RP, Marshall RM, Hale DE, et al. Diabetes risk factors in low-income Mexican-American children. American Diabetes Association. Diabetes Care. 1999. 22(2):202–207. doi:10.2337/diacare.22.2.202. http://care.diabetesjournals.org/content/22/2/202.short

309. Kirk JK, D'Agostino RBD, Bell RA, et al. Disparities in HbA1c levels between African-American and non-Hispanic white adults with diabetes. American Diabetes Association. Diabetes Care. 2006. 29(9):2130–2136. doi:10.2337/dc05-1973.

310. Flegal KM, Ezzati TM, Harris M, et al. Prevalence of diabetes in Mexican Americans, Cubans, and Puerto Ricans from the Hispanic Health and Nutrition Examination Survey, 1982–1984. American Diabetes Association. Diabetes Care. 1991. 14(7):628–638. doi:10.2337/diacare.14.7.628. http://care.diabetesjournals.org/content/14/7/628.short

311. Robbins JM, Vaccarino V, Zhang H, et al. Socioeconomic status and type 2 diabetes in African American and non-Hispanic white women and men: evidence from the Third National Health and Nutrition Examination Survey. American Journal of Public Health. 2001. 91(1):76–83. PMCID:1146485. PubMed. http://www.ncbi.nlm.nih.gov/pmc/articles/PMC1446485

312. Kington RS & Smith JP. Socioeconomic status and racial and ethnic differences in functional status associated with chronic diseases. American journal of Public Health. 1997. 87(5):805–810. PMCID:1381054. PubMed. http://www.ncbi.nlm.nih.gov/pmc/articles/PMC1381054

313. Karter AJ, Mayer-Davis EJ, Selby JV, et al. Insulin sensitivity and abdominal obesity in African-American, Hispanic, and non-Hispanic white men and women: the insulin resistance and atherosclerosis study. American Diabetes Association. Diabetes Care. 1996. 45(11):1547–1555. doi:10.2337/diab.45.11.1547. http://diabetes.diabetesjournals.org/content/45/11/1547.short

314. Drewnowski A & Darmon N. Food choices and diet costs: an economic analysis. The Journal of Nutrition. 2005. 135(4):900–904. http://jn.nutrition.org/content/135/4/900.short

315. Horowitz CR, Colson KA, Hebert PL, et al. Barriers to buying healthy foods for people with diabetes: Evidence of environmental disparities. American Journal of Public Health. 2004. 94(9):1549–1554. doi:10.2105/AJPH.94.9.1549. http://ajph.aphapublications.org/doi/abs/10.2105/AJPH.94.9.1549

316. Drewnowski A & Specter SE. Poverty and obesity: the role of energy density and energy costs. American Journal of Clinical Nutrition. 2004. 79(1):6–16. http://ajcn.nutrition.org/content/79/1/6.short

317. Morland K, Diez Roux AV, & Wing S. Supermarkets, other food stores, and obesity: the atherosclerosis risk in communities study. American Journal of Preventive Medicine. 2006. 30(4):333–339. Science Direct. http://www.sciencedirect.com/science/article/pii/S0749379705004836

318. Candib LM. Obesity and diabetes in vulnerable populations: reflection on proximal and distal Causes. Annals of family medicine. 2007. 5(6):547–556. doi:10.1370/afm.754. http://www.annfammed.org/content/5/6/547.full

319. Drewnowski A & Darmon N. The economics of obesity: dietary energy density and energy cost. American Journal of Clinical Nutrition. 2005. 82(1):265S–273S. http://ajcn.nutrition.org/content/82/1/265S.short

320. European Food Safety Authority. Bisphenol A. EFSA. Accessed February 23, 2016. http://www.efsa.europa.eu/en/topics/topic/bisphenol.htm

321. AFP RelaxNews. France bans contested chemical BPA in food packaging. December 13, 2012. NY Daily News. http://www.nydailynews.com/life-style/health/france-bans-contested-chemical-bpa-food-packaging-article-1.1219611

322. International Chemical Secretariat. Sweden to initiate a total phase out of bisphenol A. February 1, 2013. http://www.chemsec.org/news/news-2013/january-march/1117-sweden-to-initiate-a-total-phase-out-of-bisphenol-A

323. Austen I. Canada declares BPA, a chemical in plastics, to be toxic. The New York Times. October 13, 2010. http://www.nytimes.com/2010/10/14/world/americas/14bpa.html

324. Liao C, Liu F, & Kannan K. Bisphenol S, a new bisphenol analogue, in paper products and currency bills and its association with bisphenol A residues. Environmental Science & Technology. 2012. 46(12):6515–6522. PMID:22591511. PubMed. http://www.ncbi.nlm.nih.gov/pubmed/22591511

325. Cooper JE, Kendig EL, & Belcher SM. Assessment of bisphenol A released from reusable plastic, aluminium and stainless steel water bottles. Chemosphere. 2011. 85(6):943–947. Science Direct. http://www.sciencedirect.com/science/article/pii/S004565351100717X

326. Yang CZ, Yaniger SI, Jordan VC, et al. Most plastic products release estrogenic chemicals: a potential health problem that can be solved. Environmental Health Perspectives. 2011. 119(7):989–996. PMID:21367689. PubMed. http://www.ncbi.nlm.nih.gov/pubmed/21367689

327. Bittner, GD. Materials and food additives free of endocrine disruptive chemicals. Application number: 10/852,026. Publication number: US2004/0214926 A1. 2004. http://www.google.com/patents/US20040214926

328. Fassa P. Eliminate and reverse BPA toxicity. NaturalNews. October 27, 2011. http://www.natural news.com/033993_BPA_protection.html

329. Ibid.

Chemical Contaminants: Hexane

330. U.S. National Library of Medicine. Jet fuel. Haz-Map® occupational health database. Updated November 2013. U.S. National Institutes of Health. http://hazmap.nlm.nih.gov/category-details?id=698&table=copytblagents

331. U.S. Environmental Protection Agency. Hexane. Revised January 2000. Technology Transfer Network—Air Toxics. http://www.epa.gov/ttnatw01/hlthef/hexane.html

332. The National Safety Council. N-hexane chemical backgrounder. 2002. Wayback Machine. http://web.archive.org/web/20070519002303/http://www.nsc.org/ehc/chemical/N-Hexane.htm

333. Dunnick JK, Graham D, Yang RS, et al. Thirteen-week toxicity study of n-hexane in B6C3F1 mice after inhalation exposure. Toxicology. 1989. 57(2):163–72. PMID:2749745. PubMed. http://www.ncbi.nlm.nih.gov/pubmed/2749745

334. Howd RA, Rebert CS, Dickinson J, et al. A comparison of the rates of development of functional hexane neuropathy in weanling and young adult rats. Neurobehavioral Toxicology and Teratology. 1983. 5(1):63–8. PMID:6304548. PubMed. http://www.ncbi.nlm.nih.gov/pubmed/6304548

335. Cavender FL, Casey HW, Salem H, et al. A 13-week vapor inhalation study of n-hexane in rats with emphasis on neurotoxic effects. Fundamental and Applied Toxicology. 1984. 4(2 Pt 1):191–201. PMID:6724193. PubMed. http://www.ncbi.nlm.nih.gov/pubmed/6724193

336. Li H, Liu J, Sun Y, et al. N-hexane inhalation during pregnancy alters DNA promoter methylation in the ovarian granulosa cells of rat offspring. Journal of Applied Toxicology. 2013. 34(8):841–56. PMID:23740543. PubMed. http://www.ncbi.nlm.nih.gov/pubmed/23740543

337. Agency for Toxic Substances and Disease Registry (ATSDR). Toxicological profile for hexane. Public Health Service, U.S. Department of Health and Human Services. 1999. http://www.atsdr.cdc.gov/toxprofiles/tp113.pdf

338. U.S. Environmental Protection Agency. Hexane. Revised January 2000. Technology Transfer Network—Air toxics. http://www.epa.gov/ttnatw01/hlthef/hexane.html

339. U.S. Environmental Protection Agency. n-Hexane; CASRN 110-54-3. December 23, 2005. Integrated Risk Information System. http://www.epa.gov/iris/subst/0486.htm

340. Beall C, Delzell E, Rodu B, et al. Case-control study of intracranial tumors among employees at a petrochemical research facility. Journal of Environmental and Occupational Medicine. 2001. 43(12):1103–13. PMID:11765681. PubMed. http://www.ncbi.nlm.nih.gov/pubmed/11765681

341. Agency for Toxic Substances and Disease Registry (ATSDR). Toxicological profile for hexane. Public Health Service, U.S. Department of Health and Human Services. 1999. http://www.atsdr.cdc.gov/toxprofiles/tp113.pdf

342. The Cornucopia Institute. Behind the bean: the heroes and charlatans of the natural and organic soy foods industry—the social, environmental and health impacts of soy. May 8, 2009. http://www.cornucopia.org/2009/05/soy-report-and-scorecard/#more-1375

343. U.S. Environmental Protection Agency. Emission factor documentation for AP-42, section 9.11.1, vegetable oil processing final report. 1995. Office of Air Quality Planning and Standards Emission

Factor and Inventory Group. Research Triangle Park, NC. EPA Contract No. 68-D2-0159. http://www3.epa.gov/ttnchie1/ap42/ch09/bgdocs/b9s11-1.pdf

344. U.S. Environmental Protection Agency. Emission factor documentation for AP-42, section 9.11.1, vegetable oil processing final report. 1995. Office of Air Quality Planning and Standards Emission Factor and Inventory Group. Research Triangle Park, NC. EPA Contract No. 68-D2-0159. http://www3.epa.gov/ttnchie1/ap42/ch09/bgdocs/b9s11-1.pdf

345. The National Safety Council. N-hexane chemical backgrounder. 2002. Wayback Machine. http://web.archive.org/web/20070519002303/http://www.nsc.org/ehc/chemical/N-Hexane.htm

346. The Cornucopia Institute. Behind the bean: the heroes and charlatans of the natural and organic soy foods industry—the social, environmental and health impacts of soy. May 8, 2009. http://www.cornucopia.org/2009/05/soy-report-and-scorecard/#more-1375

347. Vallaeys C. DHA/ARA—replacing mother—imitating human breast milk in the laboratory, novel oils in infant formula and organic foods: safe and valuable functional food or risky marketing gimmick? The Cornucopia Institute. January 2008. http://cornucopia.org/DHA/DHA_FullReport.pdf

348. U.S. Centers for Disease Control and Prevention. n-Hexane. 1988 OSHA PEL project documentation. http://www.cdc.gov/niosh/pel88/110-54.html

349. The Cornucopia Institute. National Organic Standards Board votes on soy lecithin de-listing. May 21, 2009. http://www.cornucopia.org/2009/05/national-organic-standards-board-votes-on-soy-lecithin-de-listing

Chemical Contaminants: Pesticides

350. Plumer B. We've covered the world in pesticides. Is that a problem? Washington post. August 18, 2013. http://www.washingtonpost.com/blogs/wonkblog/wp/2013/08/18/the-world-uses-billions-of-pounds-of-pesticides-each-year-is-that-a-problem/

351. ReportLinker.com. World agriculture pesticides market. Freedonia. September 2012. 458. http://www.reportlinker.com/p0963176-summary/World-Agricultural-Pesticides-Market.html

352. Hargis C. U.S. weighing increase in herbicide levels in food supply. Inter Press Service. July 2, 2013. http://www.ipsnews.net/2013/07/u-s-weighing-increase-in-herbicide-levels-in-food-supply

353. Pesticide Action Network Europe (PANE). Is the pesticide industry really serious about their slogan? Time to change: accepting the challenge. June 26, 2013. http://www.pan-europe.info/old/Resources/Briefings/PANE%20-%202013%20-%20Is%20the%20Pesticide%20Industry%20really%20serious%20about%20their%20slogan.pdf

354. Short P & Colborn T. Pesticide use in the U.S. and policy implications: a focus on herbicides. Toxicology and Industrial Health. 1999. 15(1–2):240–75. PMID:10188206. PubMed. http://www.ncbi.nlm.nih.gov/pubmed/10188206

355. National Institutes of Environmental Health Services. Endocrine disruptors resource page. June 5, 2013. http://www.niehs.nih.gov/health/topics/agents/endocrine

356. U.S. Environmental Protection Agency. Endocrine disruptor screening program (EDSP). Updated December 5, 2013. https://www.epa.gov/endocrine-disruption

357. U.S. Environmental Protection Agency. Endocrine disruptor screening program for the 21st century: EDSP21 work plan. Office of Chemical Safety and Pollution Prevention. September 30, 2011. http://www.epa.gov/sites/production/files/2015-07/documents/edsp21_work_plan_summary_overview_final.pdf

358. U.S. Environmental Protection Agency. Endocrine Disruptor Screening Program (EDSP) Overview Updated 10.1.2015. http://www.epa.gov/endocrine-disruption/endocrine-disruptor-screening-program-edsp-overview

359. U.S. Environmental Protection Agency. Endocrine Disruptor Screening Program for the 21st Century: (EDSP21 Work Plan) September 30, 2011. http://www.epa.gov/sites/production/files/2015-07/documents/edsp21_work_plan_summary_overview_final.pdf

360. Franz US Patent #3799758A N-phosphonomethyl-glycine phytotoxicant compositions. Original Assignee Monsanto Co. Filing date: August 9, 1971. Publication date: March 26, 1974. http://www.google.com/patents/US3799758?dq=john+e+franz

361. Material Safety Data Sheet. Roundup Original Herbicide. UC Davis. November 1997. http://greenhouse.ucdavis.edu/pest/pmsds/Roundup.PDF

362. American Chemical Society. New Weapons on the Way to Battle Wicked Weeds. September 8, 2013.

363. Hoffman B. GMO crops mean more herbicide, not less. Forbes. July 2, 2013. http://www.forbes.com/sites/bethhoffman/2013/07/02/gmo-crops-mean-more-herbicide-not-less

364. Adler J. The growing menace from superweeds. Scientific American. 2011. 304:74-79. Nature. http://www.nature.com/scientificamerican/journal/v304/n5/full/scientificamerican0511-74.html

365. Behrens M, Mutlu N, Chakraborty S, et al. Dicamba resistance: enlarging and preserving biotechnology-based weed management strategies. Center for Plant Science Innovation, University of Nebraska - Lincoln. May 25, 2007. http://digitalcommons.unl.edu/cgi/viewcontent.cgi?article=1031&context=plantscifacpub

366. U.S. Environmental Protection Agency. Glyphosate; pesticide tolerances. 40 CFR Part 180. [EPA--HQ--OPP--2012--0132; FRL--9384--3]. Final rule. Federal register. May 1, 2013. 78(84):25396--25401. http://www.gpo.gov/fdsys/pkg/FR-2013-05-01/pdf/2013-10316.pdf

367. GM Watch. U.S. plans to hike allowed glyphosate levels in food supply. July 2, 2013. http://www.gmwatch.org/index.php/news/archive/2013/14769-u-s-plans-to-hike-allowed-glyphosate-levels-in-food-supply

368. Ibid.

369. GM Watch. El Salvador votes to ban glyphostate. September 19, 2013. http://gmwatch.org/index.php/news/archive/2013/15066-el-salvador-votes-to-ban-glyphostate

370. Flint JL & Barrett M. Effects of glyphosate combinations with 2,4-D or dicamba on field bindweed (Convolvulus arvensis). Weed science. 1989. 37(1):12-18 0043-1745. Food and Agriculture Organization of the United Nations. http://agris.fao.org/agris-search/search/display.do?f=1989/US/US89363.xml;US8908991

371. Ibid.

372. Bayer CropScience. Bayer CropScience announces intention to construct a state-of-the-art facility for glufosinate-ammonium herbicide. Press release. May 15, 2013. https://web.archive.org/web/20130617033346/http://www.cropscience.bayer.com/en/Media/Press-Releases/2013/Bayer-CropScience-announces-intention-construct-a-state-of-the-art-facility-glufosinate-ammonium.aspx?

373. Monsanto Company. Roundup Ready PLUS weed management solutions. Product page. 2013. http://www.roundupreadyplus.com/Pages/Home.aspx

374. Monsanto reveals plans for dicamba, glufosinate tolerant pairing. Cotton 24/7. March 1, 2012. http://www.cotton247.com/article/27603/monsanto-reveals-plans-for-dicamba-glufosinate-tolerant-pairing

375. National Pesticide Information Center. Active Ingredients for 2,4-D. Updated October 12, 2012. http://npic.orst.edu/ingred/24d.html

376. Muth N. Gene flow might turn wimps into superweeds. Nature. February 20, 2003. 421(6925):785-6. http://www.nature.com/nature/journal/v421/n6925/full/421785c.html

377. Henderson AM, Gervais JA, Luukinen B, et al. 2010. Glyphosate technical fact sheet. National Pesticide Information Center, Oregon State University Extension Services. http://npic.orst.edu/factsheets/archive/glyphotech.html#reg

378. Samsel A & Seneff S. Glyphosate's suppression of cytochrome P450 enzymes and amino acid biosynthesis by the gut microbiome: pathways to modern diseases. Entropy. 2013. 15:1416-1463. DOI:10.3390/e15041416.

379. Gillam C. Heavy use of herbicide Roundup linked to health dangers—U.S. Study. Reuters. April 25, 2013. http://www.reuters.com/article/2013/04/25/roundup-health-study-idUSL2N0DC22F20130425

380. Ibid.

381. Antoniou M, Habib ME, Howard CV, et al. Roundup and birth defects: is the public being kept in the dark? Earth Open Source. 2011. http://earthopensource.org/wp-content/uploads/RoundupandBirthDefectsv5.pdf

382. Graves L. Roundup: birth defects caused by world's top-selling weedkiller, scientists say. The Huffington post. June 24, 2011. http://www.huffingtonpost.com/2011/06/24/roundup-scientists-birth -defects_n_883578.html

383. Center for Biological Diversity. FDA to Start Testing Monsanto's Glyphosate in Food. EcoWatch, February 2016. http://ecowatch.com/2016/02/18/fda-test-food-glyphosate

384. U.S. Environmental Protection Agency. Atrazine Interim Reregistration Eligibility Decision (IRED) Q&A's — January 2003. Pesticides: Topical & Chemical Fact Sheets. Accessed March 1, 2016.

385. U.S. Environmental Protection Agency. Triazine cumulative risk assessment and atrazine, simazine, and propazine decisions. June 22, 2006. Pesticides: health and safety. May 9, 2012. http://web .archive.org/web/20150907123533/http://www.epa.gov/oppsrrd1/cumulative/triazine_fs.htm

386. Pape-Lindstrom PA & Lydy MJ. Synergistic toxicity of atrazine and organophosphate insecticides contravenes the response addition mixture model. Environmental Toxicology and Chemistry. November 1997. 16(11):2415–2420. doi:10.1002/etc.5620161130. Wiley Online Library. http:// onlinelibrary.wiley.com/doi/10.1002/etc.5620161130/abstract

387. Anderson TD & Yan Zhu K. Synergistic and antagonistic effects of atrazine on the toxicity of organophosphorodithioate and organophosphorothioate insecticides to Chironomus tentans (Diptera: Chironomidae). Pesticide Biochemistry and Physiology. 2004. 80(1):54–64. http://dx.doi .org/10.1016/j.pestbp.2004.06.003

388. U.S. Environmental Protection Agency. Basic information about atrazine in drinking water. Updated September 17, 2013. https://www.epa.gov/ingredients-used-pesticide-products/atrazine -background-and-updates

389. World Health Organization. Atrazine in drinking water: background document for development of WHO guidelines for drinking-water quality. 2003. http://www.who.int/water_sanitation_health/ dwq/atrazinerev0305.pdf

390. Berry I. Syngenta settles weedkiller lawsuit. Wall Street Journal. May 25, 2012. http://online.wsj .com/news/articles/SB10001424052702304840904577426172221346482

391. Howard C. Special report: Syngenta's campaign to protect atrazine, discredit critics. Environmental Health News. June 17, 2013. http://www.environmentalhealthnews.org/ehs/news/2013/atrazine

392. Wu M, Quirindongo M, Sass J, et al. Poisoning the well: how the EPA is ignoring atrazine contamination in surface and drinking water in the central United States. August 2009. Natural Resources Defense Council (NDRC). https://www.nrdc.org/health/atrazine/files/atrazine.pdf

393. Hayes TB, et al. 2002. Atrazine-induced hermaphroditism at 0.1 ppb in American leopard frogs (Rana pipiens): laboratory and field evidence. Environmental Health Perspectives. 111:568–575. PMCID:PMC1241446. PubMed. http://www.ncbi.nlm.nih.gov/pmc/articles/PMC1241446

394. De Noyelles F, Kettle D, & Sinn DE. The responses of plankton communities in experimental ponds to atrazine, the most heavily used pesticide in the United States. Ecology. 1982. 63(5):1285–1293. JSTOR. http://www.jstor.org/stable/1938856

395. Burken JG & Schnoor JL. Uptake and metabolism of atrazine by poplar trees. Environmental Science & Technology. 1997. 31(5):1399–1406. doi:10.1021/es960629v. American Chemical Society. http://pubs.acs.org/doi/abs/10.1021/es960629v

396. Lennartz B, Louchart X, Voltz M, et al. Diuron and simazine losses to runoff water in Mediterranean vineyards. Journal of Environmental Quality. November 13, 1996. 26(6):1493–1502. doi:10.2134/jeq1997.00472425002600060007x. Alliance of Crop, Soil, and Environmental Science Societies. https://dl.sciencesocieties.org/publications/jeq/abstracts/26/6/JEQ0260061493

397. Behki RM & Khan SU. Degradation of atrazine, propazine, and simazine by rhodococcus strain B-30. Journal of Agricultural and Food Chemistry. 1994. 42(5):1237–1241. doi:10.1021/ jf00041a036. American Chemical Society. http://pubs.acs.org/doi/abs/10.1021/jf00041a036

398. Casida JE, Gammon DW, Glickman AH, et al. Mechanisms of selective action of pyrethroid insecticides. Annual Review of Pharmacology and Toxicology. April 1983. 23:413–438. doi:10.1146/ annurev.pa.23.040183.002213. http://www.annualreviews.org/doi/abs/10.1146/annurev.pa .23.040183.002213?journalCode=pharmtox

399. Ritter L, Solomon KR, Forget J, et al. Persistent organic pollutants. United Nations Environment Programme. http://cdrwww.who.int/ipcs/assessment/en/pcs_95_39_2004_05_13.pdf

400. UNEP Stockholm Convention. Protecting human health and the environment from persistent organic pollutants. Secretariat of the Stockholm Convention. Updated 2008. http://chm.pops.int/TheConvention/Overview/tabid/3351/Default.aspx

401. Simonich SL & Hites RA. Global distribution of persistent organochlorine compounds. Science magazine. September 29, 1995. 269(5232):1851–1854. doi:10.1126/science.7569923. http://www.sciencemag.org/content/269/5232/1851.short

402. Falandysz J & Kannan K. Organochlorine pesticide and polychlorinated biphenyl residues in slaughtered and game animal fats from the northern part of Poland. Zeitschrift für Lebensmittel-Untersuchung und -Forschung. 1992. 195(1):17–21. PMID:1502854. PubMed. http://www.ncbi.nlm.nih.gov/pubmed/1502854

403. Panseri S, Biondi PA, Vigo D, et al. Occurrence of organochlorine pesticides residues in animal feed and fatty bovine tissue. Ch. 13. Food Industry. 2013. ISBN: 978-953-51-0911-2. http://dx.doi.org/10.5772/54182

404. Jukes TH. The tragedy of DDT. Rational readings on environmental concerns. Ed. JH Lehr. 1992. New York, NY: International Thomson Publishing.

405. Sharpe RM & Irvine DS. How strong is the evidence of a link between environmental chemicals and adverse effects on human reproductive health? British Medical Journal. 2004. 328(7437):447–451. PMCID:344268. PubMed. http://www.ncbi.nlm.nih.gov/pmc/articles/PMC344268

406. Safe SH. Endocrine disruptors and human health—is there a problem? An update. Environmental Health Perspectives. 2000. 108(6):487–493. PMCID:PMC1638151. PubMed. http://www.ncbi.nlm.nih.gov/pmc/articles/PMC1638151

407. Hunter DJ, Hankinson SE, Laden F, et al. Plasma organochlorine levels and the risk of breast cancer. New England Journal of Medicine. 1997. 337:1253–1258. doi:10.1056/NEJM199710303371801. http://www.nejm.org/doi/full/10.1056/nejm199710303371801

408. Cocco P, Blair A, Congia P, et al. Proportional mortality of dichloro-diphenyl-trichloroethane (DDT) workers: a preliminary report. Archives of Environmental Health: An International Journal. 1997. 2(4). DOI: 10.1080/00039899709602202. http://www.tandfonline.com/doi/abs/10.1080/00039899709602202#.Uq19QfaLrk0

409. Lear L. Rachel Carson: witness for nature. 1997. ISBN: 0-8050-3428-5. New York, NY: Henry Holt.

410. World Health Organization. DDT in indoor residual spraying: human health aspects. Environmental health criteria 241. International Programme on Chemical Safety and the Inter-Organization Programme for the Sound Management of Chemicals. 2011. http://www.who.int/ipcs/publications/ehc/ehc241.pdf

411. Whelan EM, Ross GL, & Stimola AN, eds. A summary of America's war on "carcinogens": reassessing the use of animal tests to predict human cancer risk. American Council on Science and Health. January 2005. http://www.acsh.org/wp-content/uploads/2012/04/20050126_WOCSummary.pdf

412. Dees C, Askari M, Foster JS, et al. DDT mimicks estradiol stimulation of breast cancer cells to enter the cell cycle. Molecular Carcinogenesis. 1997. 18(2):107–14. PMID: 9049186. Science. NaturalNews. http://science.naturalnews.com/1997/8807697_DDT_mimicks_estradiol_stimulation_of_breast_cancer_cells_to_enter.html

413. Hosie S, Loff S, Witt K, et al. Is there a correlation between organochlorine compounds and undescended testes? European Journal of Pediatric Surgery. 2000. 10:304–309. Mindfully. http://www.mindfully.org/Health/Organochlorine-Undescended-Testes.htm

414. Cohn BA, Cirillo PM, & Christianson RE. Prenatal DDT exposure and testicular cancer: a nested case-control study. Archives of Environmental & Occupational Health. 2010. 65(3):127–34. PMID:20705572. PubMed. http://www.ncbi.nlm.nih.gov/pubmed/20705572

415. Dich J, Zahm SH, Hanberg A, et al. Pesticides and cancer. Cancer Causes & Control. 1997. 8(3):420–443. doi:10.1023/A:1018413522959. Springer Link. http://link.springer.com/article/10.1023/A:1018413522959

416. Hileman B. Environmental estrogens linked to reproductive abnormalities, cancer. Chemical and Engineering News. January 31, 1994. http://infohouse.p2ric.org/ref/25/24379.pdf

417. Dich J, Zahm SH, Hanberg A, et al. Pesticides and cancer. Cancer Causes & Control. 1997. 8(3):420–443. doi:10.1023/A:1018413522959. Springer Link. http://link.springer.com/article /10.1023/A:1018413522959

418. Lee DH, Lee IK, Song K, et al. A strong dose-response relation between serum concentrations of persistent organic pollutants and diabetes: results from the National Health and Examination Survey 1999–2002. American Diabetes Association. Diabetes Care. 2006. 29(7):1638–1644. doi:10.2337/dc06-0543. http://care.diabetesjournals.org/content/29/7/1638.full

419. Wolff, MS, Toniolo PG, Lee WE, et al. Blood levels of organochlorine residues and risk of breast cancer. Journal of the National Cancer Institute. 1993. 85(8):648–652. doi:10.1093/ jnci/85.8.648. http://jnci.oxfordjournals.org/content/85/8/648.short

420. Toft G, Hagmar L, Giwercman A, et al. Epidemiological evidence on reproductive effects of per- sistent organochlorines in humans. Reproductive toxicity. 2004. 19(1):5–26. http://dx.doi.org /10.1016/j.reprotox.2004.05.006

421. Payne J, Scholze M, & Kortenkamp A. Mixtures of four organochlorines enhance human breast cancer cell proliferation. Environmental health perspectives. 2001. 109(4):391–397. PMCID: PMC1240280. PubMed. http://www.ncbi.nlm.nih.gov/pmc/articles/PMC1240280

422. McGlynn KA, Quraishi SM, Graubard BI, et al. Persistent organochlorine pesticides and risk of testicular germ cell tumors. Journal of the National Cancer Institute. 2008. 100(9): 663–671. doi:10.1093/jnci/djn101 http://jnci.oxfordjournals.org/content/100/9/663.short

423. Xiao P, Mori T, Kamei I, et al. Metabolism of organochlorine pesticide heptachlor and its metab- olite heptachlor epoxide by white rot fungi, belonging to genus Phlebia. FEMS Microbiology Let- ters. 2011. 314(2):140–6. PMID:21087297. Science.NaturalNews. http://science.naturalnews .com/pubmed/21087297.html

424. U.S. Environmental Protection Agency. Organophosphate pesticides: revised cumulative risk as- sessment. Updated June 10, 2002. http://nepis.epa.gov/Exe/ZyPURL.cgi?Dockey=9100BFLL.txt

425. U.S. Environmental Protection Agency. Clothianidin—registration status and related information. Updated July 27, 2012. https://web.archive.org/web/20150924062654/http://www.epa.gov /pesticides/about/intheworks/clothianidin-registration-status.html

426. Anaesth BJ. Organophosphorous insecticide poisoning. British Journal of Anaesthesia. 1989. 63(6):736–750. doi:10.1093/bja/63.6.736. http://bja.oxfordjournals.org/content/63/6/736.full .pdf+html

427. Dich J, Zahm SH, Hanberg A, et al. Pesticides and cancer. Cancer Causes & Control. 1997. 8(3):420–443. doi:10.1023/A:1018413522959. Springer Link. http://link.springer.com /article/10.1023/A:1018413522959

428. Pesticide Action Network. Parathion: identification, toxicity, use, water pollution potential, eco- logical toxicity and regulatory information. PAN Pesticides Database—Chemicals. 2010. http:// pesticideinfo.org/Detail_Chemical.jsp?Rec_Id=PC35122#Toxicity

429. Ibid.

430. Cornell University. Remaining use of pesticide ethyl parathion canceled. Press release. October 13, 2000. http://archive.is/JVZt

431. U.S. Environmental Protection Agency. Parathion (CASRN 56-38-2). Integrated Risk Information System. Updated August 22, 1988. https://cfpub.epa.gov/ncea/iris2/chemicalLanding.cfm?substance _nmbr=327

432. U.S. Environmental Protection Agency. Pesticides: reregistration—chlorpyrifos facts. EPA 738-F- 01-006. February 2002. Updated May 9, 2012. http://nepis.epa.gov/Exe/ZyPURL.cgi?Dockey =200005F3.txt

433. Goodman B. Pesticide exposure in womb linked to lower IQ: studies show kids exposed in preg- nancy may also have later problems with attention and memory. WebMD health news. April 21, 2011. http://www.webmd.com/baby/news/20110421/pesticide-exposure-in-womb-linked-to -lower-iq

434. Bouchard MF, Bellinger DC, Wright RO, et al. Attention-deficit/hyperactivity disorder and urinary metabolites of organophosphate pesticides. Pediatrics. 2010. 125(6):e1270–7. doi:10.1542/peds.2009-3058. http://pediatrics.aappublications.org/content/early/2010/05/17/peds.2009-3058.abstract

435. Klein S. Study: ADHD linked to pesticide exposure. CNN. May 17, 2010. http://www.cnn.com/2010/HEALTH/05/17/pesticides.adhd

436. Ray DE & Richard PG. The potential for toxic effects of chronic, low-dose exposure to organophosphates. Toxicology Letters. 2001. 120(1–3):343–351. http://dx.doi.org/10.1016/S0378-4274(01)00266-1

437. Jurewicz J & Hanke W. Prenatal and childhood exposure to pesticides and neurobehavioral development: review of epidemiological studies. International Journal of Occupational Medicine and Environmental Health. 2008. 21(2):121–132. doi:10.2478/v10001-008-0014-z. De Gruyter. http://www.degruyter.com/view/j/ijmh.2008.21.issue-2/v10001-008-0014-z/v10001-008-0014-z.xml

438. Engel SM, Wetmur J, Chen J, et al. Prenatal exposure to organophosphates, paraoxonase 1, and cognitive development in childhood. Environmental Health Perspectives. 2011. 119(8):1182–1188. PMCID:3237356. PubMed. http://www.ncbi.nlm.nih.gov/pmc/articles/PMC3237356

439. Rauh V, Arunajadai S, Horton M, et al. Seven-year neurodevelopmental scores and prenatal exposure to chlorpyrifos, a common agricultural pesticide. Environmental Health Perspectives. 2011. 119(8):1196–1201. PMCID:3237355. PubMed. http://www.ncbi.nlm.nih.gov/pmc/articles/PMC3237355

440. Fielder N, Kipen H, Kelly-McNeil K, et al. Long-term use of organophosphates and neuropsychological performance. American Journal of Industrial Medicine. 1997. 32(5):487–496. doi:10.1002/(SICI)1097-0274. Wiley Online Library. http://onlinelibrary.wiley.com/doi/10.1002/%28SICI%291097-0274%28199711%2932:5%3C487::AID-AJIM8%3E3.0.CO;2-P/abstract

441. Stokes L, Stark A, Marshall E, et al. Neurotoxicity among pesticide applicators exposed to organophosphates. Occupational & Environmental Medicine. 1995. 52:648–653. doi:10.1136/oem.52.10.648. http://oem.bmj.com/content/52/10/648.short

442. Bouchard MF, Chevrier J, Harley KG, et al. Prenatal exposure to organophosphate pesticides and IQ in 7-year-old children. Environmental Health Perspectives. 2011. 119(8):1189–1195. PMCID: 3237357. PubMed. http://www.ncbi.nlm.nih.gov/pmc/articles/PMC3237357

443. Sebire M, Scott A, Tyler C, et al. The organophosphorous pesticide, fenitrothion, acts as an anti-androgen and alters reproductive behavior of the male three spined stickleback, Gasterosteus aculeatus. Ecotoxicology. 2009. 18(1):122–33. PMID: 18807270. Science.NaturalNews. http://science.naturalnews.com/pubmed/18807270.html

444. Casida, JE, Gammon DW, Glickman AH, et al. Mechanisms of selective action of pyrethroid insecticides. Annual Review of Pharmacology and Toxicology. 1983. 23:413–438. doi:10.1146/annurev.pa.23.040183.002213. http://www.annualreviews.org/doi/abs/10.1146/annurev.pa.23.040183.002213?journalCode=pharmtox

445. U.S. Environmental Protection Agency. Pesticides: regulating pesticides—pyrethroids and pyrethrins. Updated April 2013. http://www.epa.gov/oppsrrd1/reevaluation/pyrethroids-pyrethrins.html

446. Ibid.

447. Casida, JE, Gammon DW, Glickman AH, et al. Mechanisms of selective action of pyrethroid insecticides. Annual Review of Pharmacology and Toxicology. 1983. 23:413–438. doi:10.1146/annurev.pa.23.040183.002213. http://www.annualreviews.org/doi/abs/10.1146/annurev.pa.23.040183.002213?journalCode=pharmtox

448. U.S. Environmental Protection Agency. Pesticides: regulating pesticides—pyrethroids and pyrethrins. Updated April 2013. http://www.epa.gov/oppsrrd1/reevaluation/pyrethroids-pyrethrins.html

449. Illinois Department of Public Health. Environmental health fact sheet: pyrethroid insecticides. June 2007. http://www.idph.state.il.us/envhealth/factsheets/pyrethroid.htm

450. Vijverberg HPM & Vanden Bercken J. Neurotoxicological effects and the mode of action of pyrethroid insecticides. Critical Reviews in Toxicology. 1990. 21(2):105–126. doi:10.3109

/10408449009089875. Informa. http://informahealthcare.com/doi/abs/10.3109
/10408449009089875

451. Soderlund DM & Bloomquist JR. Neurotoxic Actions of pyrethroid insecticides. Annual Review of Entomology. 1989. 34(1):77–96. doi:10.1146/annurev.en.34.010189.000453. http://www.annualreviews.org/doi/abs/10.1146/annurev.en.34.010189.000453?journalCode=ento

452. Weston DP & Lydy MJ. Urban and agricultural sources of pyrethroid insecticides to the Sacramento–San Joaquin Delta of California. Environmental Science & Technology. 2010. 44(5):1833–1840. doi:10.1021/es9035573. American Chemical Society. http://pubs.acs.org/doi/abs/10.1021/es9035573

453. Bhanoo S. Household pesticide is finding its way into California rivers, study suggests. The New York Times. February 3, 2010. http://green.blogs.nytimes.com/2010/02/03/household-pesticide-is-finding-its-way-into-california-rivers-study-suggests/?_r=0

454. Lu D, Wang D, Feng C, et al. Urinary concentrations of metabolites of pyrethroid insecticides in textile workers, eastern China. Environment International. 2013. 60:137–44. PMID:24056321. PubMed. http://www.ncbi.nlm.nih.gov/pubmed/24056321

455. Fukuto TR. Mechanism of action of organophosphorus and carbamate insecticides. Environmental Health Perspectives. July 1990. 87:245–254. PMCID: PMC1567830. PubMed. http://www.ncbi.nlm.nih.gov/pmc/articles/PMC1567830/

456. Roberts JR & Karr CJ. Pesticide exposure in children. Pediatrics. 1 December 2012. 130(6):e1765-e1788. DOI:10.1542/peds.2012-2758. http://pediatrics.aappublications.org/content/130/6/e1765.abstract

457. Mulla MS, Darwazeh HA, Ede L, et al. Laboratory and field evaluation of the IGR fenoxycarb against mosquitoes. Journal of the American Mosquito Control Association. 1985. 1(4):442–8. PMID:3880261. PubMed. http://www.ncbi.nlm.nih.gov/pubmed/3880261

458. McKenney CL, Cripe GM, Foss SS, et al. Comparative embryonic and larval developmental responses of estuarine shrimp (palaemonetes pugio) to the juvenile hormone agonist fenoxycarb. Archives of Environmental Contamination and Toxicology. 2004. 47(4):463–470. doi:10.1007/s00244-002-0294-4.

459. Grenier S & Grenier AM. Fenoxycarb, a fairly new insect growth regulator: review of its effects on insects. Annals of Applied Biology. 1993. 122(2):369–403. doi: 10.1111/j.1744-7348.1993.tb04042.x. Wiley Online Library. http://onlinelibrary.wiley.com/doi/10.1111/j.1744-7348.1993.tb04042.x/abstract

460. Tatarazako N & Oda S. The water flea daphnia magna (crustacea, cladocera) as a test species for screening and evaluation of chemicals with endocrine disrupting effects on crustaceans. Exotoxicology. 2007. 16:197–203. doi:10.1007/s10646-006-0120-2. Springer Link. http://link.springer.com/article/10.1007/s10646-006-0120-2#page-2

461. Dich J, Hoar Zahm S, Hanberg A, et al. Pesticides and cancer. Cancer Causes and Control. 1997. 8(3):420–443. doi:10.1023/A:1018413522959. Springer Link. http://link.springer.com/article/10.1023/A:1018413522959

462. Extension Toxicology Network. Pesticide information profiles: propoxur. Updated June 1996. Oregon State University. http://extoxnet.orst.edu/pips/propoxur.htm

463. Walsh LM & Keeney DR. Behavior and phototoxicity of inorganic arsenicals in soils. Arsenical Pesticides. Ch. 3, 35–52. doi:10.1021/bk-1975-0007.ch003. American Chemical Society. http://pubs.acs.org/doi/abs/10.1021/bk-1975-0007.ch003

464. Murphy EA & Aucott M. An assessment of the amounts of arsenical pesticides used historically in a geographical area. Science of the Total Environment. 1998. 218(2–3):89–101. http://dx.doi.org/10.1016/S0048-9697(98)00180-6

465. U.S. Environmental Protection Agency. Chromated copper arsenate (CCA) fact sheet. Pesticides: Regulating Pesticides. Updated July 2011. https://www.epa.gov/ingredients-used-pesticide-products/chromated-copper-arsenate-cca

466. Smith E, Naidu R, & Alston AM. Arsenic in the soil environment: a review. Advances in Agronomy. 1998. 64:149–195. http://dx.doi.org/10.1016/S0065-2113(08)60504-0

467. Robinson GR, Larkins P, Boughton CJ, et al. Assessment of contamination from arsenical pesticide use on orchards in the Great Valley region, Virginia and West Virginia, USA. Journal of Environmental Quality. 2007. 36(3):654–663. doi:10.2134/jeq2006.0413. https://dl.sciencesocieties.org/publications/jeq/abstracts/36/3/654

468. Hartwig J, Becker B, Erdelen C, et al. Imidacloprid—a new systemic insecticide. Pflanzenschutz-Nachrichten Bayer. http://agris.fao.org/agris-search/search.do?f=1992/DE/DE92077.xml ;DE92U0152

469. Chao SL & Casida JE. Interaction of imidacloprid metabolites and analogs with the nicotinic acetylcholine receptor of mouse brain in relation to toxicity. Pesticide Biochemistry and Physiology. 1997. 58(1):77–88. http://dx.doi.org/10.1006/pest.1997.2284

470. Elbert A, Overbeck H, Iwaya K, et al. Imidacloprid, a novel systemic nitromethylene analogue insecticide for crop protection. Brighton Crop Protection Conference, Pests and Diseases. 1990. 1:21–28. Cab Direct. http://www.cabdirect.org/abstracts/19911155517.html

471. Badgujara PC, Jaina SK, Singh A, et al. Immunotoxic effects of imidacloprid following 28 days of oral exposure in BALB/c mice. Environmental Toxicology and Pharmacology. 2013. 35(3):408–18. Science Direct. http://www.sciencedirect.com/science/article/pii/S1382668913000148

472. Di Prisco G, Cavaliere V, Annoscia D, et al. Neonicotinoid clothianidin adversely affects insect immunity and promotes replication of a viral pathogen in honey bees. Proceedings of the National Academy of Sciences of the United States of America. 2013. 110(46):18466–18471. doi:10.1073/pnas.1314923110. http://www.pnas.org/content/110/46/18466

473. Mineau P & Palmer C. Neonicotinoid Insecticides and Birds: The impact of the nation's most widely used insecticides on birds. American bird conservancy. March 2013. https://extension.entm.purdue.edu/neonicotinoids/PDF/TheImpactoftheNationsMostWidelyUsedInsecticidesonBirds.pdf

474. European Food Safety Authority. EFSA assesses potential link between two neonicotinoids and developmental neurotoxicity. Press release. December 17, 2013. http://www.efsa.europa.eu/en/press/news/131217.htm

475. Alliance for Natural Health U.S.A. Pesticides definitively linked to bee colony collapse. April 2, 2013. http://www.anh-usa.org/pesticides-definitively-linked-to-bee-colony-collapse

476. Mullin CA, Frazier M, Frazier JL, et al. High levels of miticides and agrochemicals in North American apiaries: implications for honey bee health. PLOS ONE. 2010. 5(3):e9754. PMID:20333298. Science.NaturalNews. http://science.naturalnews.com/pubmed/20333298.html

477. Wu JY, Anelli CM, & Sheppard WS. Sub lethal effects of pesticide residues in brood comb on worker honey bee (Apis mellifera) development and longevity. PLOS ONE. 2011. 6(2):e14720. PMID:21373182. Science.NaturalNews. http://science.naturalnews.com/pubmed/21373182.html

478. Girolami V, Mazzon L, Squartini A, et al. Translocation of neonicotinoid insecticides from coated seeds to seedling guttation drops: a novel way of intoxication for bees. Journal of Economic Entomology. 2009. 102(5):1808–1815. http://dx.doi.org/10.1603/029.102.0511

479. Pettis JS, Lichtenberg EM, Andree M, et al. Crop pollination exposes honey bees to pesticides which alters their susceptibility to the gut pathogen Nosema ceranae. PLOS ONE. 2013. 8(7):e70182. doi:10.1371/journal.pone.0070182.

480. Johnson RM, Evans JD, Robinson GE, et al. Changes in transcript abundance relating to colony collapse disorder in honey bees (Apis mellifera). PNAS. 2009. 106(35):14790–14795. doi:10.1073/pnas.0906970106

481. Whitehorn PR, O'Connor S, Wackers FL, et al. Neonicotinoid pesticide reduces bumble bee colony growth and queen production. Science magazine. 2012. 336(6079):351–2. doi:10.1126/science.1215025. http://www.sciencemag.org/content/336/6079/351.abstract

482. Henry M, Béguin M, Requier F, et al. A common pesticide decreases foraging success and survival in honey bees. Science magazine. 2012. 336(6079):348–350. doi:10.1126/science.1215039. http://www.sciencemag.org/content/336/6079/348.short

483. CBC Radio. Ontario beekeeper blames pesticides for staggering honeybee deaths. As it Happens with Carol Off & Jeff Douglas. July 3, 2013. www.cbc.ca/asithappens/features/2013/07/03/post-6

484. Petroff A. EU bans some pesticides to save its bees. CNN Money. April 29, 2013. http://money.cnn.com/2013/04/29/news/world/bees-ban-pesticide-europe/index.html

485. Phys.org. Syngenta challenges EU's bee-saving pesticide ban. August 27, 2013. http://phys.org/news/2013-08-syngenta-eu-bee-saving-pesticide.html

486. U.S. Environmental Protection Agency. Clothianidin—registration status and related information. 2012. http://web.archive.org/web/20150814014006/http://www.epa.gov/pesticides/about/intheworks/clothianidin-registration-status.html

487. U.S. Department of Agriculture. Report on the national stakeholders conference on honey bee health. National Honey Bee Health Stakeholder Conference Steering Committee. October 15–17, 2012. http://www.usda.gov/documents/ReportHoneyBeeHealth.pdf

488. Baker BP, Benrook CM, Groth E, et al. Pesticide residues in conventional, integrated pest management (IPM)-grown and organic foods: insights from three US data sets. Food Additives and Contaminants. 2002. 19(5):427–46. PMID:12028642. PubMed. http://www.ncbi.nlm.nih.gov/pubmed/12028642

489. Burros M. Study finds far less pesticide residue on organic produce. The New York Times. May 8, 2002. http://www.nytimes.com/2002/05/08/science/08PEST.html

490. U.S. Department of Agriculture. Organic certification: National Organic Program (NOP). Agricultural Marketing Service. Updated November4, 2013. http://www.ams.usda.gov/AMSv1.0/nop

491. U.S. Government Printing Office. Electronic Code of Federal Regulations. Title 7: agriculture. Part 205—National Organic Program. Subpart G—administrative. §205.600. Evaluation criteria for allowed and prohibited substances, methods, and ingredients. Accessed January 28, 2016. http://www.ecfr.gov/cgi-bin/text-idx?c=ecfr;sid=7f0273852439530d013eb36c05531494;rgn=div7;view=text;node=7:3.1.1.9.32.7.354;idno=7;cc=ecfr

492. Caboni P, Sherer TB, Zhang N, et al. rotenone, deguelin, their metabolites, and the rat model of Parkinson's disease. Chemical Research in Toxicology. 2004. 17(11):1540–1548. doi:10.1021/tx049867r. American Chemical Society. http://pubs.acs.org/doi/abs/10.1021/tx049867r

493. Tanner CM, Kamel F, Ross GW, et al. Rotenone, paraquat, and Parkinson's disease. Environmental Health Perspectives. 2011. 119:866–872. http://ehp.niehs.nih.gov/1002839/

494. Merchant M. Spinosad: an insecticide to make organic gardeners smile. Horticulture update. March 2004. Texas A&M University. http://aggie-horticulture.tamu.edu/newsletters/hortupdate/hortupdate_archives/2004/mar04/Spinosad.html

495. Yano BL, Bond DM, Novilla MN, et al. Spinosad insecticide: subchronic and chronic toxicity and lack of carcinogenicity in Fischer 344 rats. Toxicological Sciences. 2002. 65(2):288–298. doi:10.1093/toxsci/65.2.288. Oxford Journals. http://toxsci.oxfordjournals.org/content/65/2/288.short

Food Ingredients as Contaminants (intro)

496. Taylor MR. The de minimis interpretation of the Delaney Clause: legal and policy rationale. Journal of the American College of Toxicology. 1988. 7(4). Sage Journals. http://ijt.sagepub.com/content/7/4/529.extract

497. Verrett J & Carper J. Eating may be hazardous to your health. 1974. ISBN: 0-385-11193-2. New York, NY: Simon and Schuster.

Food Ingredients as Contaminants: Aspartame

498. Roberts HJ. Aspartame disease: an ignored epidemic. 2001. ISBN-10: 0671217224. West Palm Beach, FL: Sunshine Sentinel Press.

499. MedlinePlus. Methanol poisoning. U.S. National Library of Medicine. National Institutes of Health. Updated January 30, 2013. http://www.nlm.nih.gov/medlineplus/ency/article/002680.htm

500. Mundy W, Padilla S, Shafter T, et al. Building a database of developmental neurotoxicants: evidence from human and animal studies. U.S. Environmental Protection Agency. 2009. http://www.fluoridealert.org/wp-content/uploads/epa_mundy.pdf

501. U.S. Department of Health and Human Services. Symptoms attributed to aspartame in complaints submitted to the FDA. April 20, 1995. SweetPoison.com. http://www.sweetpoison.com/articles/0706/aspartame_symptoms_submit.html

502. Pattanaargson S, Chuapradit C, & Srisukphonraruk S. Aspartame degradation in solutions at various pH conditions. Food Chemistry and Toxicology. 2001. 66(6):808–809. Online Wiley Library. http://onlinelibrary.wiley.com/doi/10.1111/j.1365-2621.2001.tb15177.x/abstract

503. Ibid.

504. Najjar SS, Valaei N, & Mousavi SM. Effect of storage conditions on chemical compositions of diet soft drinks. Pejouhandeh, Shahid Beshti University of Medical Sciences. 2008. 13(1). http://pajoohande.sbmu.ac.ir/browse.php?a_id=660&sid=1&slc_lang=en

505. Trocho C, Pardo R, Rafecas I, et al. Formaldehyde derived from dietary aspartame binds to tissue components in vivo. Life Sciences. 1998. 63(5). PMID:9714421. PubMed. http://www.ncbi.nlm.nih.gov/pubmed/9714421

506. Soffritti M, Belpoggi F, Degli Esposti D, et al. First experimental demonstration of the multipotential carcinogenic effects of aspartame administered in the feed to Sprague-Dawley rats. Environmental Health Perspectives. 2006. 114(3):379–385. PMCID:PMC1392232. PubMed. http://www.ncbi.nlm.nih.gov/pmc/articles/PMC1392232

507. Soffritti M, Belpoggi F, Tibaldi E, et al. Life-span exposure to low doses of aspartame beginning during prenatal life increases cancer effects in rats. Environmental Health Perspectives. 2007. 115(9):1293–7. PMID:17805418. PubMed. http://www.ncbi.nlm.nih.gov/pubmed/17805418

508. Schernhammer ES, Bertrand KA, Birmann BM, et al. Consumption of artificial sweetener- and sugar-containing soda and risk of lymphoma and leukemia in men and women. American Journal of Clinical Nutrition. 2012. 96(6):1419–28. PMID:17805418. PubMed. http://www.ncbi.nlm.nih.gov/pubmed/23097267

509. Sansom W. Related studies point to the illusion of the artificial. UT Health Science Center San Antonio. June 27, 2011. http://uthscsa.edu/hscnews/singleformat2.asp?newID=3861

510. Roberts HJ. Aspartame and brain cancer. Lancet. 1997. 349(9048):362. PMID:9024408. PubMed. http://www.ncbi.nlm.nih.gov/pubmed/9024408

511. Roberts HJ. Overlooked aspartame-induced hypertension. Southern Medical Journal. 2008. 101(9):969. PMID:18708962. PubMed. http://www.ncbi.nlm.nih.gov/pubmed/18708962

512. Roberts HJ. Aspartame-induced thrombocytopenia. Southern Medical Journal. 2007. 100(5):543. PMID:17534100. PubMed. http://www.ncbi.nlm.nih.gov/pubmed/17534100

513. Roberts HJ. Perspective on aspartame-induced pseudotumor cerebri. Southern Medical Journal. 2009. 102(8):873. PMID:19593279. PubMed. http://www.ncbi.nlm.nih.gov/pubmed/19593279

514. Roberts HJ. Aspartame and headaches. Neurology. 1995. 45(8):1631. PMID:7644072. PubMed. http://www.ncbi.nlm.nih.gov/pubmed/7644072

515. Roberts HJ. Aspartame as a cause of allergic reactions, including anaphylaxis. Archives of Internal Medicine. 1996. 156(9):1027–8. PMID: 8624169. PubMed. http://www.ncbi.nlm.nih.gov/pubmed/8624169

516. Roberts HJ. Aspartame disease. Texas Heart Institute journal. 2004. 31(1):105. PMCID:PMC387446. PubMed. http://www.ncbi.nlm.nih.gov/pmc/articles/PMC387446

517. Roberts HJ. Neurological, psychiatric, and behavioral reactions to aspartame in 505 aspartame reactors. Ch. 45, pp 373-376. Dietary Phenylalanine and Brain Function. Eds. RJ Wurtman & E Ritter-Walker. 1988. ISBN-10: 1461598230. Cambridge, MA: Birkhäuser Boston.

518. Gold M. Docket no. 02P-0317. Recall aspartame as a neurotoxic drug: file #7: aspartame history. U.S. Food and Drug Administration. January 12, 2003. http://www.fda.gov/ohrms/dockets/dailys/03/Jan03/012203/02P-0317_emc-000202.txt

519. Olney JW & Ho O. Brain damage in infant mice following oral intake of glutamate, aspartate or cysteine. Nature. 1970. 227:609–611. PMID: 5464249. PubMed. http://www.ncbi.nlm.nih.gov/pubmed/5464249

520. Gross A. Letter from Dr. Andrian Gross, former FDA investigator and scientist to senator Howard Metzenbaum regarding pre-approval tests by G.D. Searle. Reprinted in U.S. Senate 1987. October 30, 1987. 430–439. http://www.dorway.com/gross.txt

521. Gordon G. NutraSweet: questions swirl. UPI investigative report. United Press International. October 13, 1987. 483–510. Retrieved from http://www.mpwhi.com/upi_nutrasweet_questions_swirl.pdf

522. Ibid.

523. U.S. Department of Health and Human Services. Docket no. 75P-0355. Aspartame: decision of the Public Board of Inquiry. 1980. Retrieved from http://www.dorway.com/pboi.txt

524. Nill A. The history of aspartame. LEDA at Harvard Law School. http://dash.harvard.edu /bitstream/handle/1/8846759/Nill,_Ashley_-_The_History_of_Aspartame.html?sequence=6

525. Stout LP. Let them eat cake? A historical analysis of FDA's decision to approve aspartame. 1997 third year paper. Harvard Law School. http://nrs.harvard.edu/urn-3:HUL.InstRepos:8846766

526. Testimony. Proceedings and debate of the 99th Congress, 1st session. Congressional record. May 7, 1985.

527. Roberts HJ. Aspartame disease: an ignored epidemic. 2001. ISBN-10: 1884243177. West Palm Beach, FL: Sunshine Sentinel Press.

528. Magnuson BA, Burdock GA, Doull J, et al. Aspartame: a safety evaluation based on current use levels, regulations, and toxicological and epidemiological studies. Critical Reviews in Toxicology. 2007. 37(8):629–727. PMID:17828671. PubMed. http://www.ncbi.nlm.nih.gov/ pubmed/17828671

529. HolisticMed.com. Aspartame and manufacturer-funded scientific reviews. 2007. http://www .holisticmed.com/aspartame/burdock/

530. Walton RG. Survey of aspartame studies: correlation of outcome and funding sources. Accessed February 24, 2016. http://www.dorway.com/peerrev.html

531. EFSA. Aspartame. European Food Safety Authority. Updated December 10, 2013. http://www. efsa.europa.eu/en/topics/topic/aspartame.htm

532. Occupational Safety & Health Administration. Methylene chloride. United States Department of Labor. Reviewed March 27, 2012. https://www.osha.gov/SLTC/methylenechloride/

533. Cong W, Wang R, Cai H, et al. Long-term artificial sweetener acesulfame potassium treatment alters neurometabolic functions in C57BL/6J mice. PLOS ONE. 2013. doi:10.1371/journal. pone.0070257. http://journals.plos.org/plosone/article?id=10.1371/journal.pone.0070257

534. Adams M. U.S. dairy industry petitions FDA to approve aspartame as hidden, unlabeled additive in milk, yogurt, eggnog and cream. NaturalNews. February 25, 2013. http://www.naturalnews .com/039244_milk_aspartame_fda_petition.html

535. Federal Register. Flavored milk; petition to amend the standard of identity for milk and 17 additional dairy products. A proposed rule by the Food and Drug Administration. February 20, 2013. https://www.federalregister.gov/articles/2013/02/20/2013-03835/flavored-milk-petition-to -amend-the-standard-of-identity-for-milk-and-17-additional-dairy-products

Food Ingredients as Contaminants: Monosodium Glutamate

536. International Glutamate Information Service. Glutamate information leaflet: glutamate—the facts. http://www.glutamate.org/pdfs/Glutamate_Information_Leaflet.pdf

537. Daley B. Has MSG gotten a bad rap? Chicago Tribune. July 17, 2013. http://articles.chicagotribune .com/2013-07-17/features/ct-food-0717-msg-20130717_1_msg-chinese-restaurant-syndrome -kikunae-ikeda

538. Win DT. MSG—flavor enhancer or deadly killer. Assumption University Journal of Technology. 2008. 12(1):43–49. http://www.journal.au.edu/au_techno/2008/jul08/journal121_article06.pdf

539. MSG Info.com. Umami: the 5th taste. Accessed February 24, 2016. http://www.msginfo.com/ about_taste_umami.asp

540. Food Insight. Everything you need to know about glutamate and monosodium glutamate. October 17, 2009. http://www.foodinsight.org/Resources/Detail.aspx?topic=Everything_You_Need_To_ Know_About_Glutamate_And_Monosodium_Glutamate

541. Jinap S & Hajeb P. Glutamate. Its applications in food and contribution to health. Appetite. August 2010. 55(1):1–10. Science Direct. http://www.sciencedirect.com/science/article/pii/ S0195666310003089

542. Sand J. A short history of MSG: Good science, bad science, and taste cultures. Gastronomica: The Journal of Food and Culture. 2005. 5(4): 38–39. JSTOR. http://www.jstor.org/stable/10.1525/ gfc.2005.5.4.38

543. Reif-Lehrer L. Adverse reactions in humans thought to be related to ingestion of elevated levels of free monosodium glutamate (MSG). Excitotoxins: proceedings of an international symposium held at the Wenner-Gren Center, Stockholm, August 26–27, 1982. Session IV. 309–330. doi:10.1007/978-1-4757-0384-9_24. Springer Link. http://link.springer.com/chapter/10.1007 /978-1-4757-0384-9_24

544. Meldrum BS. Glutamate as a neurotransmitter in the brain: review of physiology and pathology. The Journal of Nutrition. April 2000. 130(4S Supplement):1007S–15S. PMID:10736372. PubMed. http://www.ncbi.nlm.nih.gov/pubmed/10736372

545. Blaylock RL. Excitotoxins: the taste that kills. 1996. ISBN: 0-92917-32-5-2. Santa Fe, NM: Health Press.

546. Schwarcz R, Foster AC, French ED, et al. Excitotoxic models for neurodegenerative disorders. Life Sciences. 1984. 35(1):19–32. PMID:6234446. PubMed. http://www.ncbi.nlm.nih.gov/ pubmed/6234446/

547. Broberger C, Johansen J, Johansson C, et al. The neuropeptide Y/agouti gene-related protein (AGRP) brain circuitry in normal, anorectic, and monosodium glutamate-treated mice. Proceedings of the National Academy of Sciences of the United States of America. 1998. 95(25):15043–15048. http://www.pnas.org/content/95/25/15043.short

548. Rothman SM & Olney JW. Glutamate and the pathophysiology of hypoxic–ischemic brain damage. Annals of Neurology 1986. 19(2):105–111. doi:10.1002/ana.410190202. Wiley Online Library. http://onlinelibrary.wiley.com/doi/10.1002/ana.410190202/abstract

549. Raiten DJ, Talbot JM, & Fisher KD. Executive summary from the report: analysis of adverse reactions to monosodium glutamate. The Journal of Nutrition 1995. 125:2892S–2906S. PMID:7472671. http://jn.nutrition.org/content/125/11/2891S.full.pdf

550. MedlinePlus. Chinese restaurant syndrome. U.S. National Library of Medicine. Updated October 31, 2013. http://www.nlm.nih.gov/medlineplus/ency/article/001126.htm

551. Olney JW & Sharpe LG. Brain lesions in an infant rhesus monkey treated with monsodium glutamate. Science magazine. 1969. 166(3903):386–8. PMID:5812037. PubMed. http://www.ncbi .nlm.nih.gov/pubmed/5812037/

552. Olney JW. Brain lesions, obesity, and other disturbances in mice treated with monosodium glutamate. Science magazine. 1969. 164(3880):719–721. doi:10.1126/science.164.3880.719. http:// www.sciencemag.org/content/164/3880/719.short

553. Samuels A. The toxicity/safety of processed free glutamic acid (MSG): a study in suppression of information. Accountability in Research. 1999. 6(4):259–310. doi:10.1080/08989629908573933. http://www.bmartin.cc/pubs/99air/99Samuels.pdf

554. Perez VJ & Olney JW. Accumulation of glutamic acid in the arcuate nucleus of the hypothalamus of the infant mouse following subcutaneous administration of monosodium glutamate. Journal of Neurochemistry. 1972. 19(7):1777–1782. doi:10.1111/j.1471-4159.1972.tb06222.x. Wiley Online Library. http://onlinelibrary.wiley.com/doi/10.1111/j.1471-4159.1972.tb06222.x/abstract

555. Samuels A. The toxicity/safety of processed free glutamic acid (MSG): a study in suppression of information. Accountability in Research. 1999. 6:259–310. http://www.bmartin.cc/ pubs/99air/99Samuels.pdf

556. Albin RL & Greenamyre JT. Alternative excitotoxic hypotheses. Neurology. 1992. 42(4):733–8. PMID:1314341. PubMed. http://www.ncbi.nlm.nih.gov/pubmed/1314341/

557. Blaylock RL. Excitotoxins: the taste that kills. 1996. ISBN: 0-92917-32-5-2. Santa Fe, NM: Health Press.

558. Rothman SM & Olney JW. Glutamate and the pathophysiology of hypoxic–ischemic brain damage. Annals of Neurology. 1986. 19(2):105–111. doi: 10.1002/ana.410190202. Online Wiley Library. http://onlinelibrary.wiley.com/doi/10.1002/ana.410190202/abstract

559. Olney JW & Ho OL. Brain damage in infant mice following oral intake of glutamate, aspartate or cysteine. Nature. 1970. 227:609–611. doi:10.1038/227609b0. http://www.nature.com/nature/journal/v227/n5258/abs/227609b0.html

560. Reif-Lehrer L, Bergenthal J, & Hanninen L. Effects of monosodium glutamate on chick embryo retina in culture. Investigative Ophthalmology & Visual Science. 1975. 14(2):114–124. http://www.iovs.org/content/14/2/114.short

561. Blanks JC, Reif-Lehrer L, & Casper D. Effects of monosodium glutamate on the isolated retina of the chick embryo as a function of age: a morphological study. Experimental Eye Research. 32(1):105–124. Science Direct. http://dx.doi.org/10.1016/S0014-4835(81)80044-9

562. Sisk DR & Kuwabara T. Histologic changes in the inner retina of albino rats following intravitreal injection of monosodiuml-glutamate. Graefe's Archive for Clinical and Experimental Ophthalmology. 1985. 223(5): 250–258. doi:10.1007/BF02153655. Springer Link. http://link.springer.com/article/10.1007/BF02153655

563. Ohguro H, Katsushima H, Maruyama I, et al. A high dietary intake of sodium glutamate as flavoring (Ajinomoto) causes gross changes in retinal morphology and function. Experimental Eye Research. 2002. 75(3): 307–315. Science Direct. http://dx.doi.org/10.1006/exer.2002.2017

564. Sand J. A short history of MSG: Good science, bad science, and taste cultures. Gastronomica: The Journal of Food and Culture. 2005. 5(4): 38–39. JSTOR. http://www.jstor.org/stable/10.1525/gfc.2005.5.4.38

565. Ortiz GG, Bitzer-Quintero OK, Zarate CB, et al. Monosodium glutamate induced damage in liver and kidney: a morphological and biochemical approach. Biomedicine & Pharmacotherapy. 2006. 60(2):86–91. PMID:16488110. PubMed. http://www.ncbi.nlm.nih.gov/pubmed/16488110

566. Hawkins, R. The blood–brain barrier and glutamate. American Journal of Clinical Nutrition. 2009. 90(3):867S–874S. PMCID:3136011. PubMed. http://www.ncbi.nlm.nih.gov/pmc/articles/PMC3136011/

567. Price MT, Olney JW, Lowry OH, et al. Uptake of exogenous glutamate and aspartate by circumventricular organs but not other regions of brain. Journal of Neurochemistry. 1981. 36(5):1774–80. PMID: 6113269. PubMed. http://www.ncbi.nlm.nih.gov/pubmed/6113269

568. Skultétyová I, Tokarev D, & Jezová D. Stress-induced increase in blood–brain barrier permeability in control and monosodium glutamate-treated rats. Brain Research Bulletin. 1998. 45(2):175–8. PMID:9443836. PubMed. http://www.ncbi.nlm.nih.gov/pubmed/9443836

569. Thurston JH & Warren SK. Permeability of the blood–brain barrier to monosodium glutamate and effects on the components of the energy reserve in newborn mouse brain. Journal of Neurochemistry. 1971. 18(11): 2241–2244. doi:10.1111/j.1471-4159.1971.tb05084.x. Online Wiley Library. http://onlinelibrary.wiley.com/doi/10.1111/j.1471-4159.1971.tb05084.x/abstract

570. Broadwell RD & Sofroniew MV. Serum proteins bypass the blood–brain fluid barriers for extracellular entry to the central nervous system. Experimental Neurology. 1993. 120(2):245–3. PMID:8491281. PubMed. http://www.ncbi.nlm.nih.gov/pubmed/8491281

571. Gao J, Wu J, Zhao XN, et al. Transplacental neurotoxic effects of monosodium glutamate on structures and functions of specific brain areas of filial mice. Sheng Li Xue Bao. 1994. 46(1):44–51. PMID:8085168. PubMed. http://www.ncbi.nlm.nih.gov/pubmed/8085168

572. Frieder B & Grimm VE. Prenatal monosodium glutamate (MSG) treatment given through the mother's diet causes behavioral deficits in rat offspring. International Journal of Neuroscience. 1984. 23:117–126. PMID:6541212. PubMed. http://www.ncbi.nlm.nih.gov/pubmed/6541212

573. Hermanussen M, Garcia AP, Sunder M, et al. Obesity, voracity, and short stature: the impact of glutamate on the regulation of appetite. European Journal of Clinic Nutrition. 2006. 60:25–31. doi:10.1038/sj.ejcn.1602263. Nature. http://www.nature.com/ejcn/journal/v60/n1/full/1602263a.html

574. Blaylock RL. Excitotoxins: the taste that kills. 1996. ISBN-10: 0929173252. Santa Fe, NM: Health Press.

575. Schwartz GR. In bad taste: the MSG syndrome: how living without MSG can reduce head-ache, depression and asthma, and help you get control of your life. 1988. ISBN: 0929173007, 9780929173009. Santa Fe, NM: Health Press.

576. Hirata AE, Andrade IS, Vaskevicius P, et al. Monosodium glutamate (MSG)-obese rats develop glucose intolerance and insulin resistance to peripheral glucose uptake. Brazilian Journal of Medical and Biological Research. 1997. 30(5):671–674. Science Direct. http://dx.doi.org/10.1590/S0100-879X1997000500016

577. Macho L, Ficková M, Jezová, et al. Late effects of postnatal administration of monosodium glutamate on insulin action in adult rats. Physiological Research. 2000. 49(Suppl. 1): S79–85. PMID:10984075. PubMed. http://www.ncbi.nlm.nih.gov/pubmed/10984075

578. Wilson BG & Bahna SL. Adverse Reactions to Food Additives. Annals of Allergy, Asthma & Immunology. December 2005. 95(6):499-507. Nature. http://www.ncbi.nlm.nih.gov/pubmed/16400887

579. U.S. Food and Drug Administration. Re: Docket No. 02N-0434; Withdrawal of Certain Proposed Rules and Other Proposed Actions; Notice of Intent; 68 Federal register. 19766. (April 22, 2003). Dockets Management Branch (HFA-305) 21 July 2003. http://www.fda.gov/ohrms/dockets/dailys/03/jul03/072803/02N-0434_emc-000004-01.PDF

580. U.S. Food and Drug Administration. Food Additives & Ingredients: Questions and Answers on Monosodium Glutamate (MSG). 19 November 2012. http://www.fda.gov/food/ingredients packaginglabeling/foodadditivesingredients/ucm328728.htm

581. Jinap S & Hajeb P. Glutamate. Its applications in Food and Contribution to Health. Appetite. August 2010. 55(1):1-10. Science Direct. http://www.sciencedirect.com/science/article/pii/S0195666310003089

582. U.S. Food and Drug Administration. Food additives & ingredients: questions and answers on monosodium glutamate (MSG). 19 November 2012. http://www.fda.gov/food/ingredients packaginglabeling/foodadditivesingredients/ucm328728.htm

583. Food Standards Agency. Current EU Approved Additives and Their E numbers. Updated 19 November 2013. http://www.food.gov.uk/policy-advice/additivesbranch/enumberlist#Emulsifiers ,Stabilisers,ThickenersandGellingAgents

584. Truth in Labeling. E-numbers Used for Ingredients that Contain MSG. http://www.truthinlabeling .org/Enumbers.html

585. U.S. Centers for Disease Control and Prevention. Vaccine Excipient & Media Summary: excipients included in U.S. vaccines, by vaccine. Updated February 2012. http://www.cdc.gov/vaccines/pubs/pinkbook/downloads/appendices/B/excipient-table-2.pdf

586. Grabenstein JD. ImmunoFacts: Vaccines and Immunologic Drugs—2012. (37th revision). St Louis, MO: Wolters Kluwer Health. 2011. PMID:22535837. ISBN-10: 1574393448

587. U.S. Environmental Protection Agency. 63 FR 679 Federal Register: Glutamic Acid; Pesticide Tolerance Exemption. January 7, 1998.

588. Pavlista AD. Growth Regulators Increased Yield of Atlantic Potato. American Journal of Potato Research. December 2011. 88(6):479-484. doi:10.1007/s12230-011-9214-3. Springer Link. http://link.springer.com/article/10.1007/s12230-011-9214-3

589. U.S. Environmental Protection Agency. Federal register: notice of filing of pesticide petitions. October 29, 1997. 62(209):56168-56171. DOCID:fr29oc97-75. https://www.gpo.gov/fdsys/granule/FR-1997-10-29/97-28664

590. Win DT. MSG—Flavor enhancer or deadly killer. Assumption University Journal of Technology. July 2008. 12(1):43-49.

591. Truth in Labeling. MSG is being sprayed right on fruits, nuts, seeds, grains, and vegetables as they grow—even those used in baby food. Updated 14 August 2008. http://www.truthinlabeling.org/msgsprayed.html

592. Sand J. A Short History of MSG: Good Science, Bad Science, and Taste Cultures. Gastronomica: The Journal of Food and Culture. Fall 2005. 5(4):38-39. JSTOR.

593. Schwartz GR. In Bad Taste: The MSG Syndrome : How Living Without MSG Can Reduce Headache, Depression and Asthma, and Help You Get Control of Your Life. Health Press. 1988. ISBN:0929173007, 9780929173009.

594. Truth in Labeling. The real food recipeless cookbook: the secret to eating without MSG. 2012. http://www.truthinlabeling.org/CookBook_Final.pdf

595. Blaylock RL. Excitotoxins: the taste that kills. 1996. ISBN-10: 0929173252. Santa Fe, NM: Health Press.

596. Cox JA, Lysko PG, & Henneberry RC. Excitatory amino acid neurotoxicity at the N-methyl-d-aspartate receptor in cultured neurons: role of the voltage-dependent magnesium block. Brain Research. 1989. 499(2):267–272. PMID:2572299. PubMed. http://www.ncbi.nlm.nih.gov/pubmed/2572299

597. Novelli A, Reilly JA, Lysko PG, et al. Glutamate becomes neurotoxic via the N-methyl-d-aspartate receptor when intracellular energy levels are reduced. Brain Research. 1988. 451(1–2):205–212. Science Direct. http://dx.doi.org/10.1016/0006-8993(88)90765-2

Food Ingredients as Contaminants: Artificial Colors

598. International Food Information Council (IFIC) and U.S. Food and Drug Administration. Overview of food ingredients, additives & colors. FDA. Revised April 2010. http://www.fda.gov/food/ingredientspackaginglabeling/foodadditivesingredients/ucm094211.htm#coloradd

599. Red40.com. The chemistry of red 40. 2003. http://www.red40.com/pages/chemistry.html

600. Hemmessey R. Living in color: the potential dangers of artificial dyes. Forbes. August 27, 2012. http://www.forbes.com/sites/rachelhennessey/2012/08/27/living-in-color-the-potential -dangers-of-artificial-dyes

601. Kobylewski S & Jacobson MF. Food dyes: a rainbow of risks. Center for Science in the Public Interest. 2010. http://cspinet.org/new/pdf/food-dyes-rainbow-of-risks.pdf

602. Barrows JN, Lipman AL, Bailey CJ, et al. Color additives: FDA's regulatory process and historical perspectives. Food Safety Magazine. October/November 2003. U.S. Food and Drug Administration. www.fda.gov/ForIndustry/ColorAdditives/RegulatoryProcessHistoricalPerspectives/default.htm

603. Feingold BF. Biography. Updated December 25, 2011. Feingold.org. http://feingold.org/about -the-program/dr-feingold/bio

604. Severo R. Dr. Ben F. Feingold dies at 81; studied diet in hyperactivity. The New York Times. March 24, 1982. http://www.nytimes.com/1982/03/24/obituaries/dr-ben-f-feingold-dies-at-81 -studied-diet-in-hyperactivity.html

605. Feingold BF. The Feingold hypothesis. Updated January 8, 2012. Feingold.org. http://feingold.org/about-the-program/dr-feingold/theory

606. Feingold BF. Food additives in clinical medicine. International Journal of Dermatology. 1975. 14(2):113-4. PMID:1123257. PubMed. http://www.ncbi.nlm.nih.gov/pubmed/1123257

607. Feingold BF. Hyperkinesis and learning disabilities linked to the ingestion of artificial food flavors and colors. The American Journal of Nursing. 1975. 75(5):797–803. PMID:1039267. Feingold .org. http://www.feingold.org/Research/PDFstudies/Feingold75.pdf

608. Feingold BF. Hyperkinesis and learning disabilities linked to the ingestion of artificial food colors and flavors. Journal of Learning Disabilities. 1976. 9(9):19–27. PMID:1039267. Feingold.org. http://www.feingold.org/Research/PDFstudies/Feingold76.pdf

609. Feingold BF. Behavioral disturbances linked to the ingestion of food additives. Delaware Medical Journal. 1977. 49(2):89–94. PMID: 844631. PubMed. http://www.ncbi.nlm.nih.gov/pubmed/844631

610. Feingold BF. Hyperkinesis and learning disabilities linked to the ingestion of artificial food colors and flavors. Speech to American Academy of Pediatrics, New York Hilton Hotel. November 8, 1977. Feingold.org. http://www.feingold.org/Research/PDFstudies/Feingold-AAP.pdf

611. U.S. Food and Drug Administration. Federal Food, Drug, and Cosmetic Act (FD&C Act). Updated December 5, 2011. http://www.fda.gov/regulatoryinformation/legislation/federalfooddrugand cosmeticactFDCAct/default.htm

612. Chafee FH & Settipane GA. Asthma caused by FD&C approved dyes. Journal of Allergy. 1967. 40(2):65–72. Science Direct. http://www.sciencedirect.com/science/article/pii/0021870767900998

613. Feingold BF. Hyperkinesis and learning disabilities linked to the ingestion of artificial food flavors and colors. The American Journal of Nursing. 1975. 75(5):797–803. PMID:1039267. Feingold .org. http://www.feingold.org/Research/PDFstudies/Feingold76.pdf

614. Grillo CA & Dulout FN. Cytogenetic evaluation of butylated hydroxytoluene. Mutation Research/Genetic Toxicology. 1995. 345(1–2):73–78. Science Direct. http://www.sciencedirect.com/science/article/pii/0165121895900713

615. Imaida K, Fukushima S, Shirai T, et al. Promoting activities of butylated hydroxyanisole and butylated hydroxytoluene on 2-stage urinary bladder carcinogenesis and inhibition of γ-glutamyl transpeptidase-positive foci development in the liver of rats. Carcinogenesis. 1983. 4(7):895–899. doi:10.1093/carcin/4.7.895. http://carcin.oxfordjournals.org/content/4/7/895.short

616. Branen AL. Toxicology and biochemistry of butylated hydroxyanisole and butylated hydroxytoluene. Journal of the American Oil Chemists' Society. 1975. 52(2):59–63. doi:10.1007/BF02901825 BF02901825. Springer Link. http://link.springer.com/article/10.1007/BF02901825

617. U.S. National Toxicology Program. Butylated hydroxyanisole: CAS no. 25013-16-5. Report on Carcinogens, 12th ed. 2011. U.S. National Institutes of Health. http://ntp.niehs.nih.gov/ntp/roc/twelfth/profiles/ButylatedHydroxyanisole.pdf

618. Williams JI, Cram DM, Tausig FT, et al. Relative effects of drugs and diet on hyperactive behaviors: an experimental study. Pediatrics. 1978. 61(6):811–817. http://pediatrics.aappublications.org/content/61/6/811.short

619. Weiss B, Williams JH, Margen S, et al. Behavioral responses to artificial food colors. Science magazine. 1980. 207(4438):1487–1489. doi:10.1126/science.7361103. http://www.sciencemag.org/content/207/4438/1487.short

620. Schab DW & Trinh NHT. Do artificial food colors promote hyperactivity in children with hyperactive syndromes? A meta-analysis of double-blind placebo-controlled trials. Journal of Developmental & Behavioral Pediatrics. 2004. 25(6):423–434. Center for Science in the Public Interest. http://cspinet.org/new/pdf/schab.pdf

621. Center for Science in the Public Interest. CSPI urges FDA to ban artificial food dyes linked to behavior problems. June 2, 2008. http://www.cspinet.org/new/200806022.html

622. Bateman B, Warner JO, Hutchinson E, et al. The effects of a double blind, placebo controlled, artificial food colourings and benzoate preservative challenge on hyperactivity in a general population sample of preschool children. Archives of Disease in Childhood. 2004. 89:506–511. doi:10.1136/adc.2003.031435. Center for Science in the Public Interest. http://cspinet.org/new/pdf/bateman.pdf

623. Gee SN & Bigby M. Atopic dermatitis and attention-deficit/hyperactivity disorder: is there an association? Evidence-Based Dermatology: Research Commentary. 2011. 147(8):967–970. doi:10.1001/archdermatol.2011.200. http://archderm.jamanetwork.com/article.aspx?articleID=1105139

624. Rose T. The functional relationship between artificial food colors and hyperactivity. Journal of Applied Behavior Analysis. 1978. 11(4):439–446. doi:10.1901/jaba.1978.11-439. Wiley Online Library. http://onlinelibrary.wiley.com/doi/10.1901/jaba.1978.11-439/abstract

625. McCann D, Barrett A, Cooper A, et al. Food additives and hyperactive behaviour in 3-year-old and 8/9-year-old children in the community: a randomised, double-blinded, placebo-controlled trial. Lancet. 2007. 370(9598):1560–1567. doi:10.1016/S0140-6736(07)61306-3. http://www.thelancet.com/journals/lancet/article/PIIS0140673607613063/abstract

626. Weiss B. Synthetic food colors and neurobehavioral hazards: The view from environmental health research. Environmental Health Perspectives. 2012. 120(1):1–5. PMCID:3261946. PubMed. http://www.ncbi.nlm.nih.gov/pmc/articles/PMC3261946

627. Kobylewski S & Jacobson M. Food dyes: a rainbow of risks. Center for Science in the Public Interest. 2010. http://cspinet.org/new/pdf/food-dyes-rainbow-of-risks.pdf

628. Center for Science in the Public Interest. Summary of studies on food dyes. Accessed February 24, 2016. http://cspinet.org/new/pdf/dyes-problem-table.pdf

629. U.S. Government Printing Office. 21 CFR 74.340 - FD&C Red No. 40. Code of federal regulations. April 1, 2011. http://www.gpo.gov/fdsys/granule/CFR-2011-title21-vol1/CFR-2011-title21-vol1-sec74-340/content-detail.html

630. U.S. National Toxicology Program. P-Cresidine CAS No. 120-71-8. Report on Carcinogens, 12th ed. 2011. U.S. National Institutes of Health. http://ntp.niehs.nih.gov/ntp/roc/twelfth/profiles/Cresidine.pdf

631. Kobylewski S & Jacobson M. Food dyes: a rainbow of risks. Center for Science in the Public Interest. 2010. http://cspinet.org/new/pdf/food-dyes-rainbow-of-risks.pdf

632. American Cancer Society. Known and probable human carcinogens. Updated October 17, 2013. http://www.cancer.org/cancer/cancercauses/othercarcinogens/generalinformationaboutcarcinogens/known-and-probable-human-carcinogens?sitearea=PED

633. U.S. National Toxicology Program. 4-Aminobiphenyl CAS No. 92-67-1. Report on Carcinogens, 12th ed. 2011. http://ntp.niehs.nih.gov/ntp/roc/content/profiles/aminobiphenyl.pdf

634. Moutinho ILD, Bertges LC, & Assis RVC. Prolonged use of the food dye tartrazine (FD&C yellow n° 5) and its effects on the gastric mucosa of Wistar rats. Brazilian Journal of Biology. 2007. 67(1). Science Direct. http://dx.doi.org/10.1590/S1519-69842007000100019

635. U.S. National Toxicology Program. Benzidine CAS registry number: 92-87-5 toxicity effects. U.S. National Institutes of Health. http://tools.niehs.nih.gov/cebs3/ntpviews/index.cfm?action=testarticle.toxicity&cas_number=92-87-5

636. U.S. National Toxicology Program. 4-Aminobiphenyl CAS No. 92-67-1. Report on Carcinogens, 12th ed. 2011. http://ntp.niehs.nih.gov/ntp/roc/content/profiles/aminobiphenyl.pdf

637. Ershoff BH. Effects of diet on growth and survival of rats fed toxic levels of tartrazine (FD & C Yellow No. 5) and sunset yellow FCF (FD & C Yellow No. 6). Journal of Nutrition. 1977. 107(5):822–8. PMID:859044. PubMed. http://www.ncbi.nlm.nih.gov/pubmed/859044

638. U.S. National Toxicology Program. 4-Aminobiphenyl CAS No. 92-67-1. Report on Carcinogens, 12th ed. 2011. http://ntp.niehs.nih.gov/ntp/roc/content/profiles/aminobiphenyl.pdf

639. Makena P & Chung KT. Evidence that 4-aminobiphenyl, benzidine, and benzidine congeners produce genotoxicity through reactive oxygen species. Environmental and Molecular Mutagenesis. 2007. 48(5):404–13. PMID:17370336. PubMed. http://www.ncbi.nlm.nih.gov/pubmed/17370336

640. Sailstad DM, Tepper JS, Doerfler DL, et al. Evaluation of an azo and two athraquinone dyes for allergenic potential. Toxicological Sciences. 1994. 23(4):569–577. doi:10.1093/toxsci/23.4.569. Oxford Journals. http://toxsci.oxfordjournals.org/content/23/4/569.short

641. Hodgson GA. Industrial dermatitis. Postgraduate Medical Journal. 1966. 42:643–651. doi:10.1136/pgmj.42.492.643. http://pmj.bmj.com/content/42/492/643.full.pdf

642. Kobylewski S & Jacobson M. Food dyes: a rainbow of risks. Center for Science in the Public Interest. 2010. http://cspinet.org/new/pdf/food-dyes-rainbow-of-risks.pdf

643. Ibid.

644. Ibid.

645. Center for Science in the Public Interest. In Europe, dyed foods get warning label. July 20, 2010. http://cspinet.org/new/201007201.html

646. UK Food Standards Agency. Executive summary: food additives and hyperactivity. April 10, 2008. http://tna.europarchive.org/20120419000433/http://www.food.gov.uk/multimedia/pdfs/board/fsa080404a.pdf

647. U.S. Food and Drug Administration. Background document for the Food Advisory Committee: certified color additives in food and possible association with attention deficit hyperactivity disorder in children. March 30–31, 2011. Meeting materials. http://www.fda.gov/downloads/advisorycommittees/.../ucm248549.pdf

648. Schoffro Cook M. The dark side of food colors. Care2. May 10, 2013. http://www.care2.com/greenliving/the-dark-side-of-food-colors.html

649. Lucas CD, Hallagan JB, & Taylor SL. The role of natural color additives in food allergy. Advances in food and nutrition research. 2001. 43:195–216. Science Direct. http://www.sciencedirect.com/science/article/pii/S1043452601430051

650. Reddy MK, Alexander-Lindo RL, & Nair MG. Relative inhibition of lipid peroxidation, cyclooxygenase enzymes, and human tumor cell proliferation by natural food colors. Journal of Agricultural and Food Chemistry. 2005. 53(23):9268–9273. doi:10.1021/jf051399j. American Chemical Society. http://pubs.acs.org/doi/abs/10.1021/jf051399j

651. Castellar R, Obón JM, Alacid M, et al. Color properties and stability of betacyanins from Opuntia fruits. Journal of Agriculture and Food Chemistry. 2003. 51(9):2772–6. PMID:12696971. PubMed. http://www.ncbi.nlm.nih.gov/pubmed/12696971

652. Bridle P & Timberlake CF. Anthocyanins as natural food colours—selected aspects. Food Chemistry. 1997. 58(1–2):103–109. Science Direct. http://dx.doi.org/10.1016/S0308-8146(96)00222-1

653. Reddy MK, Alexander-Lindo RL, & Nair MG. Relative inhibition of lipid peroxidation, cyclooxygenase enzymes, and human tumor cell proliferation by natural food colors. Journal of Agricultural and Food Chemistry. 2005. 53(23):9268–9273. doi:10.1021/jf051399j. American Chemical Society. http://pubs.acs.org/doi/abs/10.1021/jf051399j

654. Govindarajan VS & Stahi WH. Turmeric—chemistry, technology, and quality. Critical Reviews in Food Science and Nutrition. 1980. 12(3):199–301. doi:10.1080/10408398009527278. Taylor & Francis Online. http://www.tandfonline.com/doi/abs/10.1080/10408398009527278

655. Henry BS. Natural food colours. Natural Food Colorants. 1996. 40–79. doi:10.1007/978-1-4615-2155-6_2. Springer Link. http://link.springer.com/chapter/10.1007/978-1-4615-2155-6_2

656. Lucas CD, Hallagan JB, & Taylor SL. The role of natural color additives in food allergy. Advances in Food and Nutrition Research. 2001. 43:195–216. Science Direct. http://www.sciencedirect.com/science/article/pii/S1043452601430051

657. Timberlake CF & Henry BS. Plant pigments as natural food colours. Endeavour. 1986. 10(1):31–36. Science Direct. http://dx.doi.org/10.1016/0160-9327(86)90048-7

658. Castillo M. Kraft drops Yellow Nos. 5, 6 from kid-friendly mac & cheese varieties. CBS News. October 31, 2013. http://www.cbsnews.com/news/kraft-drops-yellow-nos-5-6-from-kid-friendly -mac-cheese-varieties

659. Wilson J. Kraft removing artificial dyes from some mac and cheese. CNN Health. November 4, 2013. http://www.cnn.com/2013/11/01/health/kraft-macaroni-cheese-dyes

660. Center for Science in the Public Interest. CSPI urges FDA to ban artificial food dyes linked to behavior problems. June 2, 2008. http://www.cspinet.org/new/200806022.html

Food Ingredients as Contaminants: Chemical Preservatives

661. Feingold Association of the United States. What are the most dangerous E-numbers. http://www .feingold.org/Research/PDFstudies/E-numbers.pdf

662. Kluge H, Broz J, & Eder K. Effect of benzoic acid on growth performance, nutrient digestibility, nitrogen balance, gastrointestinal microflora and parameters of microbial metabolism in piglets. Journal of Animal Physiology and Animal Nutrition. 2006. 90(7–8):316–324. doi:10.1111/j.1439-0396.2005.00604.x. Wiley Online Library. http://onlinelibrary.wiley.com/doi/10.1111/j.1439-0396.2005.00604.x/abstract

663. The Food Commission. Additives: student pack for secondary students. January 2010. http://www .foodcomm.org.uk/images/additives_pack_secondary_student.pdf

664. McCann D, Barrett A, Cooper A, et al. Food additives and hyperactive behaviour in 3-year-old and 8/9-year-old children in the community: a randomised, double-blinded, placebo-controlled trial. Lancet. 2007. 370(9598):1560–1567. doi:10.1016/S0140-6736(07)61306-3. http://www .thelancet.com/journals/lancet/article/PIIS0140673607613063/abstract

665. Food Standards Agency. Food colours and hyperactivity. Accessed February 24, 2016. http://www .food.gov.uk/policy-advice/additivesbranch/foodcolours

666. The Food Commission. Additives: student pack for secondary students. January 2010. http://www .foodcomm.org.uk/images/additives_pack_secondary_student.pdf

667. Piper PW. Yeast superoxide dismutase mutants reveal a pro-oxidant action of weak organic acid food preservatives. Free radical biology and medicine. 1999. 27(11–12):1219–27. Science Direct. http://dx.doi.org/10.1016/S0891-5849(99)00147-1

668. Gardner LK & Lawrence GD. Benzene production from decarboxylation of benzoic acid in the presence of ascorbic acid and a transition-metal catalyst. Journal of Agricultural and Food Chemistry. 1993. 41(5):693–5. doi:10.1021/jf00029a001. American Chemical Society. http://pubs.acs.org/doi/abs/10.1021/jf00029a001

669. Patton, D. South Korea urges recall of benzene-containing drinks. BeverageDaily.com. April 18, 2006. http://www.beveragedaily.com/Markets/South-Korea-urges-recall-of-benzene-containing-drinks

670. International Council of Beverages Associations. ICBA guidance document to mitigate the potential for benzene formation in beverages. Adopted by the ICBA Council on April 29, 2006. Updated June 22, 2006. http://www.icba-net.org/files/resources/icba-benzene-guidance-english.pdf

671. Van Poucke C, Detavernier C, Van Bocxlaer JF, et al. Monitoring the benzene contents in soft drinks using headspace gas chromatography–mass spectrometry: a survey of the situation on the Belgian market. Journal of Agricultural and Food Chemistry. 2008. 56(12):4504–10. doi:10.1021/jf072580q. American Chemical Society. http://pubs.acs.org/doi/abs/10.1021/jf072580q

672. Coca-Cola Great Britain. Diet Coke ingredients & nutrition: weight loss & diabetes. Updated 2010. http://www.coca-cola.co.uk/brands/diet-coke.html

673. Fernandez C. Diet Coke to drop additive in DNA damage fear. Daily Mail.May 25, 2008. http://www.dailymail.co.uk/news/article-1021820/Diet-Coke-drop-additive-DNA-damage-fear.html

674. PepsiCo. Nutrition info for Diet Pepsi Wild Cherry. The facts about your favorite beverages. Updated January 28, 2016. http://www.pepsicobeveragefacts.com/Home/Product?formula=F0000003565&form=RTD&size=20

675. Ryan, DB. What soft drinks have sodium benzoate E211 in them? Livestrong.com. August 16, 2013. http://www.livestrong.com/article/256440-what-soft-drinks-have-sodium-benzoate-e211-in-them/

676. Newcombe R. Top E numbers to avoid. Explore E numbers. September 5, 2013. http://www.exploreenumbers.co.uk/top-10-e-numbers-try-avoid.html

677. U.S. Government Printing Office. 21 CFR 184.1490—Methylparaben. Subpart B—listing of specific substances affirmed as GRAS. 2000. http://www.gpo.gov/fdsys/granule/CFR-2000-title21-vol3/CFR-2000-title21-vol3-sec184-1490/content-detail.html

678. U.S. Government Printing Office. 21 CFR 184.1670—Propylparaben. Subpart B—listing of specific substances affirmed as GRAS. 2000. http://www.gpo.gov/fdsys/granule/CFR-2000-title21-vol3/CFR-2000-title21-vol3-sec184-1670/content-detail.html

679. Food Standards Agency. Current EU approved additives and their E numbers. Updated November 19, 2013. http://www.food.gov.uk/policy-advice/additivesbranch/enumberlist#Preservatives

680. Fujita F, Moriyama T, Higashi T, et al. Methyl p-hydroxybenzoate causes pain sensation through activation of TRPA1 channels. British Journal of Pharmacology. 2007. 151(1):134–141. PMCID:PMC2012982. PubMed. http://www.ncbi.nlm.nih.gov/pmc/articles/PMC2012982

681. Routledge EJ, Parker J, Odum J, et al. Some alkyl hydroxy benzoate preservatives (parabens) are estrogenic. Toxicology and Applied Pharmacology. 1998. 153(1):12–19. Science Direct. http://dx.doi.org/10.1006/taap.1998.8544

682. Darbre PD, Aljarrah A, Miller WR, et al. Concentrations of parabens in human breast tumours. Journal of Applied Toxicology. 2004. 24:5–13. doi:10.1002/jat.958. Research Gate. http://www.researchgate.net/publication/8900600_Concentrations_of_parabens_in_human_breast_tumours/file/79e41510c2af67d7b4.pdf

683. Scheve T. What are parabens? Discovery fit and health. October 9, 2009. http://health.howstuffworks.com/skin-care/beauty/skin-and-lifestyle/parabens1.htm

684. Ask Umbra. Ask Umbra on parabens in processed foods and personal lubricant. Grist. November 15, 2010. http://grist.org/article/2010-11-15-ask-umbra-on-parabens-in-processed-foods-and-personal-lubricant/

685. U.S. Food and Drug Administration. Select committee on GRAS substances (SCOGS) opinion: methyl paraben. Updated April 18, 2013. http://www.fda.gov/food/ingredientspackaginglabeling/gras/scogs/ucm260472.htm

686. Barr L, Metaxas G, Harbach CAJ, et al. Measurement of paraben concentrations in human breast tissue at serial locations across the breast from axilla to sternum. Journal of Applied Toxicology. 2012. 32(3):219–232. PMID:22237600. PubMed. http://www.ncbi.nlm.nih.gov/pubmed/22237600

687. Liao C, Liu F, & Kannan K. Occurrence of and dietary exposure to parabens in foodstuffs from the United States. Environmental Science & Technology. 2013. 47(8):3918–25. PMID:23506043. PubMed. http://www.ncbi.nlm.nih.gov/pubmed/23506043

688. Liao C, Liu F, & Kannan K. Occurrence of parabens in foodstuffs from China and its implications for human dietary exposure. Environment International. 2013. 57–58:68–74. PMID:23685225. PubMed. http://www.ncbi.nlm.nih.gov/pubmed/23685225

689. Ashton K & Salter Green E. The toxic consumer: living healthy in a hazardous world. 2008. ISBN-10: 1402748914. New York, NY: Sterling.

690. Hamishehkar H, Khani S, Kashanian S, et al. Geno- and cytotoxicity of propyl gallate food additive. Drug & Chemical Toxicology. 2014. 37(3)241–6. PMID:24160552. PubMed. http://www.ncbi.nlm.nih.gov/pubmed/24160552

691. Han YH, Moon HJ, You BR, et al. Propyl gallate inhibits the growth of calf pulmonary arterial endothelial cells via glutathione depletion. Toxicology in Vitro. 2010. 24(4):1183–9. PMID:20159035. PubMed. http://www.ncbi.nlm.nih.gov/pubmed/20159035

692. Foti C, Bonamonte D, Cassano N, et al. Allergic contact dermatitis to propyl gallate and pentylene glycol in an emollient cream. Australasian Journal of Dermatology. 2010. 51(2):147–8. PMID:20546226. PubMed. http://www.ncbi.nlm.nih.gov/pubmed/20546226

693. Pandhi D, Vij A, & Singal A. Contact depigmentation induced by propyl gallate. Clinical & Experimental Dermatology. 2011. 36(4):366–8. PMID:21564173. PubMed. http://www.ncbi.nlm.nih.gov/pubmed/21564173

694. Eler GJ, Peralta RM, & Bracht A. The action of n-propyl gallate on gluconeogenesis and oxygen uptake in the rat liver. Chemico-Biological Interactions. 2009. 181(3):390–9. PMID:19616523. PubMed. http://www.ncbi.nlm.nih.gov/pubmed/19616523

695. Amadasi A, Mozzarelli A, Meda C, et al. Identification of xenoestrogens in food additives by an integrated in silico and in vitro approach. Chemical Research in Toxicology. 2009. 22(1):52–63. PMID:19063592. PubMed. http://www.ncbi.nlm.nih.gov/pmc/articles/PMC2758355

696. South African National Halaal Authority. E310 (propyl gallate). SANHA. Accessed February 24, 2016. http://www.sanha.co.za/a/index.php?option=com_content&task=view&id=265&Item-id=259

697. U.S. Food and Drug Administration. Propyl gallate. Database of select committee on GRAS substances (SCOGS) reviews. Report 11, ID 121-79-9. 1973. http://www.accessdata.fda.gov/scripts/fcn/fcnDetailNavigation.cfm?rpt=scogslisting&id=260

698. Feingold.org. Some studies on BHT, BHA & TBHQ. July 29, 2012. http://www.feingold.org/Research/bht.php

699. Feingold BF. Dietary management of juvenile delinquency. International Journal of Offender Therapy and Comparative Criminology. 1979. 23(1). Feingold.org. http://www.feingold.org/Research/PDFstudies/Feingold-delinq79.pdf

700. Stokes JD & Scudder CL. The effect of butylated hydroxyanisole and butylated hydroxytoluene on behavioral development of mice. Developmental Psychobiology. 1974. 7(4):343–50. PMID:4472726. PubMed. http://www.ncbi.nlm.nih.gov/pubmed/4472726

701. Pols I. TBHQ: how butane gets in your kids' snacks. December 8, 2013. New England Health Advisory. http://nehealthadvisory.com/?p=147

702. European Food Safety Authority. Opinion of the scientific panel on food additives, flavourings, processing aids and materials in contact with food (AFC) on a request from the commission related to tertiary-butylhydroquinone (TBHQ). EFSA Journal. October 14, 2004. doi:10.2903/j.efsa.2004.84. http://www.efsa.europa.eu/en/efsajournal/pub/84.htm

703. Steinman D. Diet for a poisoned planet: how to choose safe foods for you and your family—the twenty-first century edition. ISBN-10: 1-56025-922-1. 2007. New York, NY: Thunder's Mouth Press.

704. Ibid.

705. The EXtension TOXicology NETwork. Sulfites—FDA guide to foods and drugs with sulfites. Extonet. University of California-Davis, Oregon State University. Accessed February 24, 2016. http://extoxnet.orst.edu/faqs/additive/sulf_tbl.htm

706. More than Organic. Sulphites in wine. 2013. http://www.morethanorganic.com/sulphur-in-the-bottle

707. Sulfur and sulfites. Adapted from Pure Facts. 1996. 19(10). Retrieved from Feingold.org. http://www.feingold.org/wp/PF/1995-12.pdf

708. Vally H & Thompson P. Role of sulfite additives in wine induced asthma: single dose and cumulative dose studies. Thorax. 2001. 56(10):763–769. PMCID:PMC1745927. PubMed. http://www.ncbi.nlm.nih.gov/pmc/articles/PMC1745927

709. Bai J, Lei P, Zhang J, et al. Sulfite exposure-induced hepatocyte death is not associated with alterations in p53 protein expression. Toxicology. 2013. 312:142–8. PMID:23973939. PubMed. http://www.ncbi.nlm.nih.gov/pubmed/23973939

710. U.S. Food and Drug Administration. 21CFR130.9—Sec. 130.9 Sulfites in standardized food. Revised April 1, 2013. http://www.accessdata.fda.gov/scripts/cdrh/cfdocs/cfcfr/cfrsearch.cfm?fr=130.9

711. ELISA/ACT Biotechnologies LLC. Clinical update #12: sulfites. n.d. http://www.yumpu.com/en/document/view/48121485/clud-12-sulfites-12-07-elisa-act-biotechnologies

712. Adams C. Is there any danger from sulfites in wine? The straight dope. June 26, 1988. http://www.straightdope.com/columns/read/687/is-there-any-danger-from-sulfites-in-wine

713. Ibid.

Food Ingredients as Contaminants: Emulsifiers and Thickening Agents

714. Giannini EG, Mansi C, Dulbecco P, et al. Role of partially hydrolyzed guar gum in the treatment of irritable bowel syndrome. Nutrition. 2006. 22(3):334–42. PMID:16413751. Science.NaturalNews. http://science.naturalnews.com/2006/2820846_Role_of_partially_hydrolyzed_guar_gum_in_the_treatment_of.html

715. Singh V, Kumari P, Pandey S, et al. Removal of chromium (VI) using poly(methylacrylate) functionalized guar gum. Bioresource Technology. 2009. 100(6):1977–82. PMID:19056258. Science.NaturalNews. http://science.naturalnews.com/pubmed/19056258.html

716. Kuo DC, Hsu SP, & Chien CT. Partially hydrolyzed guar gum supplement reduces high fat diet increased blood lipids and oxidative stress and ameliorates FeCl3 induced acute arterial injury in hamsters. Journal of Biomedical Science. 2009. 16:15. PMID:19272178. Science.NaturalNews. http://science.naturalnews.com/pubmed/19272178.html

717. Nakamura S, Hongo R, Moji K, et al. Suppressive effect of partially hydrolyzed guar gum on transitory diarrhea induced by ingestion of maltitol and lactitol in healthy humans. European Journal of Clinical Nutrition. 2007. 61(9):1086–93. PMID:17251924. Science.NaturalNews. http://science.naturalnews.com/pubmed/17251924.html

718. Yoon SJ, Chu DC, Raj J, et al. Chemical and physical properties, safety and application of partially hydrolized guar gum as dietary fiber. Journal of Clinical Biochemistry and Nutrition. 2008. 42:1–7. PMID:18231623. Science.NaturalNews. http://science.naturalnews.com/pubmed/18231623.html

719. Cornupcopia Institute. Executive summary—food grade carrageenan: reviewing potential harmful effects on human health. April 26, 2012. http://www.cornucopia.org/CornucopiaAnalysisofCarrageenanHealthImpacts042612.pdf

720. Jung TW, Lee SY, Hong HC, et al. AMPK activator-mediated inhibition of endoplasmic reticulum stress ameliorates carrageenan-induced insulin resistance through the suppression of selenoprotein P in HepG2 hepatocytes. Molecular and Cellular Endocrinology. 2014. 382(1):66–73. PMID:24055274. PubMed. http://www.ncbi.nlm.nih.gov/pubmed/24055274

721. Çınar BM, Çirci E, Balçık C, et al. The effects of extracorporeal shock waves on carrageenan-induced Achilles tendinitis in rats: a biomechanical and histological analysis. Acta Orthopaedica et Traumatologica Turcica. 2013. 47(4):266–72. PMID:23999515. PubMed. http://www.ncbi.nlm .nih.gov/pubmed/23999515

722. Arun O, Canbay O, Celebi N, et al. The analgesic efficacy of intra-articular acetaminophen in an experimental model of carrageenan-induced arthritis. Pain Research & Management. 2013. 18(5):e63–7. PMID: 24093120. PubMed. http://www.ncbi.nlm.nih.gov/pubmed/24093120

723. Kwon SG, Roh DH, Yoon SY, et al. Blockade of peripheral P2Y1 receptors prevents the induction of thermal hyperalgesia via modulation of TRPV1 expression in carrageenan-induced inflammatory pain rats: Involvement of p38 MAPK phosphorylation in DRGs. Neuropharmacology. 2014. 79:368–79. PMID:24333674. PubMed. http://www.ncbi.nlm.nih.gov/pubmed/24333674

724. Reis EF, Castro SB, Alves CC, et al. Lipophilic amino alcohols reduces carrageenan-induced paw edema and anti-OVA DTH in BALB/c mice. International Immunopharmacology. 2013. 17(3):727–32. PMID:24035232. PubMed. http://www.ncbi.nlm.nih.gov/pubmed/24035232

725. Yang X, Yuan L, Chen J, et al. Multitargeted protective effect of Abacopteris penangiana against carrageenan-induced chronic prostatitis in rats. Journal of Ethnopharmacology. 2014. 151(1):343–51. PMID:24211397. PubMed. http://www.ncbi.nlm.nih.gov/pubmed/24211397

726. Ekundi-Valentim E, Mesquita FP, Santos KT, et al. A comparative study on the anti-inflammatory effects of single oral doses of naproxen and its hydrogen sulfide (H2S)-releasing derivative ATB-346 in rats with carrageenan-induced synovitis. Medical Gas Research. 2013. 3(1):24. PMID: 24237604. PubMed. http://www.ncbi.nlm.nih.gov/pubmed/24237604

727. Li H, An Y, Zhang L, et al. Combined NMR and GC-MS analyses revealed dynamic metabolic changes associated with the carrageenan-induced rat pleurisy. Journal of Proteome Research. 2013. 12(12):5520–34. PMID:24131325. PubMed. http://www.ncbi.nlm.nih.gov/pubmed/24131325

728. Bhattacharyya S, Liu H, Zhang Z, et al. Carrageenan-induced innate immune response is modified by enzymes that hydrolyze distinct galactosidic bonds. Journal of Nutritional Biochemistry. 2010. 21(10):906–13. PMID:19864123. PubMed. http://www.ncbi.nlm.nih.gov/pubmed/19864123

729. Fath RB, Deschner EE, Winawer SJ, et al. Degraded carrageenan-induced colitis in CF1 mice: a clinical, histopathological and kinetic analysis. Digestion.1984. 29:197–203. doi:10.1159/000199033. Karger. http://www.karger.com/Article/Abstract/199033

730. Gong D, Chu W, Jiang L, et al. Effect of fucoxanthin alone and in combination with D-glucosamine hydrochloride on carrageenan/kaolin-induced experimental arthritis in rats. Phytotherapy Research. 2014. 28(7):1054–63.doi: 10.1002/ptr.5093. PMID:4338843. http://www.ncbi.nlm.nih.gov /pubmed/24338843

731. The Cornucopia Institute. Food-grade carrageenan: reviewing potential harmful effects on human health. 4-5. April 26, 2012. http://www.cornucopia.org/CornucopiaAnalysisofCarrageenanHealth Impacts042612.pdf

732. International Agency for Research on Cancer. IARC working group on the evaluation of the carcinogenic risk of chemicals to humans. Carrageenan. IARC monographs on the evaluation of carcinogenic risks to humans. 1983. http://monographs.iarc.fr/ENG/Monographs/vol83/mono83.pdf

733. Tobacman JK. Review of harmful gastrointestinal effects of carrageenan in animal experiments. Environmental Health Perspectives. October 2001. 109(10):983–994. PMCID:PMC1242073. PubMed. http://www.ncbi.nlm.nih.gov/pmc/articles/PMC1242073

734. U.S. Food and Drug Administration. Food additives & ingredients: food additive status list. Updated March 21, 2013. http://www.fda.gov/food/ingredientspackaginglabeling/foodadditives ingredients/ucm091048.htm

735. The Cornucopia Institute. How did carrageenan get into your organic food anyway? December 4, 2013. http://www.cornucopia.org/2013/12/carrageenan-get-organic-food-anyway

736. Marinalg International. Marinalg membership—current members. 2013. http://www.marinalg .org/our-members

737. Cohen SM & Ito N. A critical review of the toxicological effects of carrageenan and processed eucheuma seaweed on the gastrointestinal tract. Critical reviews in toxicology. 2002. 32(5):413–44. PMID:12389870. PubMed. http://www.ncbi.nlm.nih.gov/pubmed/12389870

738. The Cornucopia Institute. Food-grade carrageenan: reviewing potential harmful effects on human health. 8. April 26, 2012. http://www.cornucopia.org/CornucopiaAnalysisofCarrageenanHealth Impacts042612.pdf

739. The Cornucopia Institute. How did carrageenan get into your organic food anyway? December 4, 2013. http://www.cornucopia.org/2013/12/carrageenan-get-organic-food-anyway

740. Joanne K. Tobacman, MD, biography. University of Illinois College of Medicine at Chicago. 2013. http://chicago.medicine.uic.edu/cms/One.aspx?portalId=506244&pageId=12061305

741. Tobacman JK, Joanne K, & Tobacman MD. Citizen petition. Regulations.gov. Posted June 18, 2008. https://www.regulations.gov/#!documentDetail;D=FDA-2008-P-0347-0001

742. The Cornucopia Institute. Petition on carrageenan: gut wrenching! Tell FDA to remove toxin, making us sick, from food. Accessed February 24, 2016. http://www.cornucopia.org/carrageenan fda/#petition

743. Weil A. Q & A library: is carrageenan safe? Drweil.com. October 1, 2012. www.drweil.com/drw/u/QAA401181/Is-Carrageenan-Safe.html

744. Bhattacharyya S, Xue L, Devkota S, et al. Carrageenan-induced colonic inflammation is reduced in Bcl10 null mice and increased in IL-10-deficient mice. Mediators of Inflammation. 2013:397642. PMID:23766559. PubMed. http://www.ncbi.nlm.nih.gov/pubmed/23766559

745. Borthakur A, Bhattacharyya S, Anbazhagan AN, et al. Prolongation of carrageenan-induced inflammation in human colonic epithelial cells by activation of an NFκB-BCL10 loop. Biochimica et Biophysica Acta. 2012. 1822(8):1300–7. PMID:22579587. PubMed. http://www.ncbi.nlm.nih.gov/pubmed/22579587

746. Bhattacharyya S, Feferman L, Borthakur S, et al. Common food additive carrageenan stimulates Wnt/ β-Catenin signaling in colonic epithelium by inhibition of nucleoredoxin reduction. Nutrition and Cancer. 2014. 66(1):117–27. doi:10.1080/01635581.2014.852228. PubMed. http://www.ncbi.nlm.nih.gov/pubmed/24328990

747. The Cornucopia Institute. Shopping guide to avoiding organic foods with carrageenan. Updated December 18, 2013. http://www.cornucopia.org/shopping-guide-to-avoiding-organic-foods-with-carrageenan

748. United Soybean Board. The facts about soy lecithin: soy lecithin fact sheet. Accessed February 24, 2016. http://www.soyconnection.com/sites/default/files/Soy-Lecithin-Fact-Sheet.pdf

749. Organic Consumers Association. U.S. and Monsanto dominate global market for GM seeds. August 7, 2013. http://www.organicconsumers.org/articles/article_28059.cfm

750. Daniel KT. The whole soy story: the dark side of America's favorite health food. 2005. ISBN-10: 0967089751. Washington, DC: New Trends Publishing, Inc.

751. The Weston A. Price Foundation. Soy lecithin: from sludge to profit—excerpt from Kaayla Daniel's book: the whole soy story: the dark side of America's favorite health food. February 25, 2004. http://www.westonaprice.org/soy-alert/soy-lecithin-from-sludge-to-profit

752. Shurtleff W & Aoyagi A. History of soybeans and soyfoods: 1100 B.C. to the 1980s (unpublished manuscript). 2007. SoyInfo Center. http://www.soyinfocenter.com/HSS/lecithin1.php

753. Beal GF & Sorensen SO. Process for production of lecithin from vegetable raw materials. Original assignee: American Lecithin Co. U.S. Patent 2024398A. December 17, 1935

754. U.S. Environmental Protection Agency. Presentation abstract: nanostructured microemulsions as alternative solvents to VOCs in cleaning technologies and vegetable oil extraction. National Center for Environmental Research. Updated December 29, 2013. Wayback Machine. http://web.archive.org/web/20071013110614/http://es.epa.gov/ncer/publications/meetings/10_26_05/abstracts/do.html

755. U.S. Environmental Protection Agency. Toxicological review of n-Hexane (CAS No. 110-54-3). November 2005. https://cfpub.epa.gov/ncea/iris/iris_documents/documents/toxreviews/0486tr.pdf

756. Vallaeys C, Kastel M, Fantle W, et al. Behind the bean: the heroes and charlatans of the natural and organic soy foods industry. The Cornucopia Institute. 2009. http://www.cornucopia.org/soysurvey/OrganicSoyReport/behindthebean_color_final.pdf

757. Ema M, Hara H, Matsuomoto M, et al. Evaluation of developmental neurotoxicity of polysorbate 80 in rats. Reproductive Toxicology. January 2008. 2(1):89–99. PMID:17961976. PubMed. http://www.ncbi.nlm.nih.gov/pubmed/17961976

758. Roberts CL, Keita AV, Duncan SH, et al. Translocation of Crohn's disease Escherichia coli across M-cells: contrasting effects of soluble plant fibres and emulsifiers. Gut. 2010. 59:1331–1339. doi:10.1136/gut.2009.195370. http://gut.bmj.com/content/59/10/1331.full

759. Kucharzik T, Lügering N, Rautenberg K, et al. Role of M cells in intestinal barrier function. Annals of the New York Academy of Sciences. 2000. 915:171–83. PMID:11193574. PubMed. http://www.ncbi.nlm.nih.gov/pubmed/11193574

760. Friche E, Jensen PB, Sehested M, et al. The solvents cremophor EL and Tween 80 modulate daunorubicin resistance in the multidrug resistant Ehrlich ascites tumor. Cancer Communications. 1990. 2(9):297–303. PMID:1976341. PubMed. http://www.ncbi.nlm.nih.gov/pubmed/1976341

761. Zordan-Nudo T, Ling V, Liu Z, et al. Effects of nonionic detergents on P-glycoprotein drug binding and reversal of multidrug resistance. Journal of Cancer Research. 1993. 53:5994. http://cancerres.aacrjournals.org/content/53/24/5994.abstract?ijkey=0f84a965a96e5a749ac4be1f88daee51d5dcc2bf8&keytype2=tf_ipsecsha

762. Daher CF, Baroody GM, & Howland RJ. Effect of a surfactant, Tween 80, on the formation and secretion of chylomicrons in the rat. Food and Chemical Toxicology. 2003. 41(4):575–82. PMID:12615130. PubMed. http://www.ncbi.nlm.nih.gov/pubmed/12615130

763. Cornell University Legal Information Institute. 21 CFR 172.623—carrageenan with polysorbate 80. http://www.law.cornell.edu/cfr/text/21/172.623

764. U.S. Food and Drug Administration. Food additives & ingredients: food additive status list. Updated March 21, 2013. http://www.fda.gov/food/ingredientspackaginglabeling/foodadditives ingredients/ucm091048.htm

765. U.S. Food and Drug Administration. CFR—code of federal regulations title 21. Updated August 21, 2015. https://www.accessdata.fda.gov/scripts/cdrh/cfdocs/cfcfr/CFRSearch.cfm?fr=172.623

766. Chassaing B, Koren O, Goodrich J, et al. Dietary emulsifiers impact the mouse gut microbiota promoting colitis and metabolic syndrome. Nature. 519(7541):92–96. http://www.nature.com/nature/journal/v519/n7541/full/nature14232.html

767. Oser BL & Oser M. Nutritional studies on rats on diets containing high levels of partial ester emulsifiers. Journal of Nutrition. 1956. 60(4):489–505. http://jn.nutrition.org/content/60/4/489.short

768. Gajdová M, Jakubovsky J, & Války J. Delayed effects of neonatal exposure to Tween 80 on female reproductive organs in rats. Food and Chemical Toxicology. 1993. 31(3):183–90. PMID:8473002. PubMed. http://www.ncbi.nlm.nih.gov/pubmed/8473002

769. Goldman GS. Was there a synergistic fetal toxicity associated with the two-vaccine 2009/2010 season? Human and Experimental Toxicology. 2013. 32(5):464–475. doi:10.1177/0960327112455067. Sage Pub. http://het.sagepub.com/content/32/5/464

770. U.S. Centers for Disease Control and Prevention (CDC). Vaccine excipient & media summary: excipients included in U.S. vaccines, by vaccine. Updated February 2012. http://www.cdc.gov/vaccines/pubs/pinkbook/downloads/appendices/b/excipient-table-2.pdf

771. Coors EA, Seybold H, Merk HF, et al. Polysorbate 80 in medical products and nonimmunologic anaphylactoid reactions. Annals of Allergy, Asthma & Immunology. 2005. 95(6):593–9. PMID:16400901. Science.NaturalNews. http://science.naturalnews.com/pubmed/16400901.html

772. Dib B & Falchi M. Convulsions and death induced in rats by Tween 80 are prevented by capsaicin. International Journal of Tissue Reactions. 1996. 18(1):27–31. PMID:8880337. PubMed. http://www.ncbi.nlm.nih.gov/pubmed/8880377

773. Jayasingh P, Cornforth D, Carpenter CE, et al. Evaluation of carbon monoxide treatment in modified atmosphere packaging or vacuum packaging to increase color stability of fresh beef. Meat Science. 2001. 59(3):317–324. http://dx.doi.org/10.1016/S0309-1740(01)00086-9

774. U.S. Food and Drug Administration. Re: GRAS claim for the use of carbon monoxide in brine and modified atmosphere packaging for red meat products. March 7, 2005. http://www.fda.gov /downloads/Food/IngredientsPackagingLabeling/GRAS/NoticeInventory/ucm268850.pdf

775. Sørheima O, Aune T, & Nesbakken T. Technological, hygienic and toxicological aspects of carbon monoxide used in modified-atmosphere packaging of meat. Trends in Food Science & Technology. 1997. 8(9):307–312. Science Direct. http://dx.doi.org/10.1016/S0924-2244(97)01062-5

776. Schmit J. Carbon monoxide keeps meat red longer; is that good? USA Today. October 30, 2007. http://usatoday30.usatoday.com/money/industries/food/2007-10-30-kalsec-meat-carbon -monoxide_n.htm

777. Hunt MC, Mancini RA, Hachmeister KA, et al. Carbon monoxide in modified atmosphere packaging affects color, shelf life, and microorganisms of beef steaks and ground beef. Journal of Food Science. 2004. 69(1):FCT45–FCT52. doi:10.1111/j.1365-2621.2004.tb17854.x. Wiley Online Library. http://onlinelibrary.wiley.com/doi/10.1111/j.1365-2621.2004.tb17854.x/abstract

778. Jayasingh P, Cornforth D, Carpenter CE, et al. Evaluation of carbon monoxide treatment in modified atmosphere packaging or vacuum packaging to increase color stability of fresh beef. Meat Science. 2001. 59(3):317–324. Science Direct. http://dx.doi.org/10.1016/S0309-1740(01)00086-9

779. SilentShadow.org. Long term effects of carbon monoxide poisoning. 2004. http://www.silent shadow.org/long-term-effects-of-carbon-monoxide-poisoning.html

780. Kurokawa Y, Hayashi Y, Maekawa A, et al. Carcinogenicity of potassium bromate administered orally to F344 rats. Journal of the National Cancer Institute. 1983. 71(5):965–972. doi:10.1093/ jnci/71.5.965. http://jnci.oxfordjournals.org/content/71/5/965.short

781. Kasai H, Nishimura S, Kukokawa Y, et al. Oral administration of the renal carcinogen, potassium bromate, specifically produces 8-hydroxydeoxyguanosine in rat target organ DNA. Carcinogenesis. 1987. 8(12):1959–61. doi:10.1093/carcin/8.12.1959. http://carcin.oxfordjournals.org/content/8 /12/1959.short

782. Kurokawa Y, Maekawa A, Takahashi M, et al. Toxicity and carcinogenicity of potassium bromate—a new renal carcinogen. Environmental Health Perspectives. 1990. 87:309–335. PM-CID:PMC1567851. PubMed. http://www.ncbi.nlm.nih.gov/pmc/articles/PMC1567851

783. Perez A. 8 Foods we eat in the U.S. that are banned in other countries. BuzzFeed. June 19, 2013. http://www.buzzfeed.com/ashleyperez/8-foods-we-eat-in-the-us-that-are-banned-in-other-countries

784. California Office of Environmental Health Hazard Assessment. Bromate meets the criteria for listing as causing cancer via the authoritative bodies mechanism package 19a.1.a OEHHA Proposition 65. February 9, 2001. http://www.oehha.ca.gov/prop65/CRNR_notices/admin_listing/ intent_to_list/noilbromate.html

785. U.S. Food and Drug Administration. Food additives & ingredients: food additive status list. Updated March 21, 2013. http://www.fda.gov/food/ingredientspackaginglabeling/foodadditives ingredients/ucm091048.htm

786. Adams M. Baked goods sold in USA contain potassium bromate, a carcinogen banned in Europe but allowed in the US due to chemical loophole. Natural News. April 12, 2012. http://www .naturalnews.com/035542_potassium_bromate_baked_goods_cancer.html

787. California Office of Environmental Health Hazard Assessment. Bromate meets the criteria for listing as causing cancer via the authoritative bodies mechanism package 19a.1.a OEHHA Proposition 65. February 9, 2001. http://www.oehha.ca.gov/prop65/CRNR_notices/admin_listing/ intent_to_list/noilbromate.html

788. Iodine-Resource.com. Potassium bromate: the silent killer. Updated 2013. http://www.iodine -resource.com/potassium-bromate.html

789. Eskin BA. Iodine and mammary cancer. Advances in Experimental Medicine and Biology. 1977. 91:293–304. PMID:343535. PubMed. http://www.ncbi.nlm.nih.gov/pubmed/343535

790. Jaska E & Redfern S. Interaction of ferrous sulfate with potassium bromate and iodate in brew and dough systems. Cereal Chemistry. 975. 52(5):726–734. AACC International. http://www.aaccnet .org/publications/cc/backissues/1975/documents/chem52_726.pdf

791. Pomeranz Y & Finney KF. Protein-enriched baked products and method of making same. Patent 3,679,433. July 25, 1972.

792. Al-Hooti SN, Sidhu JS, Al-Saqer JM, et al. Effect of raw wheat germ addition on the physical texture and objective color of a designer food (pan bread). Molecular Nutrition & Food Research. 2002. 46(2):68–72. doi:10.1002/1521-3803. Wiley Online Library. http://onlinelibrary.wiley .com/doi/10.1002/1521-3803%2820020301%2946:2%3C68::AID-FOOD68%3E3.0.CO;2-W/ abstract

793. Sai K, Umemura T, Takagi A, et al. The protective role of glutathione, cysteine and vitamin C against oxidative DNA damage induced in rat kidney by potassium bromate. Cancer Science. 1992. 83(1):45–51. doi: 10.1111/j.1349-7006.1992.tb02350.x. Wiley Online Library. http:// onlinelibrary.wiley.com/doi/10.1111/j.1349-7006.1992.tb02350.x/full

794. Kurokawa Y, Maekawa A, Takahashi M, et al. Toxicity and carcinogenicity of potassium bromate—a new renal carcinogen. Environmental Health Perspectives. 1990. 87:309–335. PM-CID:PMC1567851. PubMed. http://www.ncbi.nlm.nih.gov/pmc/articles/PMC1567851

795. Watson E. FDA: Brominated vegetable oil (BVO), is safe, so removing its interim status is "not a priority." January 29, 2013. Foodnavigator-usa.com. http://www.foodnavigator-usa.com/Regulation /FDA-Brominated-vegetable-oil-BVO-is-safe-so-removing-its-interim-status-is-not-a-priority

796. Vorhees CV, Butcher RE, Wootten V, et al. Behavioral and reproductive effects of chronic developmental exposure to brominated vegetable oil in rats. Behavioral Teratology. 1983. 28(3):309–318. doi:10.1002/tera.1420280302. Wiley Library Online. http://onlinelibrary.wiley.com/doi/10.1002/ tera.1420280302/abstract

797. Munro IC, Hand B, Middleton EJ, et al. Toxic effects of brominated vegetable oils in rats. Toxicology and Applied Pharmacology. 22(3):432–439. Science Direct. http://www.sciencedirect.com/ science/article/pii/0041008X72902505

798. Horowitz BZ. Bromism from excessive cola consumption. Clinical Toxicology. 1997. 35(3):315–320. doi:10.3109/15563659709001219. Informa. http://informahealthcare.com/doi/ abs/10.3109/15563659709001219

799. Center for Science in the Public Interest. Brominated vegetable oil (BVO): emulsifier, clouding agent: soft drinks. Chemical Cuisine: Learn about Food Additives. Updated 2013. http://www .cspinet.org/reports/chemcuisine.htm#bvo

800. U.S. Food and Drug Administration. 21CFR172.175—sodium nitrite. Code of Federal Regulations. Revised April 3, 2013. http://www.accessdata.fda.gov/scripts/cdrh/cfdocs/cfCFR/CFRSearch .cfm?fr=172.175

801. Liu CY, Hsu YH, Wu MT, et al. Cured meat, vegetables, and bean-curd foods in relation to childhood acute leukemia risk: a population based case-control study. BMC Cancer. 2009. 9:15. PMID:19144145. PubMed. http://www.ncbi.nlm.nih.gov/pubmed/19144145

802. Xie TP, Zhao YF, Chen LQ, et al. Long-term exposure to sodium nitrite and risk of esophageal carcinoma: a cohort study for 30 years. International Society for Diseases of the Esophagus. 2011. 24(1):30–2. PMID:20545968. PubMed. http://www.ncbi.nlm.nih.gov/pubmed/20545968

803. Hannas BR, Das PC, Li H, et al. Intracellular conversion of environmental nitrate and nitrite to nitric oxide with resulting developmental toxicity to the crustacean Daphnia magna. PLOS ONE. 2010. 5(8):e12453. PMID:20805993. PubMed. http://www.ncbi.nlm.nih.gov/pubmed/20805993

804. de la Monte SM, Neusner A, Chu J, et al. Epidemiological trends strongly suggest exposures as etiologic agents in the pathogenesis of sporadic Alzheimer's disease, diabetes mellitus, and non-alcoholic steatohepatitis. Journal of Alzheimer's Disease. 2009. 17(3):519–29. doi:10.3233/JAD-2009-1070. http://www.ncbi.nlm.nih.gov/pmc/articles/PMC4551511

805. González CA, Jakszyn P, Pera G, et al. Meat intake and risk of stomach and esophageal adenocarcinoma within the European prospective investigation into cancer and nutrition (EPIC). Journal of the National Cancer Institute. 2006. 98(5):345–54. PMID:16507831. PubMed. http://www.ncbi .nlm.nih.gov/pubmed/16507831

806. Aschebrook-Kilfoy B, Ward MH, Gierach GL, et al. Epithelial ovarian cancer and exposure to dietary nitrate and nitrite in the NIH-AARP Diet and Health Study. European Journal of Cancer

Prevention. 2012. 21(1):65–72. PMID:21934624. PubMed. http://www.ncbi.nlm.nih.gov/pubmed/21934624

807. Jiang R, Paik DC, Hankinson JL, et al. Cured meat consumption, lung function, and chronic obstructive pulmonary disease among United States adults. American Journal of Respiratory and Critical Care Medicine. 2007. 175(8):798–804. PMCID: PMC1899290. PubMed. http://www.ncbi.nlm.nih.gov/pmc/articles/PMC1899290/

808. Wang M, Cheng G, Khariwala SS, et al. Evidence for endogenous formation of the hepatocarcinogen N-nitrosodihydrouracil in rats treated with dihydrouracil and sodium nitrite: A potential source of human hepatic DNA carboxyethylation. Chemico-Biological Interactions. 2013. 206(1):83–89. Science Direct. http://dx.doi.org/10.1016/j.cbi.2013.07.010

809. Adams M. Processed meat consumption results in 67% increase in pancreatic cancer risk, says new research. Natural News. April 20, 2005. http://www.naturalnews.com/007024.html

810. Nöthlings U, Wilkens LR, Murphy SP, et al. Meat and fat intake as risk factors for pancreatic cancer: the multiethnic cohort study. Journal of the National Cancer Institute. 2005. 97(19):1458–1465. doi:10.1093/jnci/dji292. http://jnci.oxfordjournals.org/content/97/19/1458.full

811. Ferrucci LM, Sinha R, Ward MH, et al. Meat and components of meat and the risk of bladder cancer in the NIH-AARP Diet and Health Study. Cancer. 2010. 116(18):4345–53. PMCID: PMC2936663. PubMed. http://www.ncbi.nlm.nih.gov/pmc/articles/PMC2936663

812. Khan Md.W, Arivarasu NA, Priyamvada S, et al. Protective effect of ω-3 polyunsaturated fatty acids (PUFA) on sodium nitrite induced nephrotoxicity and oxidative damage in rat kidney. Journal of Functional Foods. 2013. 5(3):956–67.Science Direct. http://dx.doi.org/10.1016/j.jff.2013.02.009

813. Aschebrook-Kilfoy B, Shu X, Gao Y, et al. Thyroid cancer risk and dietary nitrate and nitrite intake in the Shanghai women's health study. International Journal of Cancer. 2013. 132(4):897–904. doi: 10.1002/ijc.27659. Wiley Library Online. http://onlinelibrary.wiley.com/doi/10.1002/ijc.27659/abstract

814. Scanlan RA. Nitrosamines and cancer. The Linus Pauling Institute. Updated November 2000. http://margotbworldnews.com/News/Apr/Apr22/nitrosamines.html

815. Verrett J & Carper J. Eating may be hazardous to your health. 1974. ISBN: 0-385-11193-2. New York, NY: Simon and Schuster.

816. Hannas BR, Das PC, Li H, et al. Intracellular conversion of environmental nitrate and nitrite to nitric oxide with resulting developmental toxicity to the crustacean Daphnia magna. PLOS ONE. 2010. 5(8):e12453. PMID:20805993. PubMed. http://www.ncbi.nlm.nih.gov/pubmed/20805993

817. Trafton A. Nitric oxide shown to cause colon cancer. MIT News. January 19, 2009. http://web.mit.edu/newsoffice/2009/colon-cancer-0119.html

818. Gutierrez D. Sodium nitrite meat additive chemical used to poison woman in lover's spat. Natural News. August 16, 2008. http://www.naturalnews.com/023873_sodium_meat_nitrite.html

819. Lapidge S, Wishart J, Staples L, et al. Development of a feral swine toxic bait (Hog-Gone) and bait hopper (Hog-Hopper) in Australia and the USA. Proceedings of the 14th Wildlife Damage Management Conference. 2012. U.S. Department of Agriculture Animal and Plant Health Inspection Service. https://www.aphis.usda.gov/wildlife_damage/nwrc/publications/12pubs/fagerstone121.pdf

820. Kamm JJ, Dashman T, Conney AH, et al. Protective effect of ascorbic acid on hepatotoxicity caused by sodium nitrite plus aminopyrine. Proceedings of the National Academy of Sciences of the United States of America. 1973. 70(3):747–9. PMID:PMC433349. PubMed. http://www.ncbi.nlm.nih.gov/pmc/articles/PMC433349

821. Adams M. The real reason why processed meats are so dangerous to your health. NaturalNews. August 21, 2005. http://www.naturalnews.com/011148.html

822. Salama MF, Abbas A, Darweish MM, et al. Hepatoprotective effects of cod liver oil against sodium nitrite toxicity in rats. Pharmaceutical Biology. 2013. 51(11):1435–43. PMID:23862714. PubMed. http://www.ncbi.nlm.nih.gov/pubmed/23862714

Food Ingredients as Contaminants: Molecular Alteration of Food

823. Homogenization. Food science. 2009. University of Guelph. https://www.uoguelph.ca /foodscience/book-page/homogenization

824. Oster K & Ross D. The presence of ectopic xanthine oxidase in atherosclerotic plaques and myocardial tissues. Proceedings of the Society for Experimental Biology and Medicine. 1973. 144(2):523–6. PMID:4746925. PubMed. http://www.ncbi.nlm.nih.gov/pubmed/4746925

825. Ho CY & Clifford AJ. Bovine milk xanthine oxidase, blood lipids and coronary plaques in rabbits. The Journal of Nutrition. 1977. 107(5):758. PMID:870648. http://jn.nutrition.org/content/107 /5/758.full.pdf

826. Ibid.

827. Ross DJ, Sharnick SV, & Oster KA. Liposomes as a proposed vehicle for the persorption of bovine xanthine oxidase. Experimental Biology and Medicine. 1980. 163(1):141–145. Sage Journals. http://ebm.sagepub.com/content/163/1/141.short

828. Sunflower Publishing Company. Press release. Homogenized milk and atherosclerosis. PR.com. September 19, 2011. http://www.pr.com/press-release/353916

829. Enig MG. Milk homogenization & heart disease. December 13, 2003. The Weston A. Price Foundation. http://www.westonaprice.org/know-your-fats/milk-homogenization-heart-disease

830. Cardillo C, Kilcoyne CM, Cannon RO, et al. Xanthine oxidase inhibition with oxypurinol improves endothelial vasodilator function in hypercholesterolemic but not in hypertensive patients. Hypertension. 1997. 30(1):57–63. American Heart Association. http://hyper.ahajournals.org/ content/30/1/57.short

831. Houston M, Estevez A, Chumley P, et al. Binding of xanthine oxidase to vascular endothelium— kinetic characterization and oxidative impairment of nitric oxide-dependent signaling. Journal of Biological Chemistry. 1999. 274(8):4985–94. http://www.jbc.org/content/274/8/4985.short

832. Landmesser U, Spiekermann S, Dikalov S, et al. Vascular oxidative stress and endothelial dysfunction in patients with chronic heart failure—role of xanthine-oxidase and extracellular superoxide dismutase. Circulation. 2002. 106(24):3073–8. http://circ.ahajournals.org/content/106/24/3073.short

833. Miesel R & Zuber M. Elevated levels of xanthine oxidase in serum of patients with inflammatory and autoimmune rheumatic diseases. Inflammation. 1993. 17(5):551–61. PMID:8225562. PubMed. http://www.ncbi.nlm.nih.gov/pubmed/8225562

834. Moss M & Freed D. The cow and the coronary: epidemiology, biochemistry and immunology. International journal of cardiology. 2003. 87(2):203–216. PMID:12559541. PubMed. http:// www.ncbi.nlm.nih.gov/pubmed/12559541

835. Bemis S. FDA and USDA: cheese is serious! December 3, 210. Food Safety News. http://www .foodsafetynews.com/2010/12/fda-and-usda-cheese-is-serious

836. Nair MG, Mistry VV, & Oommen BS. Yield and functionality of cheddar cheese as influenced by homogenization of cream. International Dairy Journal. 2000. 10(9):647–657. Science Direct. http://www.sciencedirect.com/science/article/pii/S095869460000090X

837. Bird J. Changes in the dairy industry in the last 30 years. Journal of the Society of Dairy Technology. February 1993. 46(1):5–9. doi: 10.1111/j.1471-0307.1993.tb00850.x. Wiley Online Library. http://onlinelibrary.wiley.com/doi/10.1111/j.1471-0307.1993.tb00850.x/abstract

838. Farm-to-Consumer Legal Defense Fund. Raw milk nation. June 21, 2013. http://www.farmto consumer.org/raw_milk_map.htm

839. Johnson L. Brain shrinkage? Trans fats link to Alzheimer's. CBN News Health & Science. March 25, 2012. http://www.cbn.com/cbnnews/healthscience/2012/march/brain-shrinkage-trans-fats-link -to-alzheimers-

840. Hu FB, Stampfer MJ, Manson JE, et al. Dietary fat intake and the risk of coronary heart disease in women. New England Journal of Medicine. 1997. 337(21):1491–1499. PMID:9366580. PubMed. www.ncbi.nlm.nih.gov/pubmed/9366580

841. Stender S, Astrup A, & Dyerberg J. Ruminant and industrially produced trans fatty acids: health aspects. Food & Nutrition Research. 2008. 52. PMCID:PMC2596737. PubMed. http://www .ncbi.nlm.nih.gov/pmc/articles/PMC2596737

842. Willett WC, Stampfer MJ, Manson JE, et al. Intake of trans fatty acids and risk of coronary heart disease among women. Lancet. 1993. 341(8845):581–5. Science Direct. http://dx.doi.org/10.1016/0140-6736(93)90350-P

843. Taylor MR. Trans fat: taking the next important step. FDA voice. November 7, 2013. http://blogs.fda.gov/fdavoice/index.php/2013/11/trans-fat-taking-the-next-important-step

844. Center for Science in the Public Interest. What the food labels don't tell you. What's new—CSPI press releases. August 7, 1996. https://www.cspinet.org/new/transpr.html

845. U.S. Food and Drug Administration. Revealing trans fats. Accessed February 24, 2016. Wayback Machine. http://wayback.archive.org/web/20100311162752/http://www.pueblo.gsa.gov/cic_text/food/reveal-fats/reveal-fats.htm

846. Collier A. Deadly fats: why are we still eating them? The Independent. June 10, 2008. http://www.independent.co.uk/life-style/health-and-families/healthy-living/deadly-fats-why-are-we-still-eating-them-843400.html

847. Food and Nutrition Board. Dietary reference intakes for energy, carbohydrate, fiber, fat, fatty acids, cholesterol, protein, and amino acids (macronutrients). Institute of Medicine of the National Academies. 2005. 423. The National Academies Press. http://www.nap.edu/openbook.php?isbn=0309085373&page=423

848. Collier A. Deadly fats: why are we still eating them? The Independent. June 10, 2008. http://www.independent.co.uk/life-style/health-and-families/healthy-living/deadly-fats-why-are-we-still-eating-them-843400.html

849. Smith Y. Trans fat regulation. News Medical. Updated June 14, 2015. http://www.news-medical.net/health/Trans-Fat-Regulation.aspx

850. Ibid.

851. Norden. New Icelandic rules for trans fats. November 11, 2010. http://www.norden.org/en/news-and-events/news/new-icelandic-rules-for-trans-fats

852. European Food Safety Authority. Opinion of the Scientific Panel on Dietetic products, nutrition and allergies [NDA] related to the presence of trans fatty acids in foods and the effect on human health of the consumption of trans fatty acids. EFSA Journal. August 30, 2004. doi:10.2903/j.efsa.2004.81. http://www.efsa.europa.eu/en/efsajournal/pub/81.htm

853. Scott-Thomas C. Eu margarine industry sets stricter trans fat standards. Food Navigator. March 19, 2013. http://www.foodnavigator.com/Legislation/EU-margarine-industry-sets-stricter-trans-fat-standards

854. U.S. Food and Drug Administration. FDA news release—FDA takes step to further reduce trans fats in processed foods. November 7, 2013. http://www.fda.gov/NewsEvents/Newsroom/PressAnnouncements/UCM373939

855. Taylor MR. Trans fat: taking the next important step. FDA voice. November 7, 2013. http://blogs.fda.gov/fdavoice/index.php/2013/11/trans-fat-taking-the-next-important-step

856. Vega-López S, Ausman LM, Jalbert SM, et al. Palm and partially hydrogenated soybean oils adversely alter lipoprotein profiles compared with soybean and canola oils in moderately hyperlipidemic subjects. American Journal of Clinical Nutrition. 2006. 84(1):54–62. PMID:16825681. PubMed. http://www.ncbi.nlm.nih.gov/pubmed/16825681

857. Science Daily. Palm oil not a healthy substitute for trans fats, study finds. May 11, 2009. http://www.sciencedaily.com/releases/2009/05/090502084827.htm

858. Andrew M, de Koning L, Shannon HS, et al. A systematic review of the evidence supporting a causal link between dietary factors and coronary heart disease. Archives of Internal Medicine. 2009. 169(7):659–669. PMID:19364995. PubMed. http://www.ncbi.nlm.nih.gov/pubmed/19364995

859. Ravnskov U. The questionable role of saturated and polyunsaturated fatty acids in cardiovascular disease. Journal of Clinical Epidemiology. 1998. 51(6):443–460. Science Direct. http://www.sciencedirect.com/science/article/pii/S0895435698000183

860. Dias CB, Garg R, Wood LG, et al. Saturated fat consumption may not be the main cause of increased blood lipid levels. Medical Hypotheses. 2014. 82(2):187–95.PMID:24365276. PubMed. http://www.ncbi.nlm.nih.gov/pubmed/24365276

Animal Feed Contaminants

861. Environmental Working Group. Farm subsidy database: the United States summary information. Updated 2012. http://farm.ewg.org/region.php

862. U.S. Department of Justice. Competition and agriculture: voices from the workshops on agriculture and antitrust enforcement in our 21st century economy and thoughts on the way forward. May 2012. http://www.justice.gov/atr/public/reports/283291.pdf

863. Saxena D & Stotzky G. Bt corn has a higher lignin content than non-Bt corn. American Journal of Botany. September 2001. 88(9):1704–1706. http://www.amjbot.org/content/88/9/1704.full

864. Jung HG and Allen MS. Characteristics of plant cell walls affecting intake and digestibility of forages by ruminants. Journal of Animal Science. 1995. 73:2774–90. http://www.journalofanimal science.org/publications/jas/abstracts/73/9/2774

865. U.S. International Trade Commission. China's consumption of agricultural products increasing substantially as incomes rise, says USITC. News release 11-029. Inv. no. 332-518. March 22, 2011. http://www.usitc.gov/press_room/news_release/2011/er0322jj1.htm

866. Li YX, Li W, Wu J, et al. Contribution of additive Cu to its accumulation in pig feces: study in Beijing and Fuxin of China. Journal of Environmental Science. 2007. 19:610–615. PMID:17915692. PubMed. http://www.ncbi.nlm.nih.gov/pubmed/17915692

867. Nicholson FA, Chambers BJ, Williams JR, et al. Heavy metal contents of livestock feeds and animal manures in England and Wales. Bioresource Technologies. 1999. 70:23–31. http://www.sciencedirect.com/science/article/pii/S0960852499000176

868. Cang L, Wang Y, Zhou D, et al. Heavy metals pollution in poultry and livestock feeds and manures under intensive farming in Jiangsu Province, China. Journal of Environmental Sciences. 2004. 16(3):371–374. PMID:15272705. PubMed. http://www.ncbi.nlm.nih.gov/pubmed/15272705

869. Zhang F, Li Y, Yang M, et al. Content of heavy metals in animal feeds and manures from farms of different scales in northeast China. International Journal of Environmental Research and Public Health. 2012. 9(8):2658–68. PMCID:PMC3447579. PubMed. http://www.ncbi.nlm.nih.gov/pmc/articles/PMC3447579/

870. Nicholson FA, Chambers BJ, Williams JR, et al. Heavy metal contents of livestock feeds and animal manures in England and Wales. Bioresource Technology. 1999. 70(1):23–31. Science Direct. http://www.sciencedirect.com/science/article/pii/S0960852499000176

871. U.S. Department of Agriculture Economic Research Service. Cattle & beef. Updated May 26, 2012. http://www.ers.usda.gov/topics/animal-products/cattle-beef.aspx

872. National Cattlemen's Beef Association. Beef industry statistics—Beef USA. Directions statistics. 2013. http://www.beefusa.org/CMDocs/BeefUSA/Producer%20Ed/2013%20Directions%20Stats.pdf

873. Chicago Mercantile Exchange. Introduction to livestock and meat fundamentals. AG59.2. Accessed February 24, 2016. http://s3.amazonaws.com/zanran_storage/agmarketing.extension.psu.edu/ContentPages/15848009.pdf

874. U.S. Department of Justice and U.S. Department of Agriculture. Public workshops exploring competition issues in agriculture livestock workshop: a dialogue on competition issues facing farmers in today's agricultural marketplaces. Colorado State University. August 27, 2010. http://www.justice.gov/atr/public/workshops/ag2010/colorado-agworkshop-transcript.pdf

875. Simplot Livestock Company. Custom feeding program: Simplot. Accessed February 24, 2016. http://www.simplot.com/pdf/Simplot_Feedlot_Web_PDF.pdf

876. J.R. Simplot Company. Beef & dairy custom feeding. Cattle feedlot management. Accessed February 24, 2016. http://www.simplot.com/beef_dairy_feeding

877. J.R. Simplot Company. Cattle feedlot management. Beef & dairy custom feeding. Accessed February 24, 2016. http://www.simplot.com/beef_dairy_feeding

878. Associated Press. Simplot beefing infrastructure to increase feedlot capacity. March 12, 1998. Livestock Weekly. http://www.livestockweekly.com/papers/98/03/12/whlsimplot.asp

879. Leitner M. History of the J.R. Simplot Company. Hole notes. April 6, 2003. Michigan State University Libraries. http://archive.lib.msu.edu/tic/holen/article/2003apr6.pdf

880. Tyson Foods, Inc. Fiscal 2010 fact book. Investor relations department. 2010. http://www.tyson foods.com/Investors.aspx

881. Tyson Foods, Inc. Brands: Tyson products. 2013. http://www.tyson.com/Products/Our-Products .aspx

882. Tyson Food Service. USA Today ad: K-12—working at the heart of your school menu. 2013. http://www.tysonfoodservice.com/K-12/Tyson-Cares/Project-A.aspx

883. National Cattlemen's Beef Association. Beef industry statistics—Beef USA. Directions statistics. 2013. http://www.beefusa.org/CMDocs/BeefUSA/Producer%20Ed/2013%20Directions%20Stats .pdf

884. Krohn CC, Munsgaard L, & Jonasen B. Behaviour of dairy cows kept in extensive (loose housing/ pasture) or intensive (tie stall) environments I. Experimental procedure, facilities, time budgets— diurnal and seasonal conditions. Applied Animal Behaviour Science. 1992. 34(1–2):37–47. http:// dx.doi.org/10.1016/S0168-1591(05)80055-3

885. Rutter SM. Diet preference for grass and legumes in free-ranging domestic sheep and cattle: current theory and future application. Applied Animal Behaviour Science. 2006. 97(1):17–35. http:// dx.doi.org/10.1016/j.applanim.2005.11.016

886. Owens FN, Secrist DS, Hill WJ, et al. Acidosis in cattle: a review. Journal of Animal Science. 1998. 76(1):275–286. Animal science. https://www.animalsciencepublications.org/publications/ jas/articles/76/1/275

887. Beauchemin KA & Yang WZ. Effects of physically effective fiber on intake, chewing activity, and ruminal acidosis for dairy cows fed diets based on corn silage. Journal of Dairy Science. 2005. 88(6):2117–29. http://dx.doi.org/10.3168/jds.S0022-0302(05)72888-5

888. Nagaraja TG & Titgemeyer EC. Ruminal acidosis in beef cattle: the current microbiological and nutritional outlook. Journal of Dairy Science. 2007. 90 (Supplement):E17–E38. http://dx.doi .org/10.3168/jds.2006-478

889. Aldai N, Dugan MER, Kramer JKG, et al. Nonionophore antibiotics do not affect the trans-18:1 and conjugated linoleic acid composition in beef adipose tissue. Journal of Animal Science. 2008. 86(12):3522–32. doi:10.2527/jas.2008-0946. https://www.animalsciencepublications.org /publications/jas/abstracts/86/12/0863522

890. Kleen JL, Hooijer GA, Rehage J, et al. Subacute Ruminal Acidosis (SARA): a review. Journal of Veterinary Medicine Series A. 2003. 50(8):406–14. doi:10.1046/j.1439-0442.2003.00569.x. Wiley Online Library. http://onlinelibrary.wiley.com/doi/10.1046/j.1439-0442.2003.00569.x/abstract

891. Smith A. Cash-strapped farmers feed candy to cows. CNN Money. October 10, 2012. http://money .cnn.com/2012/10/10/news/economy/farmers-cows-candy-feed/

892. Walker DK, Titgemeyer EC, Drouillard JS, et al. Effects of ractopamine and protein source on growth performance and carcass characteristics of feedlot heifers. Journal of Animal Science. 2006. 84(10):2795–2800. doi:10.2527/jas.2005-614. https://www.animalsciencepublications.org /publications/jas/articles/84/10/0842795

893. Zuo P, Zhang Y, & Liu J. Determination of beta-adrenergic agonists by hapten microarray. Talanta. 2010. 82(1):61–66. 10.1016/j.talanta.2010.03.058. PMID:20685436. PubMed. http://www.ncbi .nlm.nih.gov/pubmed/20685436

894. Center for Food Safety. Ractopamine factsheet: lean meat = mean meat. Food safety fact sheet. February 2013. http://www.centerforfoodsafety.org/files/ractopamine_factsheet_02211.pdf

895. Consumer Reports. Consumer reports investigation of pork products finds potentially harmful bacteria? Most of which show resistance to important antibiotics. November 27, 2012. http:// pressroom.consumerreports.org/pressroom/2012/11/my-entry-4.html

896. Center for Food Safety. Public interest groups sue FDA demanding records on controversial animal growth drugs under Freedom of Information Act. Press release. October 7, 2013. http://www .centerforfoodsafety.org/press-releases/2630/public-interest-groups-sue-fda-demanding-records-on -controversial-animal-growth-drugs-under-freedom-of-information-act#

897. Merck Animal Health. Zilmax (zilpaterol hydrochloride)—Consumer Information Center. 2013. http://www.zilmax.com

898. Huffstutter PJ & Polansek T. Special report: lost hooves, dead cattle before Merck halted Zilmax sales. Reuters. December 30, 2013. http://news.yahoo.com/special-report-lost-hooves-dead-cattle -merck-halted-210140606--sector.html

899. Chan TYK. Food-borne Clenbuterol may have potential for cardiovascular effects with chronic exposure (commentary). Clinical Toxicology. 2001. 39(4):345–348. doi:10.1081/CLT-100105153. Taylor & Francis Online. http://informahealthcare.com/doi/abs/10.1081/CLT-100105153?journal Code=ctx

900. Shanghai Municipal Institute for Food and Drug Supervision. Lessons from outbreak of food poisoning caused by Clenbuterol in Shanghai in September 2006. Chinese Journal of Food Hygiene. January 2007. http://en.cnki.com.cn/Article_en/CJFDTOTAL-ZSPZ200701001.htm

901. Pulce C, Lamaison D, Keck G, et al. Collective human food poisonings by clenbuterol residues in veal liver. Veterinary and Human Toxicology. October 1991. 33(5):480–1. PMID:1746141. PubMed. http://www.ncbi.nlm.nih.gov/pubmed/1746141

902. Dalton JC. Antibiotic residue prevention in milk and dairy beef. Western Daily News. April 2006. 6(4):W-79–W80. http://www.cvmbs.colostate.edu/ilm/proinfo/wdn/2006/April%20 %20WDN06.pdf

903. Butaye P, Devriese LA, & Haesebrouck F. Antimicrobial growth promoters used in animal feed: effects of less well known antibiotics on gram-positive bacteria. Clinical Microbiology Reviews. 2003. 16(2):175–188. doi:10.1128/CMR.16.2.175-188.2003. http://cmr.asm.org/content/16/2/175

904. U.S. National Library of Medicine. Macrolide antibiotics: drug class overview. LiverTox: clinical and research information on drug-induced liver injury. November 5, 2013. http://livertox.nih.gov/ MacrolideAntibiotics.htm

905. Ireland RE, Habich D, & Norbeck DW. The convergent synthesis of polyether ionophore antibiotics: the synthesis of the monensin spiroketal. Journal of the American chemical society. 1985. 107:3271–3278. University of York. http://www.york.ac.uk/res/pac/teaching/ireland%20monensin.pdf

906. Thomas, HS. Storey's guide to raising beef cattle. 1998. 257. ISBN: 978-1-60342-454-7. North Adams, MA: Storey Publishing.

907. Dawson KA & Boling JA. Monensin-resistant bacteria in the rumens of calves on monensin-containing and unmedicated diets. Applied and Environmental Microbiology. 1983. 46(1):160–164. PMCID:PMC239282. PubMed. http://www.ncbi.nlm.nih.gov/pmc/articles/PMC239282

908. Kirkpatrick JG & Selk G. Coccidiosis in cattle. Oklahoma Cooperative Extension Service. VTMD-9129. Accessed February 24, 2016. Oklahoma State University. http://pods.dasnr.okstate .edu/docushare/dsweb/Get/Document-2677/VTMD-9129web.pdf

909. Thomas, HS. Industry reduces E. coli by 90%, but little progress with salmonella. Beef magazine. May 7, 2013. http://beefmagazine.com/beef-quality/industry-reduces-e-coli-90-little-progress -salmonella

910. CBS News/Associated Press. CDC: salmonella linked to ground beef behind 5-state outbreak. January 29, 2013. http://www.cbsnews.com/news/cdc-salmonella-linked-to-ground-beef-behind -5-state-outbreak

911. McGuirk SM & Peek S. Salmonellosis in cattle: a review. American Association of Bovine Practitioners: 36th Annual Conference. Preconvention seminar 7: dairy herd problem investigation strategies. September 15–17, 2003. University of Wisconsin School of Veterinary Medicine. http:// www.vetmed.wisc.edu/dms/fapm/fapmtools/7health/Salmorev.pdf

912. U.S. Food and Drug Administration. 21CFR589.2000, part 589—substances prohibited from use in animal food or feed. U.S. Code of Federal Regulations. Title 21, volume 6. June 5, 1997. Revised April 1, 2015. http://www.accessdata.fda.gov/scripts/cdrh/cfdocs/cfcfr/CFRSearch.cfm ?fr=589.2000

913. Washington State Department of Agriculture. Bovine spongiform encephalopathy. Updated February 4, 2013. http://agr.wa.gov/FoodAnimal/AnimalFeed/BSE.aspx

914. Washington State Department of Agriculture. Animal proteins prohibited in ruminant feed & cattle materials prohibited in all animal feed. AGR PUB 631-282. August 2013. http://agr.wa.gov/ foodanimal/animalfeed/Publications/ProhibMatDefs.pdf

915. Peterson K. Beef prices soar as cattle herds dwindle. MSN Money. June 13, 2012.

916. Lusk JL, Roosen J, & Fox JA. Demand for beef from cattle administered growth hormones or fed genetically modified corn: a comparison of consumers in France, Germany, the United Kingdom, and the United States. American Journal of Agricultural Economics. 2003. 85(1):16–29. doi:10.1111/1467-8276.00100. Oxford Journals. http://ajae.oxfordjournals.org/content/85/1/16 .short

917. Singer P & Mason J. The ethics of what we eat: why our food choices matter. 2006. ISBN: 978-1-59486-687-6. Emmaus, PA: Rodale Press.

918. U.S. Department of Justice. Tyson pleads guilty to 20 felonies and agrees to pay $7.5 million for Clean Water Act violations. June 25, 2003. http://www.justice.gov/opa/pr/2003/June/03_enrd_383.htm

919. Singer P & Mason J. The ethics of what we eat: why our food choices matter. 2006. ISBN: 978-1-59486-687-6. Emmaus, PA: Rodale Press.

920. Sierra Club. Tyson taken to task, pulls plug on fowl factory. November 27, 2007. http://web .archive.org/web/20140228162147/http://www.sierraclub.org/grassroots/stories/00027.asp

921. Singer P & Mason J. The ethics of what we eat: why our food choices matter. 2006. ISBN: 978-1-59486-687-6. Emmaus, PA: Rodale Press.

922. Attorney General of the State of Oklahoma, State of Oklahoma, ex rel. W.A. Drew Edmondson, in his capacity as Attorney General; Oklahoma Secretary of the Environment, ex rel. C. Miles Tolbert in his capacity as the Trustee for Natural Resources for the State of Oklahoma, Plaintiffs–Appellants, v. TYSON FOODS, INC.; Tyson Poultry, Inc.; Tyson Chicken, Inc.; Cobb–Vantress, Inc.; Cal–Maine Foods, Inc.; Cal–Maine Farms, Inc.; Cargill, Inc.; Cargill Turkey Production, LLC; George's, Inc.; George's Farms, Inc.; Peterson Farms, Inc.; Simmons Foods, Inc.; Willow Brook Foods, Inc., Defendants–Appellees. State of Arkansas, Amicus Curiae. No. 08-5154. United States Court of Appeals, Tenth Circuit. May 13, 2009. http://caselaw.findlaw.com/us-10th-circuit/1386118.html

923. Singer P & Mason J. The ethics of what we eat: why our food choices matter. 2006. ISBN: 978-1-59486-687-6. Emmaus, PA: Rodale Press

924. Scharbor J. Poultry litter: is it fertilizer or pollution? The Change Agent. March 2011. 42–43. Statistics for Action. http://sfa.terc.edu/materials/pdfs/CA_poultry_litter.pdf

925. Chapman HD & Johnson ZB. Use of antibiotics and roxarsone in broiler chickens in the USA: analysis for the years 1995 to 2000. Poultry science. March 2002. 81(3):356–364. http://ps.fass .org/content/81/3/356.short

926. Nachman KE, Baron PA, Raber G, et al. Roxarsone, inorganic arsenic, and other arsenic species in chicken: a U.S.-based market basket sample. Environmental Health Perspectives. May 2013. http://dx.doi.org/10.1289/ehp.1206245

927. Edmonds MS & Baker DH. Toxic effects of supplemental copper and roxarsone when fed alone or in combination to young pigs. Journal of Animal Science. 1986. 63(2):533–7. PMID:3759688. PubMed. http://www.ncbi.nlm.nih.gov/pubmed/3759688

928. Makris KC, Salazar J, Quazi S, et al. Controlling the fate of roxarsone and inorganic arsenic in poultry litter. Journal of environmental quality. 2008. 37(3):963–971. doi:10.2134/jeq2007.0416. Science Societies. https://dl.sciencesocieties.org/publications/jeq/abstracts/37/3/963

929. Wershaw RL, Garbarino JR, & Burkhardt MR. Roxarsone in natural water systems. In Wild FD, Britton LJ, Miller CV, & Kolpin DW (comps.). Effects of confined animal feeding operations (CAFOs) on hydrologic resources and the environment—proceedings of the technical meeting. August 30–September 1, 1999. U.S. Geological Survey. http://water.usgs.gov/owq/AFO /proceedings/afo/pdf/Wershaw.pdf

930. Nachman KE, Baron PA, Raber G, et al. Roxarsone, inorganic arsenic, and other arsenic species in chicken: a U.S.-based market basket sample. Environmental Health Perspectives. May 2013. Science Direct. http://dx.doi.org/10.1289/ehp.1206245

931. Schmidt CW. Arsenical association: inorganic arsenic may accumulate in the meat of treated chickens. Environmental Health Perspectives. doi:10.1289/ehp.121-a226. http://ehp.niehs.nih .gov/121-a226

932. Wang FM, Chen ZL, Zhang L, et al. Arsenic uptake and accumulation in rice (Oryza sativa L.) at different growth stages following soil incorporation of roxarsone and arsanilic acid. Plant Soil. 2006. 285(1–2):359–367. doi:10.1007/s11104-006-9021-7. Springer Link. http://link.springer .com/article/10.1007/s11104-006-9021-7

933. Duffy CF, Sims MD, & Power RF. Preliminary evaluation of dietary Natustat versus Histostat (nitarsone) for control of histomonas meleagridis in broiler chickens on infected litter. International Journal of Poultry Science. 2004. 3(12):753–757. http://docsdrive.com/pdfs/ansinet/ ijps/2004/753-757.pdf

934. Sullivan TW & Al-Timimi AA. Safety and toxicity of oietary organic arsenicals relative to performance of young turkeys: 3. Nitarsone. Poultry Science. 1972. 51(5):1582–6. doi:10.3382/ ps.0511582. Oxford Journals. http://ps.fass.org/content/51/5/1582.short

935. FDA Announces Pending Withdrawl of Approval of Nitarsone. US Food and Drug Administration. April 2015. http://www.fda.gov/AnimalVeterinary/NewsEvents/CVMUpdates/ucm440668.htm

936. Pure Salmon Campaign. Salmon farming harms other marine life. February 24, 2016. http://www .puresalmon.org/waste.html

937. Hellou J, Haya K, Steller S, et al. Presence and distribution of PAHs, PCBs and DDE in feed and sediments under salmon aquaculture cages in the Bay of Fundy, New Brunswick, Canada. Aquatic Conservation: Marine and Freshwater Ecosystems. 2005. 15(4):349–365. Wiley Online Library. http://onlinelibrary.wiley.com/doi/10.1002/aqc.678/abstract

938. Coxworth B. New "fishless" feeds could make aquaculture more sustainable. Gizmag. August 7, 2013. http://www.gizmag.com/fishless-fish-feed/28615

939. Pure Salmon Campaign. Salmon farming harms other marine life. February 24, 2016. http://www .puresalmon.org/waste.html

940. Reichhardt T. Will souped up salmon sink or swim? Nature. 2000. 406:10–12. doi:10.1038/35017657. http://www.nature.com/nature/journal/v406/n6791/full/406010a0.html

941. U.S. Food and Drug Administration. FDA has determined that the AquAdvantage salmon is as safe to eat as non-GE salmon. Updated November 24, 2015. http://www.fda.gov/ForConsumers/ ConsumerUpdates/ucm472487.htm

942. Zohar Y & Mylonas C. Endocrine manipulations of spawning in cultured fish: from hormones to genes. Aquaculture. 2001. 197(1–4):99–136. Science Direct. http://dx.doi.org/10.1016/ S00448486(01)00584-1

943. Mylonas CC, Fostier A, & Zanuy S. Broodstock management and hormonal manipulations of fish reproduction. Genercal and comparative endocrinology. 2010. 165(3):516–534. Science Direct. http://dx.doi.org/10.1016/j.ygcen.2009.03.007

944. Bye VJ & Lincoln RF. Commercial methods for the control of sexual maturation in rainbow trout (Salmo gairdneri R.). Aquaculture. 1986. 57(1–4):299–309. Science Direct. http://dx.doi.org /10.1016/0044-8486(86)90208-5

945. Thorne, PS. Environmental health impacts of concentrated animal feeding operations: anticipating hazards—searching for solutions. Environmental Health Perspectives. 2007. 115(2):296–297. PMCID:PMC1817701. PubMed. http://www.ncbi.nlm.nih.gov/pmc/articles/PMC1817701

946. Burkholder J, Libra B, Weyer P, et al. Impacts of waste from concentrated animal feeding operations on water quality. Environmental Health Perspectives. 2007. 115(2):308–312. PMCID:P-MC1817674. PubMed. http://www.ncbi.nlm.nih.gov/pmc/articles/PMC1817674

947. Kolodziej EP, Harter T, & Sedlak DL. Dairy wastewater, aquaculture, and spawning fish as sources of steroid hormones in the aquatic environment. Environmental Science & Technology. 2004. 38(23):6377–84. doi:10.1021/es049585d. American Chemical Society. http://pubs.acs.org/doi/ abs/10.1021/es049585d

948. Qu S, Kolodziej EP, Long SA, et al. Product-to-parent reversion of trenbolone: unrecognized fisks for endocrine disruption. Science magazine. 2013. 342(6156):347–351. doi:10.1126/science.1243192. http://www.sciencemag.org/content/342/6156/347

949. Peplow M. Hormone disruptors rise from the dead. Nature. September 26, 2013. http://www .nature.com/news/hormone-disruptors-rise-from-the-dead-1.13831

950. Moss M. The burger that shattered her life. The New York Times. October 3, 2009. http://www
.nytimes.com/2009/10/04/health/04meat.html

951. Villanueva L. Cargill sued over E. coli. News Inferno. December 24, 2008. http://www.news
inferno.com/cargill-sued-over-e-coli

952. Weise E. Who should pay to make ground beef safe from E. coli? USA Today. November 28, 2011.
http://usatoday30.usatoday.com/money/industries/food/story/2011-12-01/safe-meat/51447546/1

953. The Cornucopia Institute. FDA food safety rules threaten to crush the good food movement. September 19, 2013. http://www.cornucopia.org/2013/09/fda-food-safety-rules-threaten-crush-good
-food-movement

954. The Cornucopia Institute. FDA/USDA collude to eliminate true organic egg production. July 23,
2013. http://www.cornucopia.org/2013/07/fdausda-collude-to-eliminate-true-organic-egg
-production

955. U.S. Government Printing Office. §205.238. Livestock health care practice standard. Title 7:
agriculture, part 205—National organic program, subpart C—organic production and handling
requirements. Updated December 30, 2013. http://www.ecfr.gov/cgi-bin/text-idx?SID=645fd0ace-
7d274adfcc1982bb23436ed&node=7:3.1.1.9.32.3.354.12&rgn=div8

956. U.S. Government Printing Office. §205.237. Livestock feed. Title 7: agriculture, part 205—
National organic program, subpart C—organic production and handling requirements. Updated
December 30, 2013. http://www.ecfr.gov/cgi-bin/text-idx?SID=645fd0ace7d274adfcc1982bb2343
6ed&node=7:3.1.1.9.32.3.354.11&rgn=div8

957. U.S. Department of Agriculture. Grass-fed marketing claim standards: grading, certification and verifi-
cation. Agricultural Marketing Service. Accessed February 24, 2016. http://web.archive.org/web
/20150522104434/http://www.ams.usda.gov/AMSv1.0/getfile?dDocName=STELPRDC5063842

958. U.S. Government Printing Office. §205.239. Livestock living conditions. Title 7: agriculture, part
205—National Organic Program, subpart C—organic production and handling requirements.
Updated December 30, 2013. http://www.ecfr.gov/cgi-bin/text-idx?SID=645fd0ace7d274adfcc198
2bb23436ed&node=7:3.1.1.9.32.3.354.13&rgn=div8

959. The Cornucopia Institute. FDA food safety rules threaten to crush the good food movement. September 19, 2013. http://www.cornucopia.org/2013/09/fda-food-safety-rules-threaten-crush-good
-food-movement

960. The Cornucopia Institute. Organic egg scorecard. September 22, 2010. http://www.cornucopia
.org/organic-egg-scorecard

Sidebar: Zeolites and Heavy Metals

1. James F, Lonky S, & Deitsch E. Clinical evidence supporting the use of an activated clinoptilolite
suspension as an agent to increase urinary excretion of toxic heavy metals. Nutrition and Dietary
Supplements. 2009:1 11–18. Dovepress. https://www.dovepress.com/clinical-evidence-supporting
-the-use-of-an-activated-clinoptilolite-su-peer-reviewed-article-NDS

Sidebar: Toxic Elements in Fertilizers

1. Center for Media and Democracy. Biosolids. January 23, 2014. http://www.sourcewatch.org/
index.php/Biosolids

2. Grant CA & Sheppard SC. Fertilizer impacts on cadmium availability in agricultural soils and
crops. Human and ecological risk assessment: an international journal. 14(2):210–228. doi:10
.1080/10807030801934895. http://www.tandfonline.com/doi/abs/10.1080/10807030801934895

3. Aydin I, Aydin F, Saydut A, et al. Hazardous metal geochemistry of sedimentary phosphate rock used
for fertilizer (Mazıdag, SE Anatolia, Turkey). Microchemical journal. November 2010. 96(2):247–
251. Science Direct. http://www.sciencedirect.com/science/article/pii/S0026265X10000706

4. Reuss JO, Dooley HL, & Griffis W. Uptake of cadmium from phosphate fertilizers by peas, radishes, and lettuce. Journal of environmental quality. January 1978. 7(1):128–133. doi:10 .2134/
jeq1978.00472425000700010026x. Science Societies. https://dl.sciencesocieties.org/publications/
jeq/abstracts/7/1/JEQ0070010128

5. Mortvedt JJ. Heavy metal contaminants in inorganic and organic fertilizers. Fertilizer research. 1996. 43:55–61. Springer Link. http://link.springer.com/article/10.1007/BF00747683

6. Specter M. Ocean dumping is ending, but not problems; New York can't ship, bury or burn its sludge, but no one wants a processing plant. The New York times. June 29, 1992. http://www .nytimes.com/1992/06/29/nyregion/ocean-dumping-ending-but-not-problems-new-york-can -t-ship-bury-burn-its-sludge.html

7. Jerving S. New toxic sludge PR and lobbying effort gets underway. The Center for Media and De-mocracy's PR Watch. March 16, 2012. http://www.prwatch.org/news/2012/03/11273/new-toxic -sludge-pr-and-lobbying-effort-gets-underway

8. U.S. Environmental Protection Agency. U.S. Environmental fact sheet: waste-derived fertilizers. Solid waste and emergency response. EPA530-F-97-053. December 1997. http://nepis.epa.gov/ Exe/ZyPURL.cgi?Dockey=10000LTW.txt

9. Mulla DJ, Page AL, & Ganje TJ. Cadmium accumulations and bioavailability in soils from long-term phosphorus fertilization. Journal of Environmental Quality. July 1980. 9(3):408–412. doi:10.2134/jeq1980.00472425000900030016x. Science Societies. https://dl.sciencesocieties.org/ publications/jeq/abstracts/9/3/JEQ0090030408

10. Arenholt-Bindslev D & Larsen AH. Mercury levels and discharge in waste water from dental clinics. Water, air, and soil pollution. January 1996. 86(1–4):93–99. doi:10.1007/BF00279147. Springer Link. http://link.springer.com/article/10.1007/BF00279147#

11. Ibid.

12. Arenholt-Bindslev D. Dental amalgam—environmental aspects. Advances in dental research. 1992. 6(1):125–130. doi:10.1177/08959374920060010501. Sage Journals. http://adr.sagepub .com/content/6/1/125.short

13. World Health Organization. Mercury in health care. Water sanitation and health. Department of Protection of the Human Environment Policy. August 2005. http://www.who.int/water_sanitation _health/medicalwaste/mercurypolpaper.pdf

14. Hylander LD, Lindvall A, & Gahnberg L. High mercury emissions from dental clinics despite amalgam separators. Science of the total environment. 2006. 362(1–3):74–84. Science Direct. http://dx.doi.org/10.1016/j.scitotenv.2005.06.008

15. Benson J. San Francisco's free "organic biosolids compost" filled with toxic chemicals. Natural News. August 18, 2010. http://www.naturalnews.com/029504_organic_biosolids_toxic.html

16. Babish JG, Lisk DJ, Stoewsand GS, et al. Organic toxicants and pathogens in sewage sludge and their environmental effects, a special report of the subcommittee on organics in sludge. Cornell University, College of Agriculture and Life Sciences. December 1981. http://agris.fao.org/agris -search/search.do?f=1983/US/US83138.xml;US8225508

17. U.S. Environmental Protection Agency. Biosolids: targeted national sewage sludge survey report— overview. EPA 822-R-08-014. January 2009. http://nepis.epa.gov/Exe/ZyPURL.cgi?Dockey= P1005KUR.txt

18. Nicholson FA, Chambers BJ, Williams JR, et al. Heavy metal contents of livestock feeds and ani-mal manures in England and Wales. Bioresource technology. 1999. 70(1):23–31. Science Direct. http://www.sciencedirect.com/science/article/pii/S0960852499000176

19. Sharma RP & Street JC. Public health aspects of toxic heavy metals in animal feeds. Journal of the American veterinary medical association. 1980. 177(2):149–53. PMID:7429947. PubMed. http:// www.ncbi.nlm.nih.gov/pubmed/7429947

20. Webber J. Trace Metals in Agriculture. Effect of heavy metal pollution on plants: pollution moni-toring series. 1981. 2:159–184. ISBN: 978-9-40098-101-0. doi:10.1007/978-94-009-8099-0_5.

21. Zhang F, Li Y, Yang M, et al. Content of heavy metals in animal feeds and manures from farms of different scales in northeast China. Environmental research and public health. 2012. 9: 2658– 2668. doi:10.3390/ijerph9082658. http://www.mdpi.com/1660-4601/9/8/2658

22. Kastel M. The E. coli spinach contamination issue. The Cornucopia Institute. September 17, 2006. http://www.cornucopia.org/2006/09/169

Sidebar: China's Toxic Pollution Catastrophe

1. Bloomberg News. Shanghai tells children to stay inside for 7th smoggy day. December 9, 2013. http://www.bloomberg.com/news/2013-12-09/shanghai-reports-pm2-5-pollutants-13-times-who -recommended-level.html

2. Sutherland S. 3.3 million hectares of China's farmland too polluted to grow food, says official. Yahoo! news Canada. December 30, 2013. http://ca.news.yahoo.com/blogs/geekquinox/3-3 -million-hectares-china-farmland-too-polluted-230444094.html

3. Demick B. "Airpocalypse": severe pollution cripples northeastern China. Los Angeles times. October 22, 2013. http://www.latimes.com/world/worldnow/la-fg-wn-pollution-northeastern-china -20131022,0,5024464.story

4. MacKinnon E. Yangtze River runs mysteriously red. Live science. September 7, 2012. http://www .livescience.com/23038-yangtze-river-red.html

5. Kaiman J. Chinese environment official challenged to swim in polluted river. The London Guardian. February 21, 2013. http://www.theguardian.com/environment/2013/feb/21/chinese-official -swim-polluted-river

6. van Wyk B. The groundwater of 90% of Chinese cities is polluted. Danwei. February 18, 2013. http://www.danwei.com/the-groundwater-of-90-of-chinese-cities-is-polluted

7. Russia Today. China admits pollution brought about "cancer villages." RT. February 23, 2013. http://rt.com/news/china-water-pollution-cancer-346

8. Buckley C. Rice tainted with cadmium is discovered in southern china. The New York Times. May 21, 2013. http://www.nytimes.com/2013/05/22/world/asia/cadmium-tainted-rice-discovered-in -southern-china.html

9. Olesen A. China tells US to stop reporting Beijing's bad air. The Associated Press. June 5, 2012. http://bigstory.ap.org/article/china-tells-us-stop-reporting-beijings-bad-air-0

10. Gale F & Buzby J. Imports from China and food safety issues. Economic information bulletin number 52. July 2009. U.S. Department of Agriculture, Economic Research Service. http://www .ers.usda.gov/ersDownloadHandler.ashx?file=/media/156008/eib52_1_.pdf

11. Food & Water Watch. A decade of dangerous food imports from China. 2011. http://www .foodandwaterwatch.org/reports/a-decade-of-dangerous-food-imports-from-china

12. Gale F & Buzby J. Imports from China and food safety issues. Economic information bulletin number 52. July 2009. U.S. Department of Agriculture, Economic Research Service. http://www .ers.usda.gov/ersDownloadHandler.ashx?file=/media/156008/eib52_1_.pdf

13. Heyes JD. Does your Whole Foods "organic" produce come from China? July 25, 2012. Natural-News. http://www.naturalnews.com/036584_Whole_Foods_organic_produce_China.html

14. Branigan T. Chinese figures show fivefold rise in babies sick from contaminated milk. The London Guardian. December 2, 2008. http://www.theguardian.com/world/2008/dec/02/china

15. Foster P. Top 10 Chinese food scandals. The Telegraph. April 27, 2011. http://www.telegraph.co.uk /news/worldnews/asia/china/8476080/Top-10-Chinese-Food-Scandals.html

16. Kaiman J. Like some fox hair with that? China digests latest food scandals. The London Guardian. October 11, 2013. http://www.theguardian.com/world/2013/oct/11/china-food-scandals-fox-hair -animal-waste

17. Heyes JD. Does your Whole Foods "organic" produce come from China? July 25, 2012. Natural-News. http://www.naturalnews.com/036584_Whole_Foods_organic_produce_China.html

18. Liu C. Is "USDA Organic" a seal of deceit: The pitfalls of USDA certified organics produced in the United States, China and beyond. Stanford Journal of International Law. 2011. 47, 333.

19. Lavigne P. USDA does not always enforce organic label standards. The Dallas Morning News. July 25, 2006. http://lists.ibiblio.org/pipermail/livingontheland/20060727/002003.html

Sidebar: Dentistry—How Dentistry Pollutes Our Bodies with Mercury

1. Beltrán-Aguilar ED, Barker LK, Canto MT, et al. Surveillance for dental caries, dental sealants, tooth retention, edentulism, and enamel fluorosis, United States, 1988–1994 and 1999–2002.

346   **ENDNOTES**

Surveillance Summaries. U.S. Centers for Disease Control and Prevention. August 26, 2005. 54(3):1–44. http://www.cdc.gov/MMWR/preview/mmwrhtml/ss5403a1.htm

2.   U.S. National Institute of Dental and Craniofacial Research. Dental caries (tooth decay) in adults (age 20 to 64). National Institutes of Health. Updated December 10, 2013. http://www.nidcr.nih .gov/DataStatistics/FindDataByTopic/DentalCaries/DentalCariesAdults20to64.htm

3.   Reinhardt JW. Side-effects: mercury contribution to body burden from dental amalgam. Advances in Dental Research. September 1992. 6:110–3. PMID:1292449. PubMed. http://www.ncbi.nlm .nih.gov/pubmed/1292449

4.   Lorscheider FL, Vimy MJ, & Summers AO. Mercury exposure from "silver" tooth fillings: emerging evidence questions a traditional dental paradigm. The Federation of American Societies for Experimental Biology. 1995. 9(7): 504–508. http://www.fasebj.org/content/9/7/504.short

5.   World Health Organization. Elemental mercury and inorganic mercury compounds: human health aspects. Concise international chemical assessment document 50. 2003. First draft prepared by Dr. JF Risher, ATSDR. http://www.who.int/ipcs/publications/cicad/en/cicad50.pdf

6.   Stoner GE, Senti SE, & Gileadi E. Effect of sodium fluoride and stannous fluoride on the rate of corrosion of dental amalgams. Journal of Dental Research. 1979. 58:576–83. Sage Journals. http:// jdr.sagepub.com/content/50/6/1647.short

7.   Barregård L, Sällsten L, & Järvholm B. People with high mercury uptake from their own dental amalgam fillings. Occupational & Environmental Medicine. 1995. 52:124–128. doi:10.1136/ oem.52.2.124. http://oem.bmj.com/content/52/2/124.short

8.   Mortazavi SM, Daiee E, Yazdi A, et al. Mercury release from dental amalgam restorations after magnetic resonance imaging and following mobile phone use. Pakistan Journal of Biological Sciences. 2008. 11(8):1142–6. PMID:18819554. PubMed. http://science.naturalnews.com/ pubmed/18819554.html

9.   U.S. Environmental Protection Agency. Mercury in your environment. Updated July 9, 2013. http://www.epa.gov/hg/effects.htm

10.   Guzzi G, Grandi M, Cattaneo C, et al. Dental amalgam and mercury levels in autopsy tissues: food for thought. American Journal of Forensic Medicine and Pathology. 2006. 27(1):42–5. PMID: 16501347. PubMed. http://www.ncbi.nlm.nih.gov/pubmed/16501347

11.   Galic N, Prpic-Mehičic G, Prester L, et al. Dental amalgam mercury exposure in rats. Biometals. September 1999. 12(3):227–231. doi: 10.1023/A:1009267513632. http://link.springer.com /article/10.1023/A:1009267513632#

12.   Siblerud RL. A comparison of mental health of multiple sclerosis patients with silver/mercury dental fillings and those with fillings removed. Psychological Reports. 1992. 70:1139–1151. doi:10.2466/ pr0.1992.70.3c.1139. http://www.amsciepub.com/doi/abs/10.2466/pr0.1992.70.3c.1139

13.   Siblerud RL, Motl J, & Kienholz E. Psychometric evidence that mercury from silver dental fillings may be an etiological factor in depression, excessive anger, and anxiety. Psychological Reports. 1994. 74:67–80. doi:10.2466/pr0.1994.74.1.67. http://www.amsciepub.com/doi/abs/10.2466/ pr0.1994.74.1.67

14.   Echeverrial D, Aposhian HV, Woods JS, et al. Neurobehavioral effects from exposure to dental amalgam Hgo: new distinctions between recent exposure and Hg body burden. The Journal of the Federation of American Societies for Experimental Biology. 1998. 12(11):971–980. http://www .fasebj.org/content/12/11/971.short

15.   U.S. Food and Drug Administration. Appendix I: summary of changes to the classification of dental amalgam and mercury. 2009. http://www.fda.gov/MedicalDevices/ProductsandMedical Procedures/DentalProducts/DentalAmalgam/ucm171120.htm

16.   Sancho FM & Ruiz CN. Risk of suicide amongst dentists: myth or reality? International dental journal. 2010. 60(6):411–8. PMID:21302740. PubMed. http://www.ncbi.nlm.nih.gov/ pubmed/21302740

17.   McComb D. Occupational exposure to mercury in dentistry and dentist mortality. Journal of the Canadian dental association. 1997. 63(5):372–6. PMID:9170753. PubMed. http://www.ncbi.nlm .nih.gov/pubmed/9170753

18. Tezel H, Ertas OS, Ozata F, et al. Occupational health: blood mercury levels of dental students and dentists at a dental school. British dental journal. 2001. 191:449–452. doi:10.1038/sj.bdj.4801205. http://www.nature.com/bdj/journal/v191/n8/abs/4801205a.html

19. Ngim CH, Foo SC, Boey KW, et al. Chronic neurobehavioural effects of elemental mercury in dentists. British Journal of Industrial Medicine. 1992. 49(11):782–790. PMCID:PMC1039326. PubMed. http://www.ncbi.nlm.nih.gov/pmc/articles/PMC1039326

20. Ritchi KA, Gilmour WH, Macdonald EB, et al. Health and neuropsychological functioning of dentists exposed to mercury. Occupational & Environmental Medicine. 2002. 59(5):287–293. doi:10.1136/oem.59.5.287. http://oem.bmj.com/content/59/5/287.short

21. Sikorski R, Juszkiewicz T, Paszkowski T, et al. Women in dental surgeries: reproductive hazards in occupational exposure to metallic mercury. International Archives of Occupational and Environmental Health. 1987. 59(6):551–7. PMID:3679554. PubMed. http://www.ncbi.nlm.nih.gov/pubmed/3679554

22. Rowland AS, Baird DD, Weinberg CR, et al. The effect of occupational exposure to mercury vapour on the fertility of female dental assistants. Occupational & Environmental Medicine. 1994. 51:28–34. doi:10.1136/oem.51.1.28. http://oem.bmj.com/content/51/1/28.short

23. Drexler H & Schaller KH. The mercury concentration in breast milk resulting from amalgam fillings and dietary habits. Environmental Research. 1998. 77(2):124–129. Science Direct. http://dx.doi.org/10.1006/enrs.1997.3813

24. Schuurs AHB. Reproductive toxicity of occupational mercury. A review of the literature. Journal of Dentistry. 1999. 27(4):249–256. Science Direct. http://dx.doi.org/10.1016/S0300-5712(97)00039-0

25. Hörsted-Bindslev P. Amalgam toxicity—environment and occupational hazards. Journal of Dentistry. July 2004. 32(5):359–365. Science Direct. http://dx.doi.org/10.1016/j.jdent.2004.02.002

26. Pleva J. Dental mercury—a public health hazard. Reviews on Environmental Health. 2011. 10(1):1-28. doi:10.1515/REVEH.1994.10.1.1. http://www.degruyter.com/dg/viewarticle/j$002freveh.1994.10.1$002freveh.1994.10.1.1$002freveh.1994.10.1.1.xml;jsessionid=A4CA5677CF45AE161C4086325DBD0B9F

27. World Health Organization. Mercury in health care. Department of Protection of the Human Environment Policy. August 2005. http://www.who.int/water_sanitation_health/medicalwaste/mercurypolpaper.pdf

28. Arenholt-Bindslev D & Larsen AH. Mercury levels and discharge in waste water from dental clinics. Water, air, and soil pollution. January 1996. 86(1–4):93–99. doi:10.1007/BF00279147. Springer Link. http://link.springer.com/article/10.1007/BF00279147#

29. Arenholt-Bindslev D. Dental amalgam—Environmental aspects. Advances in Dental Research. September 1992. 6(1):125–130. doi:10.1177/08959374920060010501. http://adr.sagepub.com/content/6/1/125.short

30. Hylander LD, Lindvall A, & Gahnberg L. High mercury emissions from dental clinics despite amalgam separators. Science of the Total Environment. 2006. 362(1–3): 74–84. http://dx.doi.org/10.1016/j.scitotenv.2005.06.008

31. Yoshida M, Kishimoto T, Yamamura Y, et al. Amount of mercury from dental amalgam fillings released into the atmosphere by cremation. Japanese Journal of Public Health. 1994. 41(7):618–24. PMID:7919469. http://www.ncbi.nlm.nih.gov/pubmed/7919469

32. ADA Council on Scientific Affairs. Statement on dental amalgam. American Dental Association. Updated August 25, 2009. http://www.ada.org/1741.aspx

Sidebar: The Systemic, Apocalyptic Pollution of the World's Oceans

1. Ray G. The ocean is broken. Newcastle Herald. October 18, 2013. http://www.theherald.com.au/story/1848433/the-ocean-is-broken

2. Allen P. Japan tsunami: the debris trail across the Pacific—interactive. The London Guardian. May 1, 2012. http://www.theguardian.com/world/interactive/2012/may/01/japan-tsunami-debris-pacific-interactive

3. Daily Mail Online. Rubbish dump found floating in Pacific Ocean is twice the size of America. February 6, 2008. http://www.dailymail.co.uk/news/article-512424/Rubbish-dump-floating -Pacific-Ocean-twice-size-America.html

4. Mccauley L. Newly discovered "plastic island" shows global epidemic worsening. Common Dreams. January 18, 2013. https://www.commondreams.org/headline/2013/01/18-3

5. Specter M. Ocean dumping is ending, but not problems; New York can't ship, bury or burn its sludge, but no one wants a processing plant. The New York times. June 29, 1992. http://www .nytimes.com/1992/06/29/nyregion/ocean-dumping-ending-but-not-problems-new-york-can-t -ship-bury-burn-its-sludge.html

6. Casselman A. 10 biggest oil spills in history. Popular Mechanics. May 7, 2010. http://www .popularmechanics.com/science/energy/coal-oil-gas/biggest-oil-spills-in-history#slide-1

7. Richards ZT, Beger M, Pinca S, et al. Bikini Atoll coral biodiversity resilience five decades after nuclear testing. Marine Pollution Bulletin. 2008. 56:503–515. www.bikiniatoll.com /BIKINICORALS.pdf

8. Fackler M. New leaks into Pacific at Japan nuclear plant. The New York Times. August 6, 2013. http://www.nytimes.com/2013/08/07/world/asia/leaks-into-pacific-persist-at-japan-nuclear-plant .html

9. Tirone J. Tepco's claim radiation leaks confined to coast called "silly." Bloomberg News. October 6, 2013. http://www.bloomberg.com/news/2013-10-07/tepco-s-claim-radiation-leaks-confined-to -coast-called-silly-.html

10. Hotz RL. U.S. tuna has Fukushima taint. The Wall Street Journal. May 29, 2012. http://online .wsj.com/news/articles/SB10001424052702303395604577432452114613564

11. Hadlington S. Chemistry World. Waste seaweed mops up heavy metals. Chemistry World. January 20, 2006. Royal Society of Chemistry. http://www.rsc.org/chemistryworld/News/2006/January /20010601.asp

12. Besadaa V, Andradeb JM, Schultze F, et al. Heavy metals in edible seaweeds commercialised for human consumption. Journal of Marine Systems. 2009. 75(1–2):305–313. Science Direct. http:// www.sciencedirect.com/science/article/pii/S0924796308002972

13. Sindermann CJ. Ocean pollution: effects on living resources and humans. 1995. ISBN: 9780849384219. Boca Raton: CRC Press.

Sidebar: Minamata Disease: Mercury Poisoning via Industrial Pollution in Japan

1. Harada M. Minamata disease: methylmercury poisoning in Japan caused by environmental pollu-tion. Critical Reviews in Toxicology. 1995. 25(1):1–24. PMID:7734058. PubMed. http://www .ncbi.nlm.nih.gov/pubmed/7734058

2. Ministry of the Environment, Government of Japan. Minamata disease the history and measures. 2002. http://www.env.go.jp/en/chemi/hs/minamata2002/ch2.html

INDEX

A

acai superfruit powders, 84

ACAM (American College for Advancement in Medicine), 8

accreditation of laboratory, xv, xvii

accuracy of test results, xv–xvi
  from commercial labs, xvi–xvii
  for this book, xv–xviii
  up-to-date results online, xvi

acesulfame-K, 136

acid rains, 86

ACP (American College of Pediatricians), 90

activated carbon (charcoal)
  as chelating agent, 10
  in defensive eating, 207
  as toxins remover, 64

ADA (American Dental Association), 42, 48

ADD (attention-deficit disorder), 148

ADHD. *see* attention-deficit/hyperactivity disorder

ADM. *see* Archer Daniels Midland

adverse health effects. *see also* diseases
  aluminum, 88–90
  amalgam dental fillings, 43, 45–47
  in animals (*see* animal adverse effects)
  arsenic, 14
  artificial colors and flavors, 148–153
  aspartame, 131–133
  atrazine, 117
  benzoic acid and its salts, 158–160
  BPA, 100–103
  brominated vegetable oils, 175
  cadmium, 79, 83–84
  carbamates, 123
  carbon monoxide, 173
  carrageenan, 166–168
  chemical preservatives and additives, 157
  copper, 93–95
  glyphosate, 116
  hexane, 107, 110
  HFCS, 60
  homogenized milk, 179–180
  of hormones in groundwater, 198
  hydrogenated fats, 181–182
  lead, 67, 74–76
  mercury, 45–49
  MSG, 140–142
  neonicotinoids, 125
  organochlorines, 118–120
  organophosphates, 120, 121
  parabens, 160
  pesticides, 113–114
  polysorbate 80, 171
  potassium bromate, 174
  propyl gallate, 162
  pyrethroids, 122
  sodium nitrite, 175–176
  sulfites, 164–165
  TBHQ. BHA, and BHT, 163
  tin, 96–98

Agency for Toxic Substances and Disease Registry (ATSDR), xi, 97–98
  on cadmium, xi
  on hexane, 108–109

Agent Orange, 116

agricultural practices, 112. *see also* fertilizers; pesticides
  arsenic in, 16–23, 33–34
  cadmium from, 79–81
  mercury in, 41–42